Microfluidics and Nanofluidics

Microfluidics and Nanofluidics

Theory and Selected Applications

Clement Kleinstreuer

Cover image: Courtesy of the Folch Lab, University of Washington

Cover design: Anne-Michele Abbott

ISBN 978-0-470-61903-2 (cloth); ISBN 978-1-118-41527-6 (ebk); ISBN 978-1-118-41800-0 (ebk); ISBN 978-1-118-74989-0 (ebk)

Printed in the United States of America

10 9 8 7 6 5 4 3 2 1

To my family,
Christin, Nicole, and Joshua

Contents

PREFACE

"Fluidics" originated as the description of pneumatic and hydraulic control systems, where fluids were employed (instead of electric currents) for signal transfer and processing. Fluidics then broadened and now comprises the technique of handling fluid flows from the macroscale down to the nanoscale. In turn, micro-/nanofluidics is a relatively small but very important part of nanoscience and technology, as indicated by the growing number of subject-oriented engineering and physics journals.

This textbook is written primarily for mature undergraduates in engineering and physics. However, it should be of interest to first-year graduate students and professionals in industry as well. **Part A** reviews key elements of classical fluid mechanics topics, with the main focus on laminar internal flows as needed for the remaining Chapters 3 to 8. The goal is to assure the same background for all students and hence the time spent on the material of Chapter 1, "Theory," and Chapter 2, "Applications," may vary somewhat from audience to audience. **Part B,** "Microfluidics," is the heart of the book, in terms of depth and extent, because of the accessibility of the topic and its wide range of engineering applications (see Chapters 3 and 4). Dealing with the more complex transport phenomena in "Nanofluidics" (see **Part C**) is much more challenging because advanced numerical solution tools are still not readily available to undergraduate/graduate students for course assignments. Thus, Chapters 5 and 6 are more descriptive and discuss only solutions to rather simple nanoscale problems. Nevertheless, for those interested in pursuing solutions to real-world problems in micro-/nanofluidics, **Part D** provides some introductory math modeling aspects with computer applications (see Chapters 7 and 8).

When compared to current books, e.g., Tabeling (2005), Nguyen & Wereley (2006), Zhang (2007), Bruus (2007), or Kirby (2010), the present material is in content and form more transport phenomena oriented and accessible to advanced undergraduates and first-year graduate students. Most other books on microfluidics are topic-specific reviews of the exponentially growing literature. Examples include *Microfluidics* edited by S. Colin (2010) and a handbook edited by S. K. Mitra & S. Chakraborty (2011). While some of the cited books also describe elements of nanofluidics, only the recent texts by Das et al. (2007), Rogers et al. (2008), and Hornyak et al. (2008), focus exclusively on nanotechnology with chapters on nanofluids and nanofluidics. Cited references in the preface appear in the list at the end of Part A.

The main learning objectives are to gain a solid knowledge base of the fundamentals and to acquire modern application skills. Furthermore, this eight-chapter exposure should provide students with a sufficient background for advanced studies in these fascinating and very future-oriented engineering areas, as well as for expanded job opportunities. Pedagogical elements include a 50/50 physics-mathematics approach when introducing new material, illustrating concepts, showing graphical/tabulated results as well as links to flow visualizations, and, very important, providing professional problem solution steps. Specifically, the problem solution format follows strictly: System Sketch, Assumptions/Postulates, and Concept/Approach—*before* starting the solution phase which consists of symbolic math model development (see Sect. A.1 and A.2), analytic (and occasionally) numerical solution, graphs, and comments on "physical insight." After some illustrative examples, most solved text examples have the same level of difficulty as assigned homework sets listed in Section 2.6. In general, homework assignments are grouped into "concept questions" to gain physical insight, engineering problems to hone independent problem solution skills, and/or course projects. Concerning course projects, the setup, suggestions, expectations, and rewards appear at the end of Chapters 4 to 6 and 8. They are probably the most important learning experience when done right. A Solutions Manual is available for instructors adopting the textbook.

The ultimate goals after course completion are that the more serious student can solve traditional and modern fluidics problems independently, can provide physical insight, and can suggest (say, via a course project) system design improvements.

As all books, this text relied on numerous open-source material as well as contributions provided by research associates, graduate students, and former participants of the author's course "Microfluidics and Nanofluidics" at North Carolina State University (NCSU). Special thanks go to Dr. Jie Li for typing, generating the graphs and figures, checking the example solutions, formatting the text, and obtaining the cited references. The Index was generated by Zelin Xu, who also reformatted the text; the proofreading of the text was performed by Tejas Umbarkar; while Chapter 8 project results were supplied by Emily Childress and Arun Varghese Kolanjiyil, all presently PhD students in the Computational Multi-Physics Lab <http://www.mae.ncsu.edu/cmpl/> at NCSU. Some of the book manuscript was written when the author worked as a Visiting Scholar at Stanford University during summers. The engaging discussions with Prof. John Eaton and his students (Mechanical Engineering Department) and the hospitality of the Dewes, Krauskopf and Tidmarsh host families are gratefully acknowledged as well.

For critical comments, constructive suggestions, and tutorial material, please contact the author via ck@ncsu.edu.

Part A: A REVIEW OF ESSENTIALS IN MACROFLUIDICS

The review of macrofluidics repeats mostly undergraduate-level theory and provides solved examples of transport phenomena, i.e., traditional (meaning conventional macroworld) fluid mechanics, heat and mass transfer, with a couple of more advanced topics plus applications added. Internal flow problems dominate and for their solutions the differential modeling approach is preferred. Specifically, for any given problem the basic conservation laws (see Sect. A.5) are reduced based on physical understanding (i.e., system sketch plus assumptions), sound postulates concerning the dependent variables, and then solved via direct integration or approximation methods. Clearly, Part A sets the stage for most of the problems solved in Part B and Part C.

CHAPTER 1

Theory

Clearly, the general equations describing conservation of mass, momentum, and energy hold for transport phenomena occurring in all systems/devices from the macroscale to the nanoscale, outside quantum mechanics. However, for most real-world applications such equations are very difficult to solve and hence we restrict our analyses to special cases in order to understand the fundamentals and develop skills to solve simplified problems.

This chapter first reviews the necessary definitions and concepts in fluid dynamics, i.e., fluid flow, heat and mass transfer. Then the conservation laws are derived, employing different approaches to provide insight of the meaning of equation terms and their limitations.

It should be noted that Chapters 1 and 2 are reduced and updated versions of Part A chapters of the author's text *Biofluid Dynamics* (2006). The material (used with permission from Taylor & Francis Publishers) is now geared towards engineering students who already have had introductory courses in thermodynamics, fluid mechanics and heat transfer, or a couple of comprehensive courses in transport phenomena.

1.1 Introduction and Overview

Traditionally, "fluidics" referred to a technology where fluids were used as key components of control and sensing systems. Nowadays the research and application areas of "fluidics" have been greatly expanded. Specifically, *fluidics deals with transport phenomena, i.e., mass, momentum and heat transfer, in devices ranging in size from the macroscale down to the nanoscale.* As it will become evident, this modern description implies two things:

(i)   Conventional fluid dynamics (i.e., macrofluidics) forms a necessary knowledge base when solving most microfluidics and some nanofluidics problems.
(ii) Length scaling from the macroworld (in meters and millimeters) down to the micrometer or nanometer range (i.e., 1 μm = 10^{-6} m while 1 nm = 10^{-9} m) requires new considerations concerning possible changes in fluid properties, validity of the continuum hypothesis, modified boundary conditions, and the importance of new (surface) forces or phenomena.

So, to freshen up on *macrofluidics,* this chapter reviews undergraduate-level essentials in fluid mechanics and heat transfer and provides an introduction to porous media and mixture flows. Implications of *geometric scaling,* known as the "size reduction effect," are briefly discussed next.

The most important scaling impact becomes apparent when considering the area-to-volume ratio of a simple fluid conduit or an entire device:

$$\frac{\text{Device surface area}}{\text{Device volume}} = \frac{A}{\forall} \sim \frac{l^2}{l^3} = l^{-1} \tag{1.1}$$

Evidently, in the micro/nanosize limit the ratio becomes very large, i.e., $\lim_{l \to 0} l^{-1} \to \infty$, where $l \hat{=}$ system length scale such as the hydraulic diameter, channel height, or width. This implies that *in micro/nanofluidics the system's surface-area-related quantities, e.g., pressure and shear forces, become dominant.* Other potentially important micro/nanoscale forces, rightly neglected in macrofluidics, are surface tension as well as electrostatic and magnetohydrodynamic forces. To provide a quick awareness of other size-related aspects, the following tabulated summary characterizes flow considerations in macrochannels versus microchannels. Specifically, it contrasts important flow conditions and phenomena in conduits of the order of meters and millimeters vs. those in microchannels being of the order of micrometers (see Table 1.1).

Brief Comments Regarding Table 1.1. Fortunately, the *continuum mechanics assumption* holds (i.e., a fluid is homogeneous and infinitely divisible) for most microchannel flows. Hence, reduced forms of the conservation laws (see Sect. 1.3) can be employed to solve fluid flow and heat/mass transfer problems in most device geometries (see Sect. 2.1 and Chapters 3 and 4). The boundary condition of "*no velocity slip at solid walls*" is standard in macrofluidics. However, microchannels fabricated with hydrophobic material and/or having rough surfaces may exhibit liquid velocity slip at the walls. Considering *laminar* flow, the *entrance length of a conduit* can be estimated as:

$$L_{\text{entrance}} = \kappa D_h \, \text{Re} \tag{1.2}$$

where the hydraulic diameter is defined as $D_h = 4A/P$, with A being the cross-sectional area and P the perimeter, the Reynolds number $\text{Re} = u_{\text{mean}} D_h / \nu$, and $\kappa \approx 0.05$ for macroconduits and 0.5 for microchannels. For fully turbulent flow, $L_{\text{entrance}}^{\text{macro}} \approx 10 D_h$. Considering that typically $L_{\text{microchannel}} = \mathcal{O}(1\,\text{mm})$, entrance effects can be important. For example, if $L_{\text{microchannel}} \geq 0.2 L_{\text{conduit}}$ the favorite simplification "fully developed flow" cannot be assumed anymore (see Sect. 1.4). The *Reynolds number* is the most important dimensionless group in fluid mechanics. However, for microsystems employed in biochemistry as well as in biomedical and chemical engineering, the Reynolds number is usually very low, i.e., $\text{Re} \leq \mathcal{O}(1)$. In contrast, microscale cooling devices, i.e., heat exchangers, require high Reynolds numbers to achieve sufficient heat rejection. Onset to *turbulence*, mainly characterized by random fluctuations of all dependent variables, may occur earlier in microsystems than in geometrically equivalent macrosystems. In some cases, *surface roughness* over, say, 3% of the channel height may cause interesting flow phenomena near the wall, such as velocity slip and/or transition to turbulence. For microsystems with heavy liquids and high velocity gradients, energy dissipation due to *viscous heating* should be considered. The *temperature jump condition* at the wall may be applicable when dealing with convection heat transfer of rarefied gases (see Chapters 2 and 3). The last three entries in Table 1.1, i.e., *diffusion, surface tension, and electrokinetics,* are of interest almost exclusively in microfluidics and nanofluidics (see Part B and Part C).

Table 1.1 Comparison of Flows in Macrochannels vs. Microchannels

Condition/Phenomenon	Flow in Macroconduits	Flow in Microconduits
Continuum Mechanics Hypothesis	Any fluid is a continuum	Continuum assumption holds for most liquid flows when $D_h \geq 10\ \mu m$ and for gases when $D_h \geq 100\ \mu m$
Type of Fluid (i.e., liquid versus gas)	• Special considerations for compressible and/or rarefied gases • Differentiate between Newtonian and non-Newtonian liquids	May have to treat gas flow and liquid flow differently because of the impact of a given fluid's molecular structure and behavioral characteristics

Table 1.1   *Continued*

Condition/Phenomenon	Flow in Macroconduits	Flow in Microconduits
No-Slip Condition	Can generally be assumed	Liquid-solid slip may occur on hydrophobic surfaces. Velocity slip and temperature jump may occur with rarefied gases
Entrance Effects	Entrance length is usually negligible when compared to the length of the conduit	Entrance length may be on the order of a microchannel length
Reynolds Number	Important to evaluate laminar vs. transitional vs. turbulent flows	Typically $\mathrm{Re}_{D_h} \leq \mathcal{O}(1)$ justifies Stokes flow and allows for nonmechanical pumps driving fluid flow
Turbulence	Transition varies with geometry of domain, but often requires larger Re numbers than in microchannel flow. Example, $\mathrm{Re}_D\big|_{\mathrm{pipe}} \approx 2000$	Transition to turbulence may occur earlier, e.g., at $\mathrm{Re}_{D_h} \geq 1200$
Surface Roughness	Is often negligible or included in the friction factor (see Moody chart in App. B)	May need to be considered due to manufacturing limitations at this small scale; roughness may be comparable to dimensions of the system and hence causes complex flow fields near walls
Viscous Heating	Is often small/negligible	May become a major player due to high velocity gradients in tiny channels with viscous fluids
Wall Temperature Condition	Usually thermodynamic equilibrium is assumed	For rarefied gas flows, there may be a temperature jump between the solid wall and the gas
Diffusion	Present, but often very slow; therefore, often negligible	Due to the small size of channels, diffusion is important and can be used for mixing
Surface Tension	Is often negligible	May become a major contributing force, and hence is being used for small fluid volume transfer
Electrohydrodynamic effects, such as electroosmosis (EO)	Negligible	In a liquid electrolyte an electric double layer (EDL) can be formed, which is set into motion via an applied electric field $(\vec{E})$: $E_x \Rightarrow \mathrm{EDL} \rightarrow f_{\mathrm{coulomb}} \rightarrow u_{\mathrm{EO}} \approx u_{\mathrm{slip}}$

Fluidics, as treated in this book, is part of Newtonian mechanics, i.e., dealing with deterministic, or statistically averaged, processes (see Branch A in Figure 1.1).

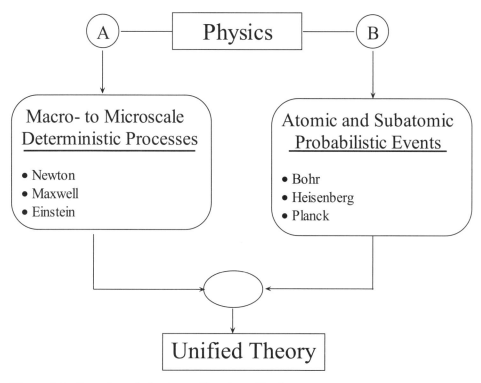

Figure 1.1   Branches of physics waiting for unification

For fluid flow in *nanoscale systems* the continuum mechanics assumption is typically invalid because the length scales of fluid molecules are on the order of nanochannel widths or heights. For example, the intermolecular distance for water molecules is 0.3-0.4 nm while for air molecules it is 3.3 nm, with a mean-free path of about 60 nm. Hence, for *rarefied gases*, not being in thermodynamic equilibrium, the motion and collision of packages of molecules have to be statistically simulated or measured. For *liquids* in nanochannels, molecular dynamics simulation, i.e., the solution of Newton's second law of motion for representative molecules, is necessary.

1.2 Definitions and Concepts

As indicated in Sect. 1.1, a solid knowledge base and good problem-solving skills in macroscale fluid dynamics, i.e., fluid flow plus heat and mass transfer, are important to model most transport phenomena in microfluidics and some in nanofluidics. So, we start out with a review of essential definitions and then revisit basic engineering concepts in macrofluidics. *The overriding goals are to understand the fundamentals and to be able to solve problems independently.*

1.2.1 Definitions

Elemental to transport phenomena is the description of fluid flow, i.e., the equation of motion, which is also called the momentum transfer equation. It is an application of Newton's second law, $\sum \vec{F}_{\text{ext.}} = m\vec{a}$, which Newton postulated for the motion of a particle. For most realistic engineering applications the equation of motion is three-dimensional (3-D) and *nonlinear*, the latter because of fluid inertia terms such as $u\,du/dx$, etc. (see App. A.5). However, it is typically independent of the *scalar* heat transfer and species mass equations, i.e., fluid properties are not measurably affected by changes in fluid temperature and species concentration, the latter in case of mixture flows. In summary, the major emphasis in Chapters 1 and 2 are on the description, solution, and understanding of the physics of fluid flow in conduits.

Here is a compilation of a few definitions:

- A *fluid* is an assemblage of gas or liquid molecules which deforms continuously, i.e., it *flows* under the application of a shear stress. Note: Solids do not behave like that; but what about borderline cases, i.e., the behavior of materials such as jelly, grain, sand, etc.?
- Key *fluid properties* are density ρ, dynamic viscosity μ, thermal conductivity k, species diffusivity $\mathcal{D}$, as well as heat capacities c_p and c_v. In general, all six are usually temperature dependent. Very important is the viscosity (see also kinematic viscosity $\nu \equiv \mu/\rho$) representing frictional (or drag) effects. Certain fluids, such as polymeric liquids, blood, food stuff, etc., are also shear rate dependent and hence called *non-Newtonian fluids* (see Sect. 2.3.4).
- *Flows, i.e., fluid in motion powered by a force or gradient,* can be categorized into:

Internal Flows	and	External Flows
- oil, air, water or steam in pipes and inside devices		- air past vehicles, buildings, and planes
- blood in arteries/veins or air in lungs		- water past pillars, submarines, etc.
- water in rivers or canals		- polymer coating on solid surfaces
- gas in pipelines		

- *Driving forces for fluid flow* include gravity, pressure differentials or gradients, temperature gradients, surface tension, electroosmotic or electromagnetic forces, etc.
- *Forces* appear either as body forces (e.g., gravity) or as surface forces (e.g., pressure). When acting on a fluid element they can be split into *normal* and *tangential forces* leading to pressure and normal/shear stresses. For example, on any surface element:

$$p \text{ or } \tau_{\text{normal}} = \frac{F_{\text{normal}}}{A_{\text{surface}}} \tag{1.3}$$

while

$$\tau_{\text{shear}} = \frac{F_{\text{tangential}}}{A_{\text{surface}}} \tag{1.4}$$

Recall: As Stokes postulated, the total stress depends on the spatial derivative of the velocity vector, i.e., $\vec{\vec{\tau}} \sim \nabla \vec{v}$ (see App. A.2). For example, shear stress τ_{shear} occurs due to relative frictional motion of fluid elements (or viscous layers). In contrast, the total pressure sums up three pressure forms, where the mechanical (or thermodynamic) pressure is experienced when moving with the fluid (and therefore labeled "static" pressure and measured with a piezometer). The dynamic pressure is due to the kinetic energy of fluid motion (i.e., $\rho \vec{v}^2 / 2$), and the hydrostatic pressure is due to gravity (i.e., $\rho g z$):

$$p_{\text{total}} = p_{\text{static}} + p_{\text{dynamic}} + p_{\text{hydrostatic}}$$
$$= p_{\text{static}} + \frac{\rho}{2} \vec{v}^2 + \rho g z = \cancel{c} \tag{1.5a,b}$$

where

$$p_{\text{thermo (or static)}} + p_{\text{dynamic}} = p_{\text{stagnation}} = p(\vec{v} = 0) \tag{1.6a,b}$$

From the fluid statics equation for a *stagnant* fluid body (or reservoir), where h is the depth coordinate, we obtain:

$$p_{\text{hydrostatic}} = p_0 + \rho g h \tag{1.7}$$

Clearly, the hydrostatic pressure due to the fluid weight appears in the momentum equation as a body force per unit volume, i.e., $\rho\vec{g}$. On the *microscopic level,* fluid molecules are randomly moving in all directions. In the presence of a wall, collisions, i.e., impulse $m\vec{v}$ per time, cause a fluctuating force on the wall. This resulting push statistically averaged over time and divided by the impact area is the pressure.

In general:

- Any fluid flow is described by its *velocity* and *pressure* fields. The velocity vector of a fluid element can be written in terms of its three scalar components:

$$\vec{v} = u\,\hat{i} + v\,\hat{j} + w\,\hat{k} \qquad \text{<rectangular coordinates>} \qquad (1.8a)$$

or

$$\vec{v} = v_r\hat{e}_r + v_\theta\hat{e}_\theta + v_z\hat{e}_z \qquad \text{<cylindrical coordinates>} \qquad (1.8b)$$

or

$$\vec{v} = v_r\hat{e}_r + v_\theta\hat{e}_\theta + v_\varphi\hat{e}_\varphi \qquad \text{<spherical coordinates>} \qquad (1.8c)$$

Its *total time derivative* is the fluid element acceleration (see Example 1.1 or Sect. A.1):

$$\left.\frac{d\vec{v}}{dt}\right|_{\substack{\text{solid}\\\text{particle}}} \hat{=} \left.\frac{D\vec{v}}{Dt}\right|_{\substack{\text{fluid}\\\text{element}}} = \vec{a}_{\text{total}} = \vec{a}_{\text{local}} + \vec{a}_{\text{convective}} = \frac{\partial\vec{v}}{\partial t} + (\vec{v}\cdot\nabla)\vec{v} \qquad (1.9)$$

where Eq. (1.9) is also known as the Stokes, material, or substantial time derivative.

- *Streamlines* for the visualization of flow fields are lines to which the local velocity vectors are tangential. In steady laminar flow streamlines and fluid-particle pathlines are identical. For example, for steady 2-D flow (see Sect. 1.4):

$$\frac{dy}{dx} = \frac{v}{u} \qquad (1.10)$$

where the 2-D velocity components $\vec{v} = (u, v, 0)$ have to be given to obtain, after integration, the streamline equation $y(x)$.

- *Dimensionless groups,* i.e., ratios of forces, fluxes, processes, or system parameters, indicate the importance of specific transport phenomena. For example, the Reynolds number is defined as (see Example 1.2):

$$\mathrm{Re}_L \equiv \frac{F_{inertia}}{F_{viscous}} := \frac{vL}{v} \tag{1.11}$$

where v is an average system velocity, L is a representative system "length" scale (e.g., the tube diameter D), and $v \equiv \mu/\rho$ is the kinematic viscosity of the fluid.

Other dimensionless groups with applications in engineering include the Womersley number and Strouhal number (both dealing with oscillatory/transient flows), Euler number (pressure difference), Weber number (surface tension), Stokes number (particle dynamics), Schmidt number (diffusive mass transfer), Sherwood number (convective mass transfer) and Nusselt number, the ratio of heat conduction to heat convection (see Sect. A.3). The most common source (i.e., derivation) of these numbers is the nondimensionalization of partial differential equations describing the transport phenomena at hand, or alternatively via scale analysis (see Example 1.2).

===

Example 1.1: Derive the material (or Stokes) derivative, D/Dt operating on the velocity vector, describing the "total time rate of change" of a fluid flow field.

Hint: For illustration purposes, use an arbitrary velocity field, $\vec{v} = \vec{v}(x,y,z;t)$, and form its total differential.
Recall: The total differential of any continuous and differentiable function, such as $\vec{v} = \vec{v}(x,y,z;t)$, can be expressed in terms of its infinitesimal contributions in terms of changes of the independent variables:

$$d\vec{v} = \frac{\partial \vec{v}}{\partial x} dx + \frac{\partial \vec{v}}{\partial y} dy + \frac{\partial \vec{v}}{\partial z} dz + \frac{\partial \vec{v}}{\partial t} dt$$

Solution:

Dividing through by dt and recognizing that $dx/dt = u$, $dy/dt = v$, and $dz/dt = w$ are the local velocity components, we have:

$$\frac{d\vec{v}}{dt} = \frac{\partial \vec{v}}{\partial x} u + \frac{\partial \vec{v}}{\partial y} v + \frac{\partial \vec{v}}{\partial z} w + \frac{\partial \vec{v}}{\partial t}$$

Substituting the "particle dynamics" differential with the "fluid element" differential yields:

$$\frac{d\vec{v}}{dt} \triangleq \frac{D\vec{v}}{Dt} = \frac{\partial \vec{v}}{\partial t} + u\frac{\partial \vec{v}}{\partial x} + v\frac{\partial \vec{v}}{\partial y} + w\frac{\partial \vec{v}}{\partial z} \equiv \frac{\partial \vec{v}}{\partial t} + (\vec{v}\cdot\nabla)\vec{v} = \vec{a}_{local} + \vec{a}_{conv.}$$

Example 1.2: Generation of Dimensionless Groups

(a) Scale Analysis

As outlined in Sect. 1.3, the Navier-Stokes equation (see Eq. (1.63)) describes fluid element acceleration due to several forces per unit mass, i.e.,

$$\vec{a}_{total} \equiv \underbrace{\frac{\partial \vec{v}}{\partial t}}_{\substack{\text{transient} \\ \text{term}}} + \underbrace{(\vec{v}\cdot\nabla)\vec{v}}_{\substack{\text{inertia force} \\ \text{term}}} = \underbrace{-\frac{1}{\rho}\nabla p}_{\substack{\sim\text{pressure} \\ \text{force}}} + \underbrace{\nu\nabla^2\vec{v}}_{\substack{\sim\text{viscous} \\ \text{force}}} + \underbrace{\vec{g}}_{\sim\text{gravity}}$$

Now, by definition:

$$Re \equiv \frac{\text{inertia force}}{\text{viscous force}} := \frac{(\vec{v}\cdot\nabla)\vec{v}}{\nu\nabla^2\vec{v}}$$

Employing the scales $\vec{v} \sim v$ and $\nabla = (\partial/\partial x,\ \partial/\partial y,\ \partial/\partial z) \sim 1/L$ where v may be an average velocity and L a system characteristic dimension, we obtain:

$$Re = \frac{\left(v\cdot\frac{1}{L}\right)v}{\nu L^{-2}v} = \frac{vL}{\nu}$$

Similarly, taking

$$\frac{\text{Local acceleration}}{\text{Convective acceleration}} \equiv \frac{\text{transient term}}{\text{inertia term}} = \frac{\partial\vec{v}/\partial t}{(\vec{v}\cdot\nabla)\vec{v}}$$

we can write with system time scale T (e.g., cardiac cycle: $T = 1$ s)

$$\frac{v/T}{vL^{-1}v} = \frac{L}{vT} = \text{Str}$$

which is the *Strouhal number*. For example, when $T \gg 1$, $\text{Str} \to 0$ and hence the process, or transport phenomenon, is quasi-steady.

(b) Nondimensionalization of Governing Equations
Taking the transient boundary-layer equations (see Sect. 1.3, Eq. (1.63)) as an example,

$$\rho \left(\frac{\partial u}{\partial t} + u \frac{\partial u}{\partial x} + v \frac{\partial u}{\partial y} \right) = -\frac{\partial p}{\partial x} + \mu \frac{\partial^2 u}{\partial y^2}$$

we nondimensionalize each variable with suitable, constant reference quantities. Specifically, approach velocity U_0, plate length ℓ, system time T, and atmospheric pressure p_0 are such quantities. Then,

$$\hat{u} = u/U_0, \ \hat{v} = v/U_0; \ \hat{x} = x/\ell, \ \hat{y} = y/\ell; \ \hat{p} = p/p_0 \ \text{and} \ \hat{t} = t/T$$

Note: Commonly, $\hat{y}$ is defined as $\hat{y} = y/\delta(x)$, where $\delta(x)$ is the varying boundary-layer thickness.
    Inserting all variables, i.e., $u = \hat{u}U_0$, $t = \hat{t}T$, etc., into the governing equation yields

$$\frac{\rho U_0}{T} \frac{\partial \hat{u}}{\partial \hat{t}} + \left[\frac{\rho U_0^2}{\ell} \right] \left(\hat{u} \frac{\partial \hat{u}}{\partial \hat{x}} + \hat{v} \frac{\partial \hat{u}}{\partial \hat{y}} \right) = -\left[\frac{p_0}{\ell} \right] \frac{\partial \hat{p}}{\partial \hat{x}} + \left[\frac{\mu U_0}{\ell^2} \right] \frac{\partial^2 \hat{u}}{\partial y^2}$$

Dividing the entire equation by, say, $\left[\rho U_0^2 / \ell \right]$ generates:

$$\underbrace{\left[\frac{\ell}{T u_0} \right]}_{\text{Strouhal \#}} \frac{\partial \hat{u}}{\partial \hat{t}} + \hat{u} \frac{\partial \hat{u}}{\partial \hat{x}} + \hat{v} \frac{\partial \hat{u}}{\partial \hat{y}} = -\underbrace{\left[\frac{p_0}{\rho U_0^2} \right]}_{\text{Euler \#}} \frac{\partial \hat{p}}{\partial \hat{x}} + \underbrace{\left[\frac{\mu}{\rho U_0 \ell} \right]}_{\substack{\text{inverse} \\ \text{Reynolds \#}}} \frac{\partial^2 \hat{u}}{\partial \hat{y}^2}$$

Comments:

    In a way three goals have been achieved:
- the governing equation is now dimensionless;
- the variables vary only between 0 and 1; and
- the overall fluid flow behavior can be assessed by the magnitude of three groups, i.e., Str, Eu, and Re numbers.

1.2.2 Flow Field Description

Any flow field can be described at either the microscopic or the macroscopic level. The *microscopic* or molecular models consider the position, velocity, and state of every molecule, or representative packages of molecules, of a fluid at all times. In contrast, averaging discrete-particle information (i.e., position, velocity, and state) over a local fluid volume yields *macroscopic* quantities, e.g., the velocity field $\vec{v}(\vec{x},t)$ in time and space of the flow field. The advantages of the molecular approach include general applicability, i.e., no need for submodels (e.g., for the stress tensor, heat flux, turbulence, wall conditions, etc.) and an absence of numerical instabilities (e.g., due to steep flow field gradients). However, considering myriads of molecules, atoms, or nanoparticles requires enormous computer resources, and hence only simple channel or stratified flows with a finite number of interacting molecules (assumed to be solid spheres) can be presently analyzed.

For example, in a 1-mm cube there are about 34 billion water molecules (about a million air molecules at STP); these high numbers make molecular dynamics simulation prohibitive but, on the other hand, intuitively validate the continuum assumption if the flow domain is sufficiently large.

Continuum Mechanics Assumption. As alluded to in Sect. 1.2.1 (see Table 1.1), fundamental to the description of all transport phenomena are the conservation laws, concerning mass, momentum, and energy, as applied to continua. In general, *solid structures and fluid flow fields are continua* as long as the local material properties can be defined as averages computed over material elements/volumes sufficiently large when compared to microscopic length scales of the solid or fluid but small relative to the (macroscopic) structure. Variations in solid-structure or fluid flow quantities can be obtained via solutions of differential equations describing the interactions between forces (or gradients) and motion. Specifically, the continuum mechanics method is an effective tool to physically explain and mathematically describe various transport phenomena without detailed knowledge of their internal molecular structures. In summary, *continuum mechanics* deals with three aspects:

- *Kinematics,* i.e., fluid element motion regardless of the cause
- *Dynamics,* i.e., the origin and impact of forces and fluxes generating fluid motion and waste heat, e.g., the stress tensor or heat flux vector, leading to entropy increase
- *Balance principles,* i.e., the mass, momentum, and energy conservation laws

Also, usually all flow properties are in *local thermodynamic equilibrium*, implying that the macroscopic quantities of the flow field can adjust swiftly to their surroundings. This local adjustment to varying conditions is rapidly achieved if the fluid has very small characteristic length and time scales of molecular collisions, when compared to the macroscopic flow variations. However, as the channel (or tube) size, typically indicated by the hydraulic diameter D_h, is reduced to the *microscale*, the surface-area-to-volume ratio becomes larger because $A/V \sim D_h^{-1}$ and wall surface effects may become important, as mentioned in Sect. 1.1.

Flow Dynamics and Fluid Kinematics.   Here, the overall goal is to find and analyze the interactions between *fluid forces*, e.g., pressure, gravity/buoyancy, drag/friction, inertia, etc., and *fluid motion*, i.e., the velocity vector field and pressure distribution from which everything else can be directly obtained or derived (see Figure 1.2). In turn, scalar transport equations, i.e., convection mass and heat transfer, can be solved based on the velocity field to obtain critical magnitudes and gradients (or fluxes) of species concentrations and temperatures, respectively.

(a) Cause-and-effect dynamics:

(b) Kinematics of a 2-D fluid element (Lagrangian frame):

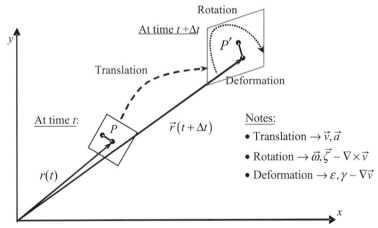

Figure 1.2   Dynamics and kinematics of fluid flow: (a) force-motion interactions; and (b) 2-D fluid kinematics

In summary, *unbalanced surface/body forces and gradients cause motion in the form of fluid translation, rotation, and/or deformation, while temperature or concentration gradients cause mainly heat or species mass transfer.* Note that flow visualization CDs

plus web-based university sources provide fascinating videos of complex fluid flow, temperature, and species concentration fields. Please check out the following links:
http://en.wikipedia.org/wiki/Flow_visualization
http://citeseerx.ist.psu.edu/viewdoc/download?doi=10.1.1.100.6782&rep=rep1&type=pdf

Balance Principles. The fundamental laws of mass, energy, and momentum conservation are typically derived in the form of mass, energy, and force balances for a differential fluid volume (i.e., a representative elementary volume) generating partial differential equations (PDEs). When a much larger *open system* is considered for mass, momentum, and/or energy balances, integral expressions result. Such balances in integral form, known as the Reynolds Transport Theorem (RTT), can be readily transformed into PDEs by employing the Divergence Theorem (see Sect. 1.3 and App. A).

Within the continuum mechanics framework, two basic flow field descriptions are of interest, i.e., the *Lagrangian viewpoint and the Eulerian (or control volume) approach* (see Figure 1.3, where C.∀. ≙ control volume and C.S. ≙ control surface).

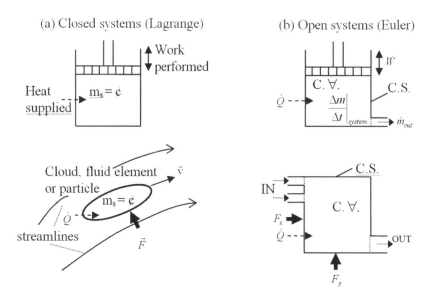

Figure 1.3   Closed versus open systems

For the *Lagrangian description* (see Figure 1.3a) one could consider first just a particle moving on a pathline with respect to a fixed Cartesian coordinate system. Initially, the particle position is at $\vec{r}_0 = \vec{r}_0(\vec{x}_0, t_0)$ and a moment later at $\vec{r} = \vec{r}(\vec{r}_0, t_0 + \Delta t)$ where based on vector algebra $\vec{r} = \vec{r}_0 + \Delta\vec{r}$. Following the particle's motion for $t > t_0$, the position vector is in general:

$$\vec{r} = \vec{r}\left(\vec{r}_0, t\right) \tag{1.12}$$

In the limit the time rate of change in location is the particle (or fluid element) velocity, i.e.,

$$\frac{d\vec{r}}{dt} = \vec{v} \tag{1.13}$$

and after a second time derivation:

$$\frac{d^2\vec{r}}{dt^2} = \frac{d\vec{v}}{dt} = \vec{a} \tag{1.14}$$

Now, the *"material point concept"* is extended to a *material volume with constant identifiable mass, forming a "closed system"* that moves and deforms with the flow but no mass crosses the material volume surface, because it is closed (see Figure 1.3a, second example). Again, the system is tracked through space, and as time expires it is of interest to know what the changes in system mass, momentum, and energy are. This can be expressed in terms of the system's extensive property B_{system} which could be mass m, momentum $m\vec{v}$, or total energy E. Thus, the key question is: "How can we express the fate of the B_{system}" or, in mathematical shorthand, what is "DB_s/Dt"? Clearly, the material time (or Stokes) derivative $D/Dt \equiv \partial/\partial t + \vec{v} \cdot \nabla$ (see Example 1.1) follows the closed system and records the total time rate of change of whatever is being tracked.

In order to elaborate on the material derivative (see Example 1.1) as employed in the Lagrangian description, a brief illustration of the various *time derivatives* is in order, i.e., $\partial/\partial t$ (local), d/dt (total of a material point or solid particle), and D/Dt (total of a fluid element). Their differences can be illustrated using acceleration (see Example 1.1 and App. A):

- $a_{x,\text{local}} = \partial u/\partial t$, where u is the fluid element velocity in the x-direction,
- $\vec{a}_{\text{particle}} = d\vec{v}/dt$ is employed in solid particle dynamics,

whereas

- $\vec{a}_{\substack{\text{fluid} \\ \text{element}}} = \dfrac{D\vec{v}}{Dt} = \underbrace{\dfrac{\partial \vec{v}}{\partial t}}_{\vec{a}_{\text{local}}} + \underbrace{\left(\vec{v} \cdot \nabla\right)\vec{v}}_{\vec{a}_{\text{convective}}}$ is the total fluid element acceleration.

In the *Eulerian frame,* an "open system" is considered where mass, momentum, and energy may readily cross boundaries, i.e., being convected across the control volume surface and local fluid flow changes may occur within the control volume over time (see Figure 1.3b). The fixed or moving control volume may be a large system/device with inlet and outlet ports, it may be small finite volumes generated by a computational mesh, or it may be in the limit a "point" in the flow field. In general, the Eulerian observer fixed to an inertial reference frame records temporal and spatial changes of the flow field at all "points" or, in case of a control volume, transient mass, momentum, and/or energy changes inside and fluxes across its control surfaces.

In contrast, the Lagrangian observer stays with each fluid element or material volume and records its basic changes while moving through space. Section 1.3 employs both viewpoints to describe mass, momentum, and heat transfer in integral form, known as the Reynolds Transport Theorem (RTT). *Thus, the RTT simply links the conservation laws from the Lagrangian to the Eulerian frame.* In turn, a surface-to-volume integral transformation then yields the conservation laws in differential form (i.e., PDEs) in the Eulerian framework, also known as the *control volume approach*.

1.2.3 Flow Field Categorization

Exact *flow problem identification*, especially in industrial settings, is one of the more important and sometimes the most difficult first task. After obtaining some basic information and reliable data, it helps to think and speculate about the physics of the fluid flow, asking:

(i)   What category does the given flow system fall into, and how does it respond to normal as well as extreme changes in operating conditions? Figure 1.4 may be useful for categorization of real fluids and types of flows.

(ii) What variables and system parameters play an important role in the observed transport phenomena, i.e., linear or angular momentum transfer, fluid mass or species mass transfer, and heat transfer?

(iii) What are the key dimensionless groups and what are their expected ranges?

Answers to these questions assist in grouping the flow problem at hand. For example, with the exception of "superfluids," all others are viscous, some more (e.g., syrup) and some less (e.g., rarefied gases). However, with the advent of *Prandtl's boundary-layer concept* the flow field, say, around an airfoil has been traditionally divided into a very thin (growing) viscous layer and beyond that an unperturbed *inviscid region* (see Schlichting & Gersten, 2000). This paradigm helped to better understand actual fluid mechanics phenomena and to simplify velocity and pressure as well as drag and lift calculations. Specifically, at sufficiently high approach velocities a fluid layer adjacent to a submerged body experiences steep gradients due to the "no-slip" condition and hence constitutes a *viscous flow region*, while outside the boundary layer frictional effects are negligible (see Prandtl equations versus Euler equation in Sect. 1.3.3.3). Clearly, with the prevalence of powerful

CFD software and engineering workstations, such a fluid flow classification is becoming more and more superfluous for practical applications.

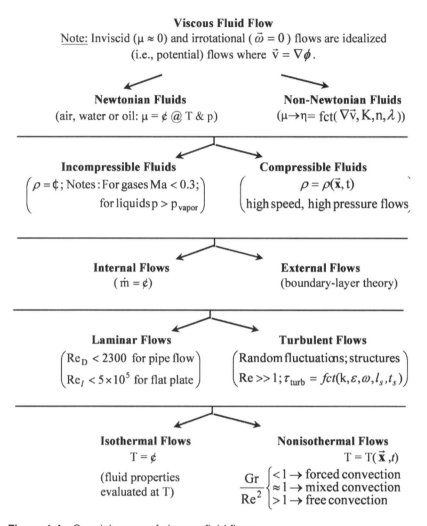

Figure 1.4 Special cases of viscous fluid flows

While, in addition to air, water and almost all oils are *Newtonian*, some synthetic motor oils are shear rate dependent and that holds as well for a variety of new (fluidic) products. This implies that modern engineers have to cope with the analysis and computer modeling of *non-Newtonian fluids* (see Sect. 2.3.4). For example, Latex paint is shear thinning, i.e., when painting a vertical door rapid brush strokes induce high shear rates ($\dot{\gamma} \sim dw/dz$) and the paint viscosity/resistance is very low.

When brushing stops, locally thicker paint layers (due to gravity) try to descend slowly; however, at low shear rates the paint viscosity is very high and hence "teardrop" formation is avoided and a near-perfect coating can dry on the vertical door.

All natural phenomena change with time and hence are *unsteady* (i.e*., transient*) while in industry it is mostly desirable that processes are steady, except during production line start-up, failure, or shut-down. For example, turbines, compressors, and heat exchangers operate continuously for long periods of time and hence are labeled "steady-flow devices"; in contrast, pacemakers, control systems, and drink dispensers work in a time-dependent fashion. In some cases, like a heart valve, devices change their orientation periodically and the associated flows oscillate about a mean value. In contrast, it should be noted that the term *uniform* implies "no change with system *location*," as in uniform (i.e., constant over a cross section) velocity or uniform particle distribution, which all could still vary with time.

Mathematical flow field descriptions become complicated when *laminar flow* turns unstable due to high speed and/or geometric irregularities ranging from surface roughness to complex conduits. The deterministic laminar flow turns *transitional* on its way to become *fully turbulent*, i.e., chaotic, transient 3-D with random velocity fluctuations, which help in mixing but also induce high apparent stresses. As an example of "flow transition," picture a group (on bikes or skis) going faster and faster down a mountain while the terrain gets rougher. The initially quite ordered group of riders/skiers may change swiftly into an unbalanced, chaotic group. So far no *universal model for turbulence*, let alone for the transitional regime from laminar to turbulent flow, has been found. Thus, major efforts focus on direct numerical simulation (DNS) of turbulent flows which are characterized by relatively high Reynolds numbers and chaotic, transient 3-D flow pattern.

Basic Flow Assumptions and Their Mathematical Statements.   Once a given fluid dynamics problem has been categorized (Figure 1.4), some justifiable assumptions have to be considered in order to simplify the equations describing the flow system's transport phenomena. The three most important ones are time dependence, dimensionality, and flow (or Reynolds number) regime. Especially, if justifiable, steady laminar 1-D (*parallel or unidirectional or fully developed*) flow simplifies a given problem analysis (see Sect. 1.3.3.3 with Table 1.2 as well as Table 2.1).

1.2.4 Thermodynamic Properties and Constitutive Equations

Thermodynamic Properties.   Examples of thermodynamic properties are mass and volume (extensive properties) as well as velocity, pressure, and temperature (intensive properties), all essential to characterize a general system, process, or device. In addition, there are *transport properties*, such as viscosity, diffusivity, and thermal conductivity, which are all temperature dependent and may greatly influence, or even largely determine, a fluid flow field. Any extensive, i.e., mass-dependent, property divided by a unit mass is called a *specific property*, such as the specific volume $v = \forall/m$ (where its inverse is the fluid density) or the specific energy $e = E/m$ (see Sect. 1.3). An *equation of state* is a correlation of independent intensive properties, where for a simple compressible system just two describe the state of such a system. A famous example is the ideal-gas relation, $p\forall = mRT$, where $m = \rho\forall$ and R is the gas constant.

At the *microscopic level* (based on kinetic theory), the fluid temperature is directly proportional to the kinetic energy of the fluid's molecular motion (Probstein, 1994). Specifically, $3kT = \langle m\vec{v} \cdot \vec{v} \rangle$, where k is the Boltzmann constant, m is the molecular mass, and $\vec{v}$ is the fluctuating velocity vector. The pressure, as indicated, is the result of molecular bombardment. The density depends macroscopically on both pressure and temperature and microscopically on the number of molecules per unit volume; for example, there are 2.69×10^{19} air molecules in 1 cm^3. Comparing the compressibility of liquids vs. gases, it takes $\Delta p_{water}/\Delta p_{air} = 2.15 \times 10^4$ to achieve the same fractional change in density.

Constitutive Equations.   Looking ahead (see Sect. 1.3), when considering the conservation laws for fluid flow and heat transfer, it is apparent that additional relationships must be found in order to solve for the variables: velocity vector $\vec{v}$, fluid pressure p, fluid temperature T, and species concentration c as well as stress tensor $\vec{\vec{\tau}}$, heat flux vector $\vec{q}$, and species flux $\vec{j}_c$. Thus, this is necessary for reasons of (i) mathematical closure, i.e., a number of unknowns require the same number of equations, and (ii) physical evidence, i.e., additional material properties other than the density ρ are important in the description of system/material/fluid behavior. These additional relations, or *constitutive equations*, are fluxes which relate via "material properties" to gradients of the principal unknowns. Specifically, for basic *linear* proportionalities we recall:

Stokes' postulate, i.e., the fluid stress tensor

$$\vec{\vec{\tau}} = \mu\left(\nabla\vec{v} + \nabla\vec{v}^T\right)\ \left[\text{N}/\text{m}^2\right] \tag{1.15}$$

where μ is the dynamic viscosity;
    Fourier's law, i.e., the heat conduction flux

$$\vec{q} = -k\nabla T\ \left[\text{J}/\text{m}^2\cdot\text{s}\right] \tag{1.16}$$

where k is the thermal conductivity; and
    Fick's law for the species mass flux

$$\vec{j}_c = -\mathcal{D}_{AB} \nabla c \; \left[\text{kg}/\text{m}^2 \cdot \text{s} \right] \tag{1.17}$$

where $\mathcal{D}_{AB}$ is the binary diffusion coefficient.

Of these three constitutive equations, the total stress tensor is in macrofluidics the most important and complex one (see Sect. 1.3.3.2 for more details). For, say, one-dimensional (1-D) cases to move fluid elements relative to each other, a shear force $F_{\text{tangential}} = \tau_{\text{shear}} A_{\text{interface}}$ is necessary. Thus, for simple shear flows (see Figure 1.5) $\tau_{yx} = \mu \, du/dy$. In general, the shear stress is proportional to $\nabla \vec{v}$ and the dynamic viscosity is just temperature dependent for Newtonian fluids (e.g., air, water, and oil) or shear rate dependent for polymeric liquids, paints, blood (at low shear rates), food stuff, etc.

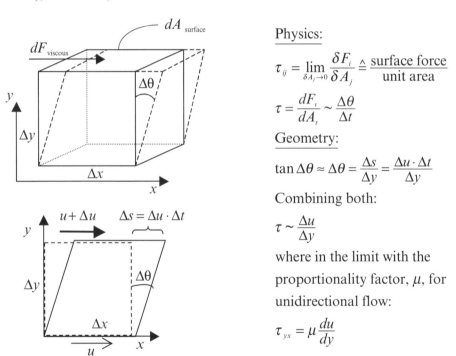

Physics:

$$\tau_{ij} = \lim_{\delta A_j \to 0} \frac{\delta F_i}{\delta A_j} \triangleq \frac{\text{surface force}}{\text{unit area}}$$

$$\tau = \frac{dF_v}{dA_s} \sim \frac{\Delta \theta}{\Delta t}$$

Geometry:

$$\tan \Delta \theta \approx \Delta \theta = \frac{\Delta s}{\Delta y} = \frac{\Delta u \cdot \Delta t}{\Delta y}$$

Combining both:

$$\tau \sim \frac{\Delta u}{\Delta y}$$

where in the limit with the proportionality factor, μ, for unidirectional flow:

$$\tau_{yx} = \mu \frac{du}{dy}$$

Figure 1.5 Illustration of the shear stress derivation for simple shear flow

1.3 Conservation Laws

After proper problem recognition and classification (see Sect. 1.2.3), central to engineering analysis are the tasks of realistic, accurate, and manageable modeling followed by analytical (or numerical) solution. While in most cases a given system's conservation laws are known, to solve the equations, often subject to complex boundary conditions and closure models, is quite a different story. So, after highlighting different approaches to derive conservation laws for mass, linear momentum, and energy, several examples in this section as well as in Chapter 2 illustrate basic phenomena and solution techniques.

1.3.1 Derivation Approaches

Derivation of the conservation laws describing all essential transport phenomena is very important because they provide a deeper understanding of the underlying physics and implied assumptions, i.e., the power and limitations of a particular mathematical model. Of course, derivations are regarded by most as boring and mathematically quite taxing; however, for those, it's time to become a convert for the two beneficial reasons stated. Furthermore, one should not forget the power of *dimensional analysis* (DA) which requires only simple algebra when nondimensionalizing governing equations and hence generating dimensionless groups. Alternatively, *scale analysis* (SA) is a nifty way of deriving dimensionless groups as demonstrated in this chapter (see Example 1.2). Both DA and SA are standard laboratory/computational tools for estimating dominant transport phenomena, graphing results, to evaluate engineering systems, and to test kinematic/dynamic similarities between a physical model and the actual prototype.

Outside the cutting-edge research environment, fluid mechanics problems are solved as special cases, i.e., the conservation equations are greatly reduced based on justifiable assumptions on a case-by-case basis (see Sect. 1.3.3.3). Clearly, the simplest case is *fluid statics* where the fluid mass forms a "whole body," either stationary or moving without any *relative* velocities (see Eq. (1.7)). The popular (because very simple) *Bernoulli equation,* for frictionless fluid flow along a representative streamline, balances kinetic energy ($\sim \rho \bar{v}^2$), flow work ($\sim \Delta p$), and potential energy ($\sim \rho g z$) and hence in some cases provides useful pressure-velocity-elevation correlations. The most frequently used equations in macrofluidics and microfluidics are the *Navier-Stokes (N-S) equations*, describing momentum and heat transfer for constant-property fluids, assuming that the flow is a continuum. After some basic fluid flow applications, relatively new material is introduced to broaden the student's knowledge base and provide a higher skill level to cope with today's engineering problems encountered in industry or graduate school.

There are basically four ways of obtaining specific equations expressing the conservation laws:

(i) *Molecular Approach:* Fluid properties and transport equations can be obtained from kinetic theory and the Boltzmann equation, respectively, employing statistical means. Alternatively, $\sum \vec{F} = m\vec{a}$ is solved for each molecule using direct numerical integration.

(ii) *Integral Approach:* Starting with the RTT for a fixed open control volume (Euler), specific transport equations in integral form can be obtained (see Sect. 1.3.2).

(iii) *Differential Approach:* Starting with 1-D balances over an REV (representative elementary volume) and then expanding them to 3-D, the mass, momentum, and energy transfer equations in differential form can be formulated. Alternatively, the RTT is transformed via the divergence theorem, where in the limit the field equations in differential form are obtained (see Sect. 1.3.3).

(iv) *Phenomenological Approach:* Starting with balance equations for an open system, transport phenomena in complex transitional, turbulent, or multiphase flows are derived largely based on empirical correlations and dimensional analysis considerations. A very practical example is the description of transport phenomena with fluid compartment models. These "compartments" are either well-mixed, i.e., *transient lumped-parameter models* without any spatial resolution, or transient with a one-dimensional resolution in the axial direction.

Especially for the (here preferred) differential approach (iii), the system-specific fluid flow assumptions have to be carefully stated and justified.

1.3.2 Reynolds Transport Theorem

Consider $\boldsymbol{B}_{\text{system}} \equiv \boldsymbol{B}$ to be an arbitrary extensive quantity of a *closed system,* say, a moving material volume. In general, such a system could be an ideal piston-cylinder device with enclosed (constant) gas mass, a rigid tank without any fluid leaks, or an identifiable pollutant cloud—all subject to forces and energy transfer (see Figure 1.3a). In any case, $\boldsymbol{B}$ represents the system's mass, momentum, or energy.

Task 1 is to express in the Lagrangian frame the fate of $\boldsymbol{B}$ in terms of the material derivative, $D\boldsymbol{B}/Dt$, i.e., the total time rate of change of $\boldsymbol{B}$ (see Sect. 1.2.2 reviewing the two system approaches and Example 1.1 discussing the operator D/Dt). Specifically, based on:

- Conservation of mass

$$\boldsymbol{B} \equiv m_{\text{system}} = \cancel{c} \rightarrow \frac{Dm}{Dt} = 0 \qquad (1.18a)$$

- Conservation of momentum (or Newton's second law)

$$\boldsymbol{B} \equiv \left(m\vec{v}\right)_{\text{system}} \rightarrow m\frac{D\vec{v}}{Dt} = m\vec{a}_{\text{total}} = \sum \vec{F}_{\text{external}} = \sum \vec{F}_{\text{surface}} + \sum \vec{F}_{\text{body}} \qquad (1.18b)$$

- Conservation of energy or first law of thermodynamics

$$\boldsymbol{B} \equiv E_{\text{system}} \rightarrow \frac{DE}{Dt} = \dot{Q} - \dot{W} \qquad (1.18c)$$

In *Task 2* the conservation laws, in terms of $D\boldsymbol{B}/Dt$, are related to an *open system*, i.e., in the Eulerian frame. Here, for a fixed control volume (C.∀.) with material streams flowing across the control surface (C.S.), and possibly accumulating inside C.∀., we observe with specific quantity $\beta \equiv \boldsymbol{B}/m$, or $\rho\beta = \boldsymbol{B}/\forall$:

$$\begin{bmatrix} Total \text{ time rate of} \\ \text{change of system} \\ \text{property } \boldsymbol{B} \end{bmatrix} = \begin{bmatrix} Local \text{ time rate of} \\ \text{change of } \boldsymbol{B}/\forall \equiv \rho\beta \\ \text{within the C.∀.} \end{bmatrix} + \begin{bmatrix} Net \text{ efflux of } (\rho\beta), \\ \text{i.e., net material} \\ \text{convection across} \\ \text{control surface C.S.} \end{bmatrix}$$

or in mathematical shorthand:

$$\left.\frac{D\boldsymbol{B}}{Dt}\right|_{\text{closed system}} = \frac{\partial}{\partial t} \iiint\limits_{\text{C.∀.}} (\rho\beta)d\forall + \iint\limits_{\text{C.S.}} (\rho\beta)\vec{v} \cdot d\vec{A} \qquad (1.19)$$

Equation (1.19), which is formally derived in any undergraduate fluids text, is the RTT for a fixed control volume. Clearly, the specific quantity β can be expressed as:

$$\beta \triangleq \frac{\boldsymbol{B}}{m} := \begin{cases} 1 & \text{mass per unit mass} \\ \vec{v} & \text{momentum per unit mass} \\ e & \text{energy per unit mass} \end{cases} \qquad (1.20a\text{-}c)$$

Extended Cases. For a *moving* control volume the fluid velocity $\vec{v}$ is replaced by $\vec{v}_{\text{relative}} = \vec{v}_{\text{fluid}} - \vec{v}_{\text{C.∀.}}$ (see Example 1.6). The operator $\partial/\partial t$, acting on the first term on the right-hand-side (RHS) of Eq. (1.19), has to be replaced by d/dt when the control volume is *deformable*, i.e., the C.S. moves with time (see Example 1.5).

For a *noninertial* coordinate system, for example, when tracking an accelerating system such as a rocket, $\sum \vec{F}_{external}$ of Eq. (1.18b) is expressed as:

$$m\vec{a}_{abs} = m\left(\frac{d\vec{v}}{dt} + \vec{a}_{rel}\right) \tag{1.21a}$$

where $m\vec{a}_{rel}$ accounts for noninertial effects (e.g., arbitrary C.∀. acceleration):

$$\vec{a}_{rel} = \frac{d^2\vec{R}}{dt^2} \tag{1.21b}$$

In case of C.∀.. rotation,

$$\vec{a}_{rel} = \frac{d\vec{w}}{dt} \times \vec{r} \tag{1.21c}$$

Specifically, for a *rotating* material volume, the fluid angular momentum per unit volume $\beta \equiv \rho(\vec{r} \times \vec{v})$ has to be considered. The law of *conservation of angular momentum* states that the rate of change of angular momentum of a material volume is equal to the resultant moment on the volume (see any undergraduate fluids text for more details).

Setting Up the Reynolds Transport Theorem.   There are a few sequential steps necessary for tailoring the general RTT toward a specific flow system description and solving the resulting integral equations:

(1) Identify the extensive quantity $\boldsymbol{B}_{system}$, e.g., the mass of the identifiable material, linear (or angular) momentum, or total energy. As a result, the specific property of the closed system $\beta = \boldsymbol{B}/m_{system}$ is known (see Eqs. (1.18 and 1.20)).
(2) Determine $DB/Dt\big|_{system}$ for each conservation case, i.e., mass, momentum, or energy (see Eq. (1.18)).
(3) Select a "smart" control volume and determine *if*:

- the control volume is fixed, or moving at $\vec{v}_{C.\forall.} = ¢$, or accelerating at $\vec{a}_{rel} = ¢$, or accelerating and rotating (see "Extended Cases" above);
- the flow problem is steady or transient, i.e., is $\partial/\partial t = 0$ or $\partial/\partial t \neq 0$ (note, $\partial/\partial t \iiint_{C.\forall.} (\rho\beta)d\forall \cong 0$ when the rate of change of $(\rho\beta)$ inside the C.∀. is negligible);
- a control surface moves, i.e., we have a deformable C.∀. where $\partial/\partial t \rightarrow d/dt$;
- the fluid properties are constant or variable;

- the inflow/outflow velocity fields are constant, i.e., uniform, or a function of inlet/exit space variables; and
- the resulting integral balance equations for mass and momentum (or energy) are decoupled or not.

(1) Set up the momentum, i.e., force balance, equation for each coordinate direction.
(2) Solve the volume and/or surface integrals (use integration tables, if necessary).
(3) Follow the inflow/outflow sign convention (see Figure 1.6), i.e., $IN \hat{=} " - "$ and $OUT \hat{=} " + "$
(4) Check the results for correctness, i.e., apply common sense.

Inflow: $\vec{v} \; d\vec{A} = -v_n dA$

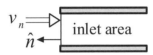

Outflow: $\vec{v} \; d\vec{A} = v_n dA$

Figure 1.6   Sign convention for the "net efflux" RTT term (Recall: $\cos 180° = -1$; and $\cos 0° = 1$)

1.3.2.1 Fluid Mass Conservation in Integral Form

Conservation of mass is very intuitive and standard in daily-life observations. A given mass of a fluid may change its thermodynamic state, i.e., liquid or gaseous, but it can neither be destroyed nor created. This, as the other two conservation laws, can be expressed in *integral form* for a control volume of any size and shape or derived in *differential form*.

In order to track within the Lagrangian frame an identifiable constant mass of fluid, we set (see Figure 1.3a and Eq. (1.20a))

$$\boldsymbol{B}_{\text{system}} \equiv m \text{ and hence } \beta \equiv 1$$

The conservation principle requires that, with $m = \text{¢}$, $Dm/Dt = 0$ and hence Eq. (1.19) reads:

$$0 = \frac{\partial}{\partial t} \iiint_{C.\forall.} \rho \, d\forall + \iint_{C.S.} \rho \vec{v} \cdot d\vec{A} \qquad (1.22)$$

Thus, we just completed Steps (i) and (ii) of the "**setting-up-the-RTT**" procedure. Aspects of Step (iii) are best illustrated with a couple of examples.

Example 1.3: Volumetric Flow Rate

Consider a liquid-filled tank (depth H) with a horizontal slot outlet (height $2h$ and width w) where the short-tube locally varying outlet velocity can be expressed as:

$$u \approx \sqrt{2g(H-z)}$$

A constant fluid mass flow rate, $\dot{m}_{in}$, is added to maintain the liquid depth H. The z-coordinate indicates the location of the center of the outlet. Find Q_{outlet} as a function of H and h.

Sketch:	Assumptions:	Concepts:
	• Steady incompressible flow • Outflow velocity as $u(z) = \sqrt{2g(H-z)}$ based on Torricelli's law	• Mass RTT • Fixed, nondeforming C.$\forall$., i.e., H=const. • Torricelli's law

Solution:

The given $u(z)$-equation is a special case of Bernoulli's equation. Specifically, with $\partial/\partial t \iiint \rho \, d\forall = 0$ (steady state because no system parameter changes with time), we have with $\rho = \phi$ (incompressible fluid):

$$0 = \iint_{C.S.} \vec{v} \cdot d\vec{A} \qquad (E1.3\text{-}1)$$

Fluid mass crosses the control surface at two locations (see graph). Recalling that "inflow" is negative and "outflow" positive (see Figure 1.6), Eq. (E1.3-1) reads, with $\dot{m}_{in}/\rho = Q_{in}$,

Graph:

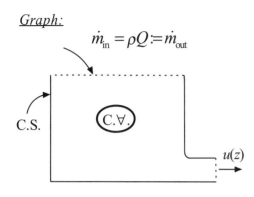

$$\dot{m}_{in} = \rho Q := \dot{m}_{out}$$

C.S.

C.∀.

$u(z)$

$$Q_{in} + \iint_{A_{slot}} \vec{v} \cdot d\vec{A} = 0$$

or

$$Q_{in} = \int_A v_n \, dA \qquad (E1.3\text{-}2)$$

Here $Q_{in} = Q_{outlet}$ because H = constant;

$v_n = u(z) = \sqrt{2g(H-z)}$ and $dA = w\,dz$ so that

$$Q_{outlet} = w\sqrt{2g} \int_{-h}^{h} \sqrt{H-z} \, dz$$

which yields

$$Q_{outlet} = K\left[(H+h)^{3/2} - (H-h)^{3/2}\right] \qquad (E1.3\text{-}3)$$

where $K \equiv 2/3\,w\sqrt{2g}$.

For $h \ll H$, as it is often the case, Eq. (E1.3-3) can be simplified to (see Sect. 1.4):

$$Q_{outlet} \approx 2\,wh\sqrt{2gH} \qquad (E1.3\text{-}4)$$

Comment: Equation (E1.3-4) is known as _Torricelli's law._

What can be deduced from Example 1.3 is that for _incompressible_ fluid flow through a conduit,

$$\iint_{C.S.} \vec{v} \cdot d\vec{A} = 0 \qquad (1.23)$$

or $-Q_{in} + \int_{A_{outlet}} \vec{v} \cdot d\vec{A} = 0$, i.e., in general:

$$\boxed{Q = \int_A \vec{v} \cdot d\vec{A} = \int_A v_n \, dA = \bar{v} A} \qquad (1.24)$$

where $\bar{v} = Q/A$ is the cross-sectionally averaged velocity. We also recall that the mass flow rate at any point in a conduit is:

$$\boxed{\dot{m} = \rho \bar{v} A = \rho Q = \text{constant}} \qquad (1.25)$$

which holds for any fluid and is a key "internal flow" condition, reflecting conservation of mass. Clearly, for a C.∀. with multiple inlets and outlets:

$$\sum Q_{\text{in}} = \sum Q_{\text{out}} \qquad (1.26)$$

as illustrated in Example 1.4.

Example 1.4: Multiport Flow Junction

Consider a feed pipe (A, v_1) bifurcating into two outlet pipes (A_2, v_2 and A_3, v_3) where a small hole (A_4) has been detected in the junction area. Develop an equation for the leak Q_4.

Sketch:	_Assumptions:_	_Approach:_
	• Steady incompressible flow • Fixed, nondeformable C.∀. • Constant velocities	• Reduced mass RTT, i.e., direct use of Eq. (1.26) • Sign convention (Figure 1.6b)

Solution:

The fact that the inlet/outlet velocities are all constant simplifies Eq. (1.23) to Eq. (1.26), i.e.,

$$-Q_1 - Q_2 + Q_3 + Q_4 = 0 \text{ where } Q_i = v_i A_i$$

Thus,

$$Q_{\text{leak}} \equiv Q_4 = v_1 A_1 + v_2 A_2 - v_3 A_3$$

The next example considers a *deforming* control volume inside a tank due to a single outflow in terms of a variable velocity. Thus, Eq. (1.22) has to be rewritten as:

$$0 = \frac{d}{dt} \iiint_{C.\forall.} \rho \, d\forall + \iint_{C.S.} \rho \vec{v} \cdot d\vec{A} \tag{1.27a}$$

For incompressible fluid flow, we have:

$$\left.\frac{d\forall}{dt}\right|_{C.\forall.} = -\iint_{C.S.} \vec{v} \cdot d\vec{A} \tag{1.27b}$$

Example 1.5: Draining of a Tank: A "deformable C.∀." because the fluid level decreases and hence we have a shrinking C.∀. with moving C.S.

Consider a relatively small tank of diameter D and initially filled to height h_0. The fluid drains through a pipe of radius (r_0) according to (see Example 2.1):

$$u(r) = 2\bar{u}\left[1 - \left(r/r_0\right)^2\right]$$

where $\bar{u} = \sqrt{2gh}$ (see Example 1.3), r_0 is the outlet pipe radius, and r is its variable radius, $0 \le r \le r_0$. The fluid depth was h_0 at time $t = 0$. Find $h(t)$ for a limited observation time Δt.

Sketch:	Assumptions:	Control Volume:
	• Transient incompressible flow • Stationary tank but "deforming" liquid volume C.∀.	

Solution:

Equation (1.27b) can be expanded for this problem with $\forall_{C.\forall.} = \left(D^2\pi/4\right)h$ to

$$\left(\frac{D^2\pi}{4}\right)\frac{dh}{dt} = -\int_0^{r_0} u(r)\, dA \tag{E1.5-1}$$

where $dA = 2\pi r dr$ *is the variable ring element* as part of the cross-sectional area of the outlet pipe. Thus,

$$\frac{dh}{dt} = -\frac{16\sqrt{2gh}}{D^2} \int_0^{r_0} \left[1 - \left(\frac{r}{r_0} \right)^2 \right] r dr \qquad \text{(E1.5-2a)}$$

or

$$\frac{dh}{dt} = -4\sqrt{2g} \left(\frac{r_0}{D} \right)^2 \sqrt{h} \qquad \text{(E1.5-2b)}$$

subject to $h(t=0) = h_0$. Separation of variables and integration yield:

$$\sqrt{h_0} - \sqrt{h} = \sqrt{8g} \left(\frac{r_0}{D} \right)^2 t \qquad \text{(E1.5-3a)}$$

or

$$\frac{h(t)}{h_0} = \left[1 - \sqrt{\frac{8g}{h_0}} \left(\frac{r_0}{D} \right)^2 t \right]^2 \qquad \text{(E1.5-3b)}$$

Graph:

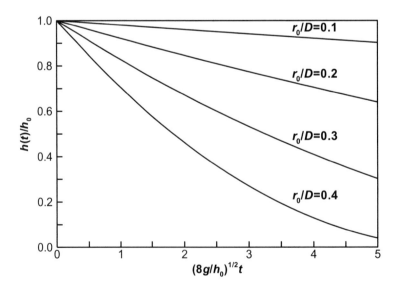

Comments:

- The standard assumption that $h = ¢$, i.e., being a reservoir, is only approximately true when $r_0/D < 0.1$.
- A variable speed dh/dt, i.e., accelerated tank draining, occurs when $r_0/D > 0.1$.

1.3.2.2 Momentum Conservation in Integral Form

Focusing on linear momentum transfer (in contrast to angular momentum transfer) the momentum conservation law is again derived via the integral.

Forces acting on an identifiable *fluid* element accelerate it, as it is well known from Newton's second law of motion $(m\vec{a})_{\text{fluid}} = \Sigma \vec{F}_{\text{external}}$. Specifically, we set (see Eq. 1.18b)):

$$B_{\text{system}} \equiv m\vec{v} \text{ and hence } \beta \equiv \vec{v}.$$

As previously indicated, with $m = ¢$, $m D\vec{v}/Dt = \Sigma\vec{F}_{\text{ext}}$ and hence Eq. (1.19) reads:

$$m\frac{D\vec{v}}{Dt} = m\vec{a}_{\text{total}} = \sum\vec{F}_{\text{body}} + \sum\vec{F}_{\text{surface}} = \frac{\partial}{\partial t}\int\limits_{\text{C.∀.}}\rho\vec{v}d\forall + \int\limits_{\text{C.S.}}\vec{v}\rho\vec{v}\cdot d\vec{A} \qquad (1.28a)$$

As discussed, for control volumes *accelerating* without rotation relative to inertial coordinates, an additional force $\vec{F}_{\text{inertia}} \sim \vec{a}_{\text{relative}}$ appears, where $\vec{a}_{\text{rel}}$ is the C.∀. acceleration relative to the fixed frame of reference X-Y-Z. Thus, with

$$\vec{v}_{XYZ} = \vec{v}_{xyz} + \vec{v}_{\text{relative}} \text{ or } \vec{a}_{XYZ} = \vec{a}_{\text{absolute}} = \frac{d\vec{v}}{dt} + \vec{a}_{\text{rel}}$$

we can express the sum of all forces as:

$$\sum\vec{F}_{\text{total}} = m\vec{a}_{\text{abs}} = m\left(\frac{D\vec{v}}{Dt} + \vec{a}_{\text{rel}}\right)$$

i.e., for an accelerating and deforming control volume:

$$m\frac{D\vec{v}}{Dt} = \Sigma\vec{F}_B + \Sigma\vec{F}_S - \int\limits_{\text{C.∀.}}\vec{a}_{\text{rel}}dm = \frac{d}{dt}\int\limits_{\text{C.∀.}}\rho\vec{v}d\forall + \int\limits_{\text{C.S.}}\vec{v}\rho\vec{v}_{\text{rel}}\cdot dA \qquad (1.28b)$$

Example 1.6: Force on a Disk Moving into an Axisymmetric Jet

A steady uniform round jet impinges upon an approaching conical disk as shown. Find the force exerted on the disk, where v_{jet}, A_{jet}, v_{disk}, diameter D, angle θ, and fluid layer thickness t are given.

Sketch:	*Control volume:*
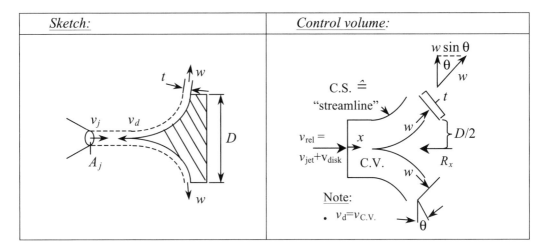	

Assumptions: as stated; constant averaged velocities and properties
Approach: RTT (mass balance and 1-D force balance)
Solution:

(a) Mass Conservation:

$$0 = \int_{C.S.} \rho \vec{v} \cdot d\vec{A} = -\int_{A_{jet}} \rho v_{rel} dA + \int_{A_{exit}} \rho w dA \qquad \text{(E1.6-1)}$$

where $v_{rel} = v_{jet} - \left(-v_{disk}\right) = v_j + v_d$ and $A_{exit} \approx \pi Dt$.

$$\therefore -\rho A_j \left(v_j + v_d\right) + \rho(\pi Dt)w = 0$$

Hence,

$$w = \frac{\left(v_j + v_d\right)A_j}{\pi Dt} \qquad \text{(E1.6-2)}$$

(b) Momentum Conservation:

$$B = \left(m\vec{v} \right)_s \, ; \, \beta = \vec{v}; \; DB/Dt := \vec{F}_{surf} = -R_x$$

$$\Sigma \vec{F}_{surf} = \int_{C.S.} \vec{v} \rho \vec{v}_{rel} \cdot d\vec{A} \xrightarrow{\text{x-component}} -R_x = \int_{C.S.} u \rho v_{rel} dA \qquad \text{(E1.6-3a.b)}$$

Thus,

$$-R_x = -v_{rel} \rho v_{rel} A_{jet} + w \, \sin \, \theta \rho w A_{exit}$$

where with Eq. (E1.6-2):

$$R_x = \rho A_j \left(v_j + v_d \right)^2 \left[1 - \frac{A_j \sin\theta}{\pi t D} \right] \qquad \text{(E1.6-4)}$$

Comment:

The resultant fluid structure force (E1.6-4) can be rewritten as:

$$R_x = \left(\dot{m} v_x \right)_{out} - \left(\dot{m} v_x \right)_{in} \qquad \text{(E1.6-5)}$$

which is the result of a change in fluid flow momentum. If the disk would move away with $v_d = v_j$, i.e., escaping the jet, $v_{rel} = 0$ and hence $R_x = 0$.

Expanding on Eq. (E1.6-5), we can generalize that the net momentum flux due to applied forces is:

$$\Sigma \vec{F} = \dot{m} \left(\alpha_2 \vec{v}_2 - \alpha_1 \vec{v}_1 \right) \qquad \text{(1.29)}$$

where the correction factor α accounts for the variation of v^2 across the inlet ① or outlet ② duct section. Specifically,

$$\rho \int_A v^2 dA = \alpha \dot{m} v_{average} = \alpha \rho A v_{av}^2$$

so that

$$\alpha \equiv \frac{1}{A} \int \left(\frac{v}{v_{av}}\right)^2 dA \tag{1.30}$$

Example 1.7: Force on a Submerged Body

Find the drag on a submerged elliptic rod of characteristic thickness h, based on a measured velocity profile downstream from the body, say,

$$u(y) = u_\infty / 4 [1 + y/h]; \ 0 \le y \le 3h$$

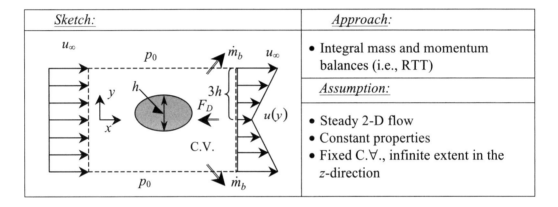

Sketch:	*Approach:*
	• Integral mass and momentum balances (i.e., RTT)
	Assumption:
	• Steady 2-D flow • Constant properties • Fixed C.∀., infinite extent in the z-direction

Solution:

(a) Mass Conservation:

$$\dot{m}_{in} = \dot{m}_{boundary} + \dot{m}_{out} \tag{E1.7-1}$$

$$0 = \int_{C.S.} \rho \vec{v} \cdot dA = \rho u_\infty 2(3h) + \dot{m}_{boundary} + 2\rho \frac{u_\infty}{4} \int_0^{3h} \left(1 + \frac{y}{h}\right) dy \tag{E1.7-2a}$$

$$\therefore \ \dot{m}_{boundary} = 6\rho h u_\infty - \frac{15}{4} \rho h u_\infty = \frac{9}{4} \rho h u_\infty \tag{E1.7-2b}$$

(b) Momentum Conservation (*x*-momentum):

$$-F_D = \int_{C.S.} v_x \rho \vec{v} \cdot d\vec{A} := \left(\dot{m}v_x\big|_{exit} + \dot{m}v_x\big|_b\right) - \dot{m}v_x\big|_{inlet} \tag{E1.7-3a}$$

$$-F_D = -u_\infty \rho u_\infty (6h) + u_\infty \left(\frac{9}{4}\rho h u_\infty\right) + 2\rho \int_0^{3h} \left[u(y)\right]^2 dy \qquad \text{(E1.7-3b)}$$

$$F_D = \frac{9}{8}\rho h u_\infty^2 \qquad \text{(E1.7-3c)}$$

Comment:

The fluid flow field inside the C.∀., especially behind the submerged body, is very complex. The RTT treats it as a "black box" and elegantly obtains $F_D = F_{fixation}$ via "velocity defect" measurements, indicated with the given velocity profile $u(y)$. Note that the given system is 2-D with $z \to \infty$; but, after integration the result, $F_x = -F_D$, is obtained in 1-D.

1.3.2.3 Conservation Laws of Energy and Species Mass

Although many natural and industrial fluid flow problems are nonisothermal, fluid mechanics education typically only deals with constant-temperature flows, leaving thermal flows for separate thermodynamics and convection heat transfer courses. Species mass transfer is almost entirely left for chemical and biomedical engineers. Being more comprehensive, we highlight the energy RTT plus the resulting heat transfer equation with an analogy to mass transfer, to lay the groundwork for some interesting engineering applications in remaining sections plus Chapters 3, 5 and 6.

Two approaches are considered which highlight *global energy balance* for closed and open systems, as well as convection heat transfer described by the *energy RTT*.

Global Energy Balance.   The first law of thermodynamics states that energy forms can be converted but the total energy is constant, i.e., conserved:

$$E_{total} = \underbrace{E_{kinetic} + E_{potential} + E_{internal}}_{=E_{system}} + \underbrace{E_{mass\ flow} + E_{\substack{work \\ performed}} + E_{\substack{heat \\ transfered}}}_{=E_{surrounding}} \ldots = \cent \qquad (1.31)$$

Taking the total time derivative in an "engineering approach" yields:

$$\frac{dE_{total}}{dt} \approx \frac{\Delta E_{total}}{\Delta t} = 0 = \left.\frac{\Delta E}{\Delta t}\right|_{system} + \sum \dot{E}_{net\ exchange} \qquad (1.32a)$$

Expressing the net energy rate efflux as $\sum \dot{E}_{out} - \sum \dot{E}_{in}$, we have:

$$\left.\frac{\Delta E}{\Delta t}\right|_{\text{system}} = \sum \dot{E}_{\text{in}} - \sum \dot{E}_{\text{out}} \qquad (1.32b)$$

For *observation time* Δt the *global energy balance* for any system can be written as:

$$\underbrace{\sum E_{\text{in}} - \sum E_{\text{out}}}_{\substack{\text{Net energy transfer by heat,} \\ \text{work, and/or mass flow in/out}}} = \Delta E_{\text{system}} := \underbrace{\left[\Delta kE + \Delta pE + \Delta U\right]_{\text{system}}}_{\substack{\text{Change of total energy inside the system} \\ \text{(e.g., kinetic, potential, and internal energies)}}} \qquad (1.33)$$

where the internal energy U is the sum of all microscopic energy forms, i.e., mainly due to molecular vibration.

For example, for a *closed system* with heat transferred to the system and work done by the system (see Figure 1.7):

$$\dot{Q} - \dot{W} = \frac{\Delta E}{\Delta t} \qquad (1.34)$$

or during Δt considering only an internal energy change from State 1 <initial> to State 2 <final>:

$$Q - W = \Delta E \approx \Delta U = U_2 - U_1 \qquad (1.35)$$

Heat can be transferred via conduction, convection, and/or thermal radiation, while work done by the closed system, such as a piston-cylinder device, is typically boundary or electric work:

(a) Closed System   **(b) Open System**

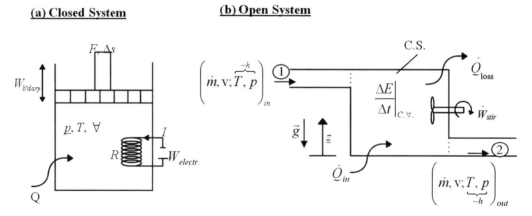

Figure 1.7   Energy transfer for closed and open systems

$$W_b = \int_1^2 F ds := \int_1^2 p d\forall \quad \text{while} \quad W_{el} = \int_1^2 VI dt \qquad (1.36\text{a,b})$$

For an *open system*, i.e., control volume, energy forms flowing in and out of the system have to be accounted for. Thus, Eq. (1.33) can be written for a fixed control volume with uniform streams entering and leaving (see Figure 1.7b):

$$\frac{\Delta E}{\Delta t}\Big|_{C.\forall.} = \Sigma\left[\dot{Q} + \dot{W}\right]_{in} - \Sigma\left[\dot{Q} + \dot{W}\right]_{out} + \Sigma\left[\dot{m}\left(h + \frac{v^2}{2} + gz\right)\right]_{in}$$
$$- \Sigma\left[\dot{m}\left(h + \frac{v^2}{2} + gz\right)\right]_{out} \qquad (1.37)$$

where for constant fluid properties $h \equiv \tilde{u} + pv$ is the enthalpy per unit mass; h combines internal energy and flow work due to pressure p moving specific volume v, i.e., pv; $e_{kin} = v^2/2$ is the specific kinetic energy, and $e_{pot} = gz$ is the potential energy per unit mass.

As an aside, for steady, single-inlet/outlet, frictionless flows without internal energy changes, heat transferred, and work done, the *Bernoulli equation* appears:

$$\left[\frac{p}{\rho} + \frac{v^2}{2} + gz\right]_{in} = \left[\frac{p}{\rho} + \frac{v^2}{2} + gz\right]_{out} \qquad (1.38)$$

Energy Conservation in Integral Form.   The global energy balance (see Eq. (1.33) or Eq. (1.37)) is a special case of the energy RTT (see Eq. (1.40)). Specifically, taking $B_{system} \equiv E_{total}$ and hence $\beta \equiv e_t$, the energy RTT reads [*Recall*: Eqs. (1.18) and (1.34)]:

$$\frac{DE_t}{Dt}\Big|_{\substack{close \\ system}} = \Sigma\dot{W} + \Sigma\dot{Q} = \frac{\partial}{\partial t}\int_{C.V.} \rho e_t d\forall + \int_{C.S.} \rho e_t \vec{v} \cdot d\vec{A} \qquad (1.39)$$

Typically, $E_{total} = E_{kinetic} + E_{internal} := (m/2)|\vec{v}|^2 + m\tilde{u}$, i.e., the left-hand-side (LHS) has to be separately expressed. The physical meaning of each term in Eq. (1.39) can be summarized as:

$$\begin{Bmatrix} \text{Time rate of change} \\ \text{of } E_{total} \text{ in material} \\ \text{volume moving with} \\ \text{the flow} \end{Bmatrix} = \begin{Bmatrix} \text{Rate of work } \dot{W} \\ \text{done on the C.}\forall.\text{ by} \\ \text{surface forces plus} \\ \text{body forces} \end{Bmatrix} + \begin{Bmatrix} \text{Net heat flux } \vec{q}_t \\ \text{across the C.S.} \\ \text{of the} \\ \text{control volume} \end{Bmatrix}$$

where the total heat flux $\vec{q}_t = \vec{q}_{cond.} + \vec{q}_{conv.} + \vec{q}_{rad.}$. In some cases, in order to complete the energy conservation law, a distributed internal heat source term, e.g., $\int_V \rho \hat{q}_{int} d\forall$, may have to be added. Finally, using Eq. (1.39), the energy RTT reads for a stationary control volume:

$$\int_{C.\forall.} \rho\left(\vec{v}\cdot\vec{f}_b\right)d\forall + \int_{C.S.} \left(\vec{v}\cdot\vec{\vec{\tau}}\right)dA + \int_{C.S.} \vec{q}_t\cdot d\vec{A} + \int_{C.\forall.} \rho\hat{q}_{int}d\forall$$
$$= \frac{\partial}{\partial t}\int_{C.\forall.} \rho e_t d\forall + \rho e_t \vec{v}\cdot d\vec{A} \tag{1.40}$$

Employing the *Divergence Theorem* (see App. A), the RTTs (i.e., Eqs. (1.22), (1.28), and (1.40)) can be transformed into PDEs as shown in Sects. 1.3.3.1 to 1.3.3.4. To connect Eqs. (1.37) and (1.40) is left as a homework assignment.

1.3.3 Conservation Equations in Differential Form

As discussed and illustrated in Sect. 1.3.2, the RTT provides integral system quantities such as flow rate, fluxes, and forces. In microfluidics and nanofluidics, high-resolution results in terms of *local* velocities, pressures, temperatures, concentrations as well as shear rates and fluxes are most desirable. Hence, starting with the RTT, fluid mass continuity, linear momentum, heat transfer, and species mass transfer equations are derived in *differential form*. Alternatively, the conservation laws are then also derived starting with mass, momentum, and energy balances for a fluid element.

1.3.3.1 Fluid Mass Conservation

Continuity Equation in Differential Form Based on the RTT. As discussed, the RTT is great for computing global (or integral) quantities, such as flow rates and mass fluxes or forces and energies without knowledge of the *detailed* fluid flow field *inside* the open system (i.e., the control volume). However, if it is necessary to find point-by-point density variations as well as velocity and pressure distributions in order to analyze fluid flow patterns, the conservation laws in *differential* rather than integral form have to be solved.

The conservation equations can be readily derived from the RTT, e.g., Eq. (1.19), by considering an infinitesimally small control volume $d\forall$ and then expressing each term in the form of a volume integral. For example, Eq. (1.22) contains a surface integral which has to be transformed into a volume integral, employing *Gauss' Divergence Theorem* (App. A):

$$\iint_S \vec{v} \cdot dS \equiv \iiint_\forall \nabla \cdot \vec{v} d\forall \tag{1.41}$$

where $\vec{v}$ is a vector in surface S, ∇ is the del operator (see App. A), and $(\nabla \cdot \vec{v})$ is the divergence of the vector field.

Using Eq. (1.41) to express the surface integral in Eq. (1.22) as a volume integral, Eq. (1.22) can be written as:

$$\frac{\partial}{\partial t}\iiint \rho d\forall + \iiint \nabla \cdot (\rho\vec{v}) d\forall = 0$$

or, following Leibniz's rule (App. A) and switching the operation for the first term, we have

$$\iiint \left[\frac{\partial \rho}{\partial t} + \nabla \cdot (\rho\vec{v}) \right] d\forall = 0$$

Clearly, either $d\forall = 0$ (not physical) or

$$\frac{\partial \rho}{\partial t} + \nabla \cdot (\rho\vec{v}) = 0 \tag{1.42}$$

Equation (1.42) is known as the *continuity equation*, stating fluid mass conservation on a differential basis. Note the special cases:

- For steady flow $\left(\frac{\partial}{\partial t} \equiv 0 \right)$: $\nabla \cdot (\rho\vec{v}) = 0$ (1.43)

- For incompressible fluids $(\rho = \cancel{c})$: $\nabla \cdot \vec{v} = 0$ (1.44)

It should be noted that the widely applicable Eq. (1.44) holds for transient flow as well.

Continuity Derived from a Mass Balance. In order to gain more physical insight, Eq. (1.42) is now derived based on a 3-D mass balance. A fluid mass balance over an open system, say, a cube of volume $\Delta\forall = \Delta x \Delta y \Delta z$, yields:

$$\sum \dot{m}_{in} - \sum \dot{m}_{out} = \left. \frac{\partial m}{\partial t} \right|_{\Delta\forall} \tag{1.45}$$

from a global perspective. Using Eq. (1.45) in 1-D on a differential basis (see Figure 1.8):

$$\left[(\rho u)\big|_x - (\rho u)\big|_{x+\Delta x} \right] \Delta y \Delta z = \frac{\partial \rho}{\partial t} \Delta \forall \tag{1.46}$$

Now, Taylor's truncated series expansion (see App. A) states:

$$f\big|_{x+\Delta x} = f\big|_x + \frac{\partial f}{\partial x} \Delta x + \cdots \tag{1.47}$$

so that

$$-\frac{\partial (\rho u)}{\partial x} \Delta x \Delta y \Delta z = \frac{\partial \rho}{\partial t} \Delta x \Delta y \Delta z \tag{1.48}$$

Adding the other two net fluxes ρv and ρw in the y- and z-direction, respectively, and dividing by the arbitrary volume, $\Delta \forall$, yield:

$$-\frac{\partial (\rho u)}{\partial x} - \frac{\partial (\rho v)}{\partial y} - \frac{\partial (\rho w)}{\partial z} = \frac{\partial \rho}{\partial t} \tag{1.49}$$

Thus,

$$\frac{\partial \rho}{\partial t} + \nabla \cdot (\rho \vec{v}) = 0 \tag{1.50}$$

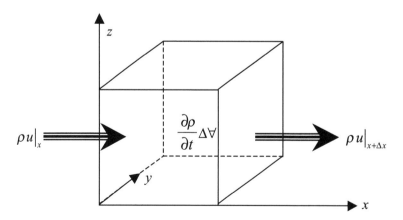

Figure 1.8 One-dimensional fluid mass balance for a 3-D control volume

Example 1.8: Use of the Continuity Equation (Two Problems: A&B)

(a) For steady laminar fully developed pipe flow of an incompressible fluid, the axial flow is:

$$v_z(r) = v_{max}\left[1 - \left(\frac{r}{r_0}\right)^2\right] \tag{E1.8-1}$$

Show that the radial (or normal) velocity $v_r = 0$.

Sketch:	*Assumptions:*
$r = r_0$ r $r = 0$ $-v_z(r)$ z v_{max} <u>Note</u>: $v_{max} = 2v_{average}$ (see Example 2.2)	• Steady implies $\partial/\partial t \equiv 0$ • Incompressible fluid: $\rho = \cancel{c}$ • Axisymmetric pipe: $\partial/\partial\Theta \equiv 0$ • Fully developed flow: $\partial/\partial z \equiv 0$

Solution:

Based on the assumptions, Eq. (1.42) is appropriate and reads in cylindrical coordinates (App. A):

$$\frac{1}{r}\frac{\partial(rv_r)}{\partial r} + \frac{1}{r}\frac{\partial v_\Theta}{\partial\Theta} + \frac{\partial v_z}{\partial z} = 0 \tag{E1.8-2}$$

Clearly, the given velocity profile $v_z(r)$ (see Eq. (E1.8-1)) is not a function of z, i.e., $\partial v_z/\partial z = 0$, which implies with $\partial/\partial\Theta = 0$ (axisymmetry) that Eq. (E1.8-2) reduces to:

$$\frac{\partial(rv_r)}{\partial r} = 0 \tag{E1.8-3}$$

Partial integration yields:

$$v_r = f(z)/r \tag{E1.8-4}$$

where $0 \leq r \leq r_0$, i.e., r could be zero. That fact and the boundary condition $v_r(r = r_0) = 0$, i.e., no fluid penetrates the pipe wall, force the physical solution $v_r \equiv 0$.

Indeed, if $v_r \neq 0$, such a radial velocity component would alter the axial velocity profile to $v_z = v_z(r, z)$, which implies *developing* flow; that happens, for example, in the pipe's entrance region or due to a porous pipe wall through which fluid can escape or is being injected.

(b) Consider 2-D steady laminar symmetric flow in a smooth converging channel where axial velocity values were measured at five points (see sketch). Estimate the fluid element acceleration a_x at point C as well as the normal velocity v at point B'. All distances are 2 cm and the centerline velocities are 5 m/s at A; 7 m/s at B; 10 m/s at C; and 12 m/s at D.

Sketch:	Concept:
	Approximate via finite differencing the reduced acceleration and continuity equations.

Solution:

From Sect. 1.2.2 (Eq. (1.14)) the axial acceleration can be written as:

$$a_x = \frac{\partial u}{\partial t} + u\frac{\partial u}{\partial x} + v\frac{\partial u}{\partial y} + w\frac{\partial u}{\partial z} \qquad \text{(E1.8-5)}$$

Based on the stated assumptions $\left(\partial/\partial t = 0, \partial/\partial y = 0, w = 0\right)$:

$$a_x = u\frac{\partial u}{\partial x} \approx u\frac{\Delta u}{\Delta x} = u_c\frac{u_D - u_B}{x_D - x_B} = 10\frac{12-7}{0.04}$$

$$\therefore \qquad \boxed{a_x|_C \approx 1250 \text{ m/s}^2}$$

In order to find $v|_{B'}$, we employ the 2-D continuity equation in rectangular coordinates:

$$\frac{\partial u}{\partial x} + \frac{\partial v}{\partial y} = 0 \qquad \text{(E1.8-6)}$$

which can be approximated as:

$$\frac{\Delta v}{\Delta y} = -\frac{\Delta u}{\Delta x}$$

or

$$\Delta v = -\frac{10-5}{0.04}0.02 = -2.5 \text{ m/s}$$

Recall that $\Delta v = v|_{B'} - v|_B$ where $v_B \equiv 0$ <symmetry> so that

$$\boxed{v|_{B'} = -2.5 \text{ m/s}}$$

Comment:

This is a very simple example of finite differencing where derivatives are approximated by finite differences of all variables. Discretization of the governing equations describing the conservation laws is the underlying principle of CFD (computational fluid dynamics) software (see Chapter 7).

1.3.3.2 Linear Momentum Conservation

Momentum Equation in Differential Form Based on the RTT.   In order to obtain the equation of motion describing any *point* in a fluid flow field, all terms in the RTT have to be again converted to volume integrals, employing Gauss' Divergence Theorem (see Eq. (1.39)).

The Equation of Motion. First, body forces in Eq. (1.28) are logically expressed in terms of volume integrals, i.e.,

$$\vec{F}_B = \iiint\limits_{\text{C.}\forall.} \rho \vec{f}_B d\forall$$

and surface forces in terms of surface integrals, i.e.,

$$\vec{F}_S = \iint\limits_{\text{C.S.}} \vec{\vec{T}} \cdot d\vec{A}$$

where $\vec{f}_B$ is a body force per unit mass and $\vec{\vec{T}}$ is the total (or Cauchy) stress tensor. Now, for a stationary control volume the linear momentum equation in integral form reads:

$$\int\limits_{\text{C.V.}} \rho \vec{f}_B d\forall + \int\limits_{\text{C.S.}} \vec{T} \cdot d\vec{A} = \frac{\partial}{\partial t} \int\limits_{\text{C.V.}} \rho \vec{v} d\forall + \int\limits_{\text{C.S.}} \vec{v} \rho \vec{v} \cdot d\vec{A} \qquad (1.51)$$

<u>Recall</u>: This is a (3-component) vector equation in principle for the velocity field $\vec{v}$. It contains $\vec{T} \equiv -p\vec{I} + \vec{\tau}$, i.e., the (9-component) total stress tensor (see App. A), as an additional unknown because in most cases $\vec{f}_B$ is simply weight ($\vec{g}$) per unit mass. Thus, in order to solve this closure problem, we have to know the *thermodynamic pressure p* and an expression for the *stress tensor* $\vec{\tau}$. Recall that $\vec{I}$ is the unit tensor, i.e., only ones on the diagonal and zeros everywhere else in the 3×3 matrix, elevating the product $p\vec{I}$ to a "pseudotensor" because p is just a scalar.

Now, converting all surface integrals of Eq. (1.51) into volume integrals yields:

$$\iiint\limits_{\text{C.V.}} \left[\frac{\partial(\rho\vec{v})}{\partial t} + \nabla \cdot (\rho\vec{v}\vec{v}) + \nabla \cdot \left(p\vec{I} - \vec{\tau} \right) - \rho\vec{g} \right] d\forall = 0 \qquad (1.52)$$

or

$$\frac{\partial(\rho\vec{v})}{\partial t} + \nabla \cdot (\rho\vec{v}\vec{v}) = -\nabla p + \nabla \cdot \vec{\tau} + \rho\vec{g} \qquad (1.53)$$

Equation (1.53) is the Cauchy equation of motion (or linear momentum equation) for any fluid and with gravity as the body force. In order to reduce it in complexity and provide some physical meaning, let's consider *constant fluid properties* and express the unknown stress tensor in terms of the principal variable. Employing Stokes' hypothesis, we have in vector notation:

$$\vec{\tau} = \mu \left(\nabla\vec{v} + \nabla\vec{v}^{\text{tr}} \right) \equiv \mu\vec{\dot{\gamma}} \qquad (1.54a)$$

and in index (or tensor) notation:

$$\tau_{ij} = \mu \left(\frac{\partial v_i}{\partial x_j} + \frac{\partial v_j}{\partial x_i} \right) \equiv \mu\dot{\gamma}_{ij} \qquad (1.54b)$$

where $\dot{\gamma}_{ij}$ is the shear rate tensor.

Stress Tensors and Stress Vectors.   Physically τ_{ij} represents a force field per unit area (see Sect. 1.2.4 and Example 1.9) as a result of the resistance to the rate of deformation of fluid elements, i.e., internal friction. This insight leads for Newtonian fluids, such as air, water, and typical oils, to the postulate:

$$\tau_{ij} = \text{fct}\left(\varepsilon_{ij}\right) \tag{1.55}$$

where $\varepsilon_{ij} = \left(v_{i,j} + v_{j,i}\right)/2$ is the strain rate tensor, familiar from solid mechanics. Now, Stokes suggested that $\vec{\vec{\tau}}$ is a *linear function* of $\vec{\vec{\varepsilon}}$, which is not the case for non-Newtonian fluids and rarefied gases, as well as some fluid flow in microscale devices. Specifically, for Newtonian fluids:

$$\vec{\vec{\tau}} = \lambda(\nabla \cdot \vec{v})\vec{\vec{I}} + 2\mu\vec{\vec{\varepsilon}} \tag{1.56}$$

where the viscosity coefficients λ and μ depend only on the thermodynamic state of the fluid. For incompressible flow $\nabla \cdot \vec{v} = 0$ (see Eq. (1.44)) and the total stress tensor reduces to

$$\tau_{ij} = 2\mu\varepsilon_{ij} \tag{1.57}$$

where again

$$\varepsilon_{ij} = \frac{1}{2}\left(\frac{\partial v_i}{\partial x_j} + \frac{\partial v_j}{\partial x_i}\right) = \frac{1}{2}\dot{\gamma}_{ij} \tag{1.58}$$

so that

$$\tau_{ij} = \mu\left(\nabla\vec{v} + \nabla\vec{v}^{|\text{tr}}\right) \equiv \mu\dot{\gamma}_{ij} \tag{1.59}$$

The Cauchy or *total stress tensor* $\vec{\vec{T}} = -p\vec{\vec{I}} + \vec{\vec{\tau}}$ being an unknown in Eq. (1.53) constitutes a closure problem, i.e., $\vec{\vec{T}}$ has to be related to the principal variable $\vec{v}$ or its derivatives. As mentioned, p is the *thermodynamic pressure*, $\vec{\vec{I}}$ is the necessary unit tensor for homogeneity, and $\vec{\vec{\tau}}$ is the stress tensor. For any coordinate system, the stress vector $\vec{\tau}$ relates to the *symmetric* second-order tensor $\vec{\vec{T}}$ as:

$$\vec{\tau} = \hat{n} \cdot \vec{\vec{T}} = \vec{\vec{T}} \cdot \hat{n} \tag{1.60}$$

where $\hat{n}$ is the normal (unit) vector. Note, without tensor symmetry, i.e., $T_{ij} \neq T_{ji}$, angular momentum would not be conserved (see Batchelor, 1967). It is more insightful to write $\vec{\vec{T}} = -p\vec{\vec{I}} + \vec{\vec{\tau}}$ in *tensor (or index) notation* so that the total stress tensor reads:

$$T_{ij} = -p\delta_{ij} + \tau_{ij} \tag{1.61}$$

where $-p\delta_{ij}$ is interpreted as the isotropic part (e.g., fluid statics and inviscid flow) and τ_{ij} is the deviatoric part for which a constitutive equation has to be found. Physically, $\tau_{ij} \triangleq \tau_{(i=\text{surface-normal})(j=\text{stress-direction})}$ represents a force field per unit area as a result of the resistance to the rate of deformation of fluid elements, i.e., $\tau_{ij} \sim \varepsilon_{ij}$ (see Figure 1.5).

The relation between τ_{ij} and ε_{ij} (plus vorticity tensor ζ_{ij}) can be more formally derived, starting with a fluid element displacement from point P (with $\vec{v}$ at t) to point P' ($\vec{v} + d\vec{v}$ at $t + dt$) a distance ds apart. Expanding the total derivative in Cartesian coordinates:

$$d\vec{v} = \begin{bmatrix} \dfrac{\partial u}{\partial x} & \dfrac{\partial u}{\partial y} & \dfrac{\partial u}{\partial z} \\[2mm] \dfrac{\partial v}{\partial x} & \dfrac{\partial v}{\partial y} & \dfrac{\partial v}{\partial z} \\[2mm] \dfrac{\partial w}{\partial x} & \dfrac{\partial w}{\partial y} & \dfrac{\partial w}{\partial z} \end{bmatrix} d\vec{s} = \nabla \vec{v} d\vec{s} \tag{1.62}$$

The spatial changes, or deformations, the fluid element is experiencing can be expressed as the "rate-of-deformation" tensor (see dyadic product $\nabla \vec{v}$ in App. A):

$$\frac{d\vec{v}}{ds} = \nabla \vec{v} := \frac{\partial v_i}{\partial x_j} \tag{1.63a,b}$$

Equation (1.63) can be decomposed into the strain rate tensor ε_{ij} (symmetrical part) and the vorticity (or rotation tensor) ζ_{ij}:

$$\frac{\partial v_i}{\partial x_j} = \varepsilon_{ij} + \zeta_{ij} \tag{1.64}$$

It can be readily shown that:

- $\zeta_{yx} = -\zeta_{xy} = \omega_z,$ $\qquad \zeta_{xz} = -\zeta_{zx} = \omega_y,$ $\qquad$ and $\qquad \zeta_{zy} = -\zeta_{yz} = \omega_x,$ $\qquad$ where $\omega_z = \frac{1}{2}(\partial v/\partial x - \partial u/\partial y)$, etc.; thus, the tensor ζ_{ij} collapses to the *vorticity vector:*

$$2\vec{\omega} = \nabla \times \vec{v} = \vec{\zeta} \tag{1.65a,b}$$

- $\varepsilon_{ii} := \dot{\gamma}_{ii} \equiv \partial v_i / \partial x_i$ indicates volume change (dilation)
- $\varepsilon_{ij} := \dot{\gamma}_{ij}/2, i \neq j$, represents element distortion

- The shear rate tensor $\dot{\gamma}_{ij} \equiv \partial v_i / \partial x_j + \partial v_j / \partial x_i$ and hence the $1/2$ in $\varepsilon_{ij} = \dot{\gamma}_{ij}/2$ is mathematically necessary in order to match Eq. (1.64).

As mentioned, Stokes suggested that $\vec{\bar{\tau}}$ is a *linear* function of $\vec{\bar{\varepsilon}}$, which is not the case for non-Newtonian fluids, rarefied gases, and some fluid flows in microscale devices, e.g., bioMEMS. Specifically, for Newtonian fluids, i.e., air, water, and most oils:

$$\vec{\bar{\tau}} = \lambda(\nabla \cdot \vec{v})\vec{\bar{I}} + 2\mu\vec{\bar{\varepsilon}} \qquad (1.66)$$

where the viscosity coefficients λ and μ depend only on the thermodynamic state of the fluid. For incompressible flow $\nabla \cdot \vec{v} = 0$ and the total stress tensor reduces to:

$$T_{ij} = -p\delta_{ij} + 2\mu\varepsilon_{ij} \qquad (1.67)$$

where

$$2\mu\varepsilon_{ij} \equiv \tau_{ij} = \mu\left(\frac{\partial v_i}{\partial x_j} + \frac{\partial v_j}{\partial x_i}\right) \equiv \mu\dot{\gamma}_{ij} \qquad (1.68)$$

Here, $\dot{\gamma}_{ij} \equiv 2\varepsilon_{ij}$ is called the shear rate tensor (see App. A for all stress and shear rate components in rectangular and cylindrical coordinates).

Of great importance is the wall shear stress vector ($\vec{\tau}_{wall} \equiv$ WSS) as a result of frictional (or viscous) effects and the *no-slip condition* for macroscale systems, i.e., at any solid surface:

$$\vec{v}_{fluid} = \vec{v}_{wall} \qquad (1.69)$$

Typically, $\vec{v}_{wall} = 0$, i.e., the wall is stationary and impermeable. The experimentally verified no-slip condition in macrofluidics generates velocity gradients normal to the wall at all axial flow speeds. As illustrated in Figure 1.5,

$$\tau_{wall} \equiv \text{WSS} \sim \begin{cases} \partial u/\partial n & \text{(due to the no-slip condition)} \\ \mu & \text{(due to viscous fluid effects)} \\ F_{tang}/A & \text{(tangential force per unit surface area)} \end{cases} \qquad (1.70\text{a-c})$$

Very high or low WSS values have been related to device malfunctions in mechanical and arterial diseases in biomedical engineering (Kleinstreuer, 2006).

Integration of $\vec{\tau}_{wall}$ over the entire surface of a submerged body or inside a conduit yields the frictional drag:

$$F_{viscous} = \int_A \tau_{wall} \, dA \qquad (1.71)$$

Viscous drag (frictional effect) plus form drag (pressure effect) make up the total drag:

$$F_{drag} = \int_A (\tau_w + p) \, dA \qquad (1.72)$$

which, for most cases, would require elaborate CFD (computational fluid dynamics) analysis to evaluate the WSS and pressure distributions on the submerged body surface and then integrating.

Force Balance Derivation.   A more physical approach for deriving the (linear) momentum equation starts with a force balance for a representative elementary volume (REV). Employing rectangular coordinates and an incompressible fluid, external surface and body forces accelerate an REV of mass m, so that we can write Newton's second law of motion per unit volume as (cf. Figure 1.9):

$$\rho \frac{D\vec{v}}{Dt} = \sum \vec{f}_{surface} + \sum \vec{f}_{body} \qquad (1.73)$$

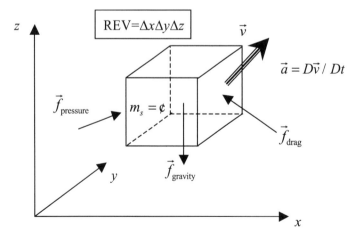

Figure 1.9   Closed system, i.e., accelerating material volume (REV)

Applying Newton's second law of motion, the REV is a control volume (i.e., fluid element) for which we record local and convective momentum changes due to *net* pressure, viscous, and gravitational forces per unit volume, viz.:

$$\rho\left[\frac{\partial \vec{v}}{\partial t}+(\vec{v}\cdot\nabla)\vec{v}\right]=\vec{f}_{\substack{net\\pressure}}+\vec{f}_{\substack{net\\viscous}}+\vec{f}_{\substack{net\\buoyancy}} \tag{1.74}$$

$$f_{\substack{net\\pressure}}=f_p\big|_x-f_p\big|_{x+\Delta x}=-\frac{\partial f_p}{\partial x}\Delta x \tag{1.75}$$

and with

$$f_p\equiv\frac{p\Delta A}{\Delta V},\ f_{\substack{net\\pressure}}=-\frac{\partial p}{\partial x}\frac{\Delta y\Delta z}{\Delta x\Delta y\Delta z}\Delta x=-\frac{\partial p}{\partial x},\ \text{etc.}$$

In 3-D:

$$\vec{f}_{\substack{net\\pressure}}=-\nabla p=-\left(\frac{\partial p}{\partial x}\vec{i}+\frac{\partial p}{\partial y}\vec{j}+\frac{\partial p}{\partial z}\vec{k}\right) \tag{1.76}$$

Similarly, the net viscous force per unit volume in the *x*-direction reads (see Figure 1.10):

$$f_{\substack{net\\viscous}}=f_v\big|_z-f_v\big|_{z+\Delta z}=-\frac{\partial f_v}{\partial z}\Delta z \tag{1.77}$$

and with

$$f_v\equiv\frac{\tau\Delta A}{\Delta V},\ f_{\substack{net\\viscous}}=-\frac{\partial \tau}{\partial z}\frac{\Delta x\Delta z}{\Delta x\Delta y\Delta z}\Delta z=-\frac{\partial \tau}{\partial z},\ \text{etc.}$$

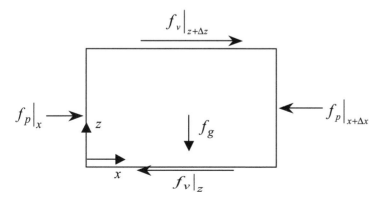

Figure 1.10 Control volume for 1-D force balances

In 3-D, the net frictional force can be expressed as:

$$\underset{\substack{\text{viscous}}}{\vec{f}_{net}} = \nabla \cdot \vec{\vec{\tau}} =$$

$$\left(\frac{\partial \tau_{xx}}{\partial x} + \frac{\partial \tau_{yx}}{\partial y} + \frac{\partial \tau_{zx}}{\partial z} \right)\vec{i} + \left(\frac{\partial \tau_{xy}}{\partial x} + \frac{\partial \tau_{yy}}{\partial y} + \frac{\partial \tau_{zy}}{\partial z} \right)\vec{j} + \left(\frac{\partial \tau_{xz}}{\partial x} + \frac{\partial \tau_{yz}}{\partial y} + \frac{\partial \tau_{zz}}{\partial z} \right)\vec{k} \qquad (1.78)$$

As discussed, with Stokes' hypothesis for incompressible Newtonian fluids, we have (see Eq. (1.59)):

$$\vec{\vec{\tau}} = \mu \left(\nabla \vec{v} + \nabla \vec{v}^{\text{tr}} \right) \qquad (1.79)$$

Taking the divergence of the tensor field, i.e., $\nabla \cdot \vec{\vec{\tau}} = \mu \nabla^2 \vec{v}$ (see App. A) allows expressing Eq. (1.73) as:

$$\rho \left[\frac{\partial \vec{v}}{\partial t} + (\vec{v} \cdot \nabla)\vec{v} \right] = -\nabla p + \mu \nabla^2 \vec{v} + \rho \vec{g} \qquad (1.80)$$

This linear momentum equation and the continuity equation $\nabla \cdot \vec{v} = 0$ are the N-S equations.

At the *molecular level for gas flow*, momentum transfer can be explained via the net shear stress as the result of mean momentum flux in the normal direction of flow (see Wilcox, 2007):

$$\tau_{yx} = 0.5 \rho v_{\text{thermal}}, \; \lambda_{\text{mfp}} \, du/dy = \mu \, du/dy \qquad (1.81\text{a,b})$$

where v_{thermal} (approximately 4/3 times the speed of sound in air) is the average molecular velocity and λ_{mfp} is the mean free path, i.e., the average distance traveled by a gas molecule before collision.

==

Example 1.9: Wall Boundary Conditions and Shear Stress in Simple Couette Flow

Note: This detailed solution illustrates the differential approach for solving the reduced Navier-Stokes equations.

Consider Couette flow, i.e., a viscous fluid between two parallel plates a small gap h apart, where the upper plate moves with a constant velocity u_o, in general due to an external tangential force, F_{pull}.

Note: The experimentally observed boundary condition for a conventional fluid at any solid surface demands that:

$$\vec{v}_{\text{fluid}} = \vec{v}_{\text{wall}}$$

where in rectangular coordinates $\vec{v} = (u,v,w)$ or $\vec{v} = u\hat{i} + v\hat{j} + w\hat{k}$.

Applying for the present case (see system sketch), we have for a stationary solid wall $(y = 0)$:

$$u_{\text{fluid}} = 0 \qquad <\text{``no-slip'' condition}>$$

For the moving wall $(y = h)$:

$$u_{\text{fluid}} = u_0 \qquad <\text{``no-slip'' condition}>$$

and the normal velocity component at both walls is:

$$v_{\text{fluid}} = 0 \qquad <\text{``no-penetration'' condition}>$$

Sketch:	*Assumptions:*	*Approach:*
	• Steady laminar fully developed (unidirectional) flow • Constant fluid properties	• Reduced N-S equations based on assumptions and boundary conditions

Solution:

Translating the problem statement plus assumptions into mathematical shorthand, we have (see Sect. 1.2.3):

Movement of the upper plate (u_0 = constant keeps the viscous fluid between the plates in motion via frictional effects propagating normal to the plate; hence, the usual "driving force" $\partial p / \partial x \equiv 0$.

Steady flow⇒all time derivatives are zero, i.e., $\partial / \partial t \equiv 0$.

Laminar <u>unidirectional flow</u>⇒only one velocity component dependent on one dimension (1-D) is nonzero, i.e., $\vec{v} = (u,0,0)$, where $u = u(y)$ only. This implies <u>parallel</u> or <u>fully developed flow</u> where $\partial / \partial x \equiv 0$ and hence $v = 0$.

In summary, we can postulate that

$$u = u(y), \ v = w = 0$$

$$\frac{\partial u}{\partial x} = 0; \ \frac{\partial p}{\partial x} = 0; \ g_x = 0$$

Checking Eqs. (1.82a-c) with these postulates, we realize the following using Sect. A.5:

$$\underline{\text{Continuity equation confirms:}} \ 0 + \frac{\partial v}{\partial y} = 0 \succ \underline{\underline{v = 0}} \qquad (E1.9\text{-}1)$$

or better, $v = 0 \Rightarrow \partial u / \partial x = 0$; i.e., fully developed flow:

$$\underline{x\text{-Momentum yields:}} \ 0 + 0 = 0 + v\left(0 + \frac{\partial^2 u}{\partial y^2}\right) + 0 \qquad (E1.9\text{-}2)$$

and

$$\underline{y\text{-Momentum collapses to:}} \ \frac{\partial p}{\partial y} = \rho g_y \ \text{<fluid statics>} \qquad (E1.9\text{-}3)$$

Thus, Eq. (E1.9-2) can be written as

$$\frac{d^2 u}{dy^2} = 0 \qquad (E1.9\text{-}4a)$$

subject to the "no-slip" conditions

$$u(y = 0) = 0 \ \text{ and } \ u(y = h) = u_0 \qquad (E1.9\text{-}4b,c)$$

Double integration of (E1.9-4a) and inserting the boundary conditions (BCs) (E1.9-4b,c) yields:

$$u(y) = u_0 \frac{y}{h} \qquad (E1.9\text{-}5)$$

Of the stress tensor (Eq. (1.54b)), $\tau_{ij} = \mu\left(\partial u_i / \partial x_j + \partial u_j / \partial x_i\right)$, only τ_{yx} is nonzero, i.e.,

$$\tau_{xy} = \mu\left(\frac{\partial u}{\partial y} + \frac{\partial v}{\partial x}\right) \tag{E1.9-6}$$

With $v = 0$ and Eq. (E1.9-5)

$$\tau_{xy} = \frac{\mu u_0}{h} = \cancel{c}$$

Of the vorticity tensor $\omega_{ij} = \partial u_i/\partial x_j - \partial u_j/\partial x_i$ (Sect. A5), only ω_{yx} is nonzero, i.e.,

$$\omega_{yx} = \frac{\partial u}{\partial y} - \frac{\partial v}{\partial x} := \frac{u_0}{h} = \cancel{c} \tag{E1.9-7}$$

which implies that the fluid elements between the plates rotate with constant angular velocity ω_{yx}, while translating with $u(y)$.

Note: The wall shear stress at the upper (moving) plate is also constant, i.e.,

$$\tau_w = \mu\frac{\partial u}{\partial y}\bigg|_{y=h} = \frac{\mu u_0}{h}$$

so that

$$F_{drag} = -\int \tau_w dA_{plate} = F_{pull} = \frac{\mu u_0}{h} A_{surface} = \text{constant}$$

Profiles:

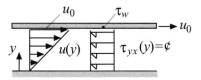

Comments:

In the absence of a pressure gradient, only viscous effects set the fluid layer into (linear) motion. The necessary "pulling force" is inversely proportional to the gap height, i.e., the thinner the fluid layer, the larger is the shear stress and hence F_{pull}.

1.3.3.3 Reduced Forms of the Momentum Equation

Returning to Eq. (1.53), which is generally known as Cauchy's *equation of motion*, we now introduce simplifications of increasing magnitude. Fluid properties, i.e., density ρ and dynamic viscosity μ, are typically constant; but, in general, ρ and μ are functions of temperature T, pressure p, and species concentration c. Thus, the underlying assumptions for $\rho = ¢$ and $\mu = ¢$ are that:

- only relatively small temperature variations occur;
- the Mach number $M \equiv v_{\text{fluid}} / a_{\text{sound}} < 0.3$, which may be only violated by gases;
- pressure drops in gas flow are relatively small, and cavitation in liquid flow is avoided;
- concentration variations of components in mixture flows are small.

(i) The Navier-Stokes Linear Momentum Equation:
    Dividing Eq. (1.53) through by the constant density ρ and recalling that $\mu / \rho \equiv v$, the kinematic viscosity, we have with Stokes' hypothesis (Eq. (1.59)):

$$\underbrace{\frac{\partial \vec{v}}{\partial t} + (\vec{v} \cdot \nabla) \vec{v}}_{\frac{D\vec{v}}{Dt} = \vec{a}_{\text{total}}} = \underbrace{- \frac{1}{\rho} \nabla p}_{\vec{f}_{\text{net pressure}}} + \underbrace{v \nabla^2 \vec{v}}_{\vec{f}_{\text{net viscous}}} + \underbrace{\vec{g}}_{\vec{f}_{\text{body}}} \tag{1.80}$$

Clearly, Eq. (1.80) is Newton's particle dynamics equation applied to fluid elements.
    For example, for steady 2-D flows, the N-S equations read in rectangular coordinates (see App. A):

- (Continuity) $\dfrac{\partial u}{\partial x} + \dfrac{\partial v}{\partial y} = 0$

- (x-Momentum) $u\dfrac{\partial u}{\partial x} + v\dfrac{\partial u}{\partial y} = -\dfrac{1}{\rho}\dfrac{\partial p}{\partial x} + v\left(\dfrac{\partial^2 u}{\partial x^2} + \dfrac{\partial^2 u}{\partial y^2}\right) + g_x$ (1.82a-c)

- (y-Momentum) $u\dfrac{\partial v}{\partial x} + v\dfrac{\partial v}{\partial y} = -\dfrac{1}{\rho}\dfrac{\partial p}{\partial y} + v\left(\dfrac{\partial^2 v}{\partial x^2} + \dfrac{\partial^2 v}{\partial y^2}\right) + g_y$

On a professional level this set of four PDEs, subject to appropriate boundary conditions, is now being routinely solved for the four unknowns, u, v, w, and p, using numerical software packages on desktop workstations, HPC clusters, and supercomputers (see Chapter 8). In a classroom environment, only reduced forms of Eqs. (1.82a-c) can be solved.

(ii) Prandtl's Boundary-Layer Equations:
    As indicated in Example 1.9, the fluid velocity is zero at a stationary wall. Now, considering relatively high-speed fluid flow past a (horizontal) solid surface, the

quasi-uniform high velocity suddenly has to reduce, within a narrow region, to zero at the stationary wall. This region of high-velocity gradients is called a "thin shear layer," or more generally a *boundary layer*. For example, Figure 1.11 depicts such a (laminar) boundary layer of thickness $\delta(x)$, formed along a horizontal stationary flat plate (e.g., a giant razor blade) which is approached by a uniform fluid stream of velocity u_∞ with $10^3 < \mathrm{Re} = u_\infty \ell / v < 10^5$. It can be readily demonstrated that the $v \partial^2 u / \partial x^2$ term (axial momentum diffusion) of Eq. (1.82b) is negligible and that the y-momentum equation collapses to $-(1/\rho)\partial p / \partial y = 0$; i.e., $p = p(x)$ only. As a result, Eqs. (1.82a-c) reduce to:

$$\frac{\partial u}{\partial x} + \frac{\partial v}{\partial y} = 0 \tag{1.83}$$

and

$$u\frac{\partial u}{\partial x} + v\frac{\partial u}{\partial y} = -\frac{1}{\rho}\frac{\partial p}{\partial x} + v\frac{\partial^2 u}{\partial y^2} \tag{1.84}$$

inside the boundary layer $0 \le y \le \delta(x)$ and $0 \le x \le \ell$.

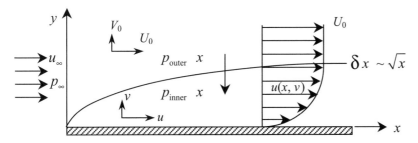

Figure 1.11   Laminar flat-plate boundary layer

(iii) Stokes' Creeping Flow Equation:

When the viscous forces are dominant, the Reynolds number (Re = inertial forces/viscous forces) is very small, i.e., the term $(\vec{v} \cdot \nabla)\vec{v}$ in Eq. (1.80) is negligible. As a result, the *Stokes equation* is *obtained* which holds for "creeping" flows Eq. (1.80) (see Example 2.13 for an application):

$$\rho\frac{\partial \vec{v}}{\partial t} = -\nabla p + \mu \nabla^2 \vec{v} \tag{1.85}$$

(iv) Euler's Inviscid Flow Equation:

For frictionless flows $(\mu \equiv 0)$, Eq. (1.80) reduces to:

$$\rho\frac{D\vec{v}}{Dt} = -\nabla p + p\vec{g} \tag{1.86}$$

which is the *Euler equation*. Although ideal fluids, i.e., inviscid flows, hardly exist, the second-order term also vanishes when $\nabla^2\vec{v} \approx 0$; for example, outside boundary layers as indicated with the velocity profile in Figure 1.11. In fact, aerodynamics people employ Eq. (1.86) to find the pressure field around airfoils (see p_{outer} in Figure 1.11).
(v) Bernoulli's Equation:
    Equation (1.86) applied in 2-D to a representative streamline along coordinate "*s*" yields (see Figure 1.12):

$$\frac{\partial v_s}{\partial t} + v_s\frac{\partial v_s}{\partial s} + \frac{1}{\rho}\frac{\partial p}{\partial s} + g\underbrace{\sin\theta}_{\partial z/\partial s} = 0 \tag{1.87a}$$

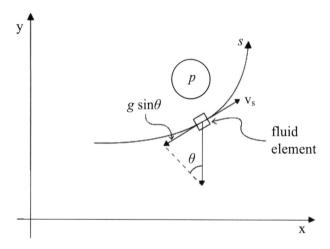

Figure 1.12   A fluid element along a representative streamline

which leads to the *Bernoulli equation*. Multiplying Eq. (1.87a) through by ∂s and integrating yield for *steady incompressible inviscid flows*:

$$\frac{v^2}{2} + \frac{p}{\rho} + gz = C \tag{1.87b}$$

where v and p are locally averaged quantities along a representative streamline and the *z*-coordinate extends against the direction of gravity. Thus, for two points on a representative streamline the total energy per unit mass is balanced, i.e. (see also Eq. (1.38)):

$$\frac{v_1^2}{2} + \frac{p_1}{\rho} + gz_1 = \frac{v_2^2}{2} + \frac{p_2}{\rho} + gz_2 \tag{1.88}$$

For example, for a given system where $v_2 = 0$ (e.g., point 2 of a streamline is on the front of a submerged body) and $g\Delta z \approx 0$, we have:

$$p_2 = p_1 + \frac{\rho v_1^2}{2} \tag{1.89}$$

where p_2 is the total or *stagnation point pressure*, p_1 is the *thermodynamic* (or a *static*) *pressure* at point 1, and $\rho v_1^2/2$ is the *dynamic pressure* at point 1. One application of Eq. (1.89) is the Pitot-static tube, which measures $\Delta p = p_2 - p_1$, so that $v_1 = \sqrt{2(p_2 - p_1)/\rho}$ (see Figure 1.13).

An *extended, i.e., more realistic form of Bernoulli's equation* adds a frictional loss term to the RHS of Eq. (1.88). For example, multiplying Eq. (1.88) through by ρ yields an energy balance per unit volume:

$$\underbrace{\frac{\rho}{2}v_1^2 + p_1 + \rho gz_1}_{\sim E_{total}} = \underbrace{\frac{\rho}{2}v_2^2}_{\sim E_{kin.}} + \underbrace{p_2}_{\sim \tilde{E}_{pr.}} + \underbrace{\rho gz_2}_{\sim \tilde{E}_{pot.}} + \underbrace{H_f}_{E_{loss}} \tag{1.90}$$

where H_f represents an energy loss between stations ① and ②.

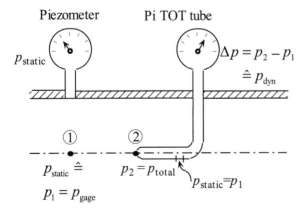

Figure 1.13   Different manometers to measure different pressures

Naturally, Eq. (1.90) can also be expressed in terms of heights (or "heads"):

$$\frac{v_1^2}{2g} + \frac{p_1}{\gamma} + z_1 = \frac{v_2^2}{2g} + \frac{p_2}{\gamma} + z_2 + h_f \tag{1.91}$$

The specific weight $\gamma \equiv \rho g$ and the frictional loss term $h_f \sim \tau_{wall} \sim \Delta p$ is usually expressed as a portion of the kinetic energy. Clearly, while Eq. (1.91) is based on the conservation of energy, Eq. (1.88) is based on the conservation of linear momentum.

1.3.3.4 Energy and Species Mass Conservation

Employing again the divergence theorem (see Eq. (1.41)), we can rewrite Eq. (1.40) as:

$$\rho \frac{De_t}{Dt} \equiv \rho \left[\frac{\partial e_t}{\partial t} + (\vec{v} \cdot \nabla)e_t \right] = \rho \left(\vec{f}_b \cdot \vec{v}E \right) + \nabla \cdot \left(\vec{\vec{T}} \cdot \vec{v} \right) + \nabla \cdot q_t \tag{1.92}$$

where the specific total energy is simply $e_t = |\vec{v}|^2/2 + \tilde{u} \cong |\vec{v}|^2/2 + c_v T$ and $\vec{q}_t = -k\nabla T + ...$ Clearly, T is now the temperature.

Another derivation of the energy equation, resulting in a directly applicable form, starts with $h = \tilde{u} + p/\rho$ <enthalpy per unit mass> as the principal unknown, and considering $\vec{q}$ <diffusive heat flux> and $v\Phi$ <energy dissipation due to viscous stress>, we obtain:

$$\frac{\partial}{\partial t}(\rho h) + \nabla \cdot (\rho \vec{v} h) = -\nabla \cdot \vec{q} + \mu \Phi \tag{1.93}$$

With $dh \equiv c_p dT$, or simplified to $h = c_p T$ when $c_p = $ constant, $\vec{q} = -k\nabla T$ after Fourier, and $\mu\Phi = \tau_{ij} \partial v_i / \partial x_j$, we obtain for thermal flow with constant fluid properties the *heat transfer equation*:

$$\frac{\partial T}{\partial t} + (\vec{v} \cdot \nabla)T = \alpha \nabla^2 T + \frac{\mu}{\rho c_p} \Phi \tag{1.94}$$

where $\alpha \equiv k/(\rho c_p)$ is the thermal diffusivity and Φ is given in Sect. A.5.

It is interesting to note that when contracting Eq. (1.94),

$$\frac{DT}{Dt} = a\nabla^2 T + S_T \tag{1.95}$$

has the same form as the *species mass transfer* equation:

$$\frac{Dc}{Dt} = \mathcal{D}\nabla^2 c + S_c \tag{1.96}$$

where $\mathcal{D}$ is the binary diffusion coefficient and S_c denotes possible species sinks or sources. Clearly, momentum diffusivity v (Eq. (1.80)), thermal diffusivity a (Eq. (1.94)), and mass diffusivity $\mathcal{D}$ (Eq. (1.96)) have the same dimensions [length^2/time].

Section 1.3 is summarized in compact form via Table 1.2 and Figure 1.14. Specifically, Table 1.2 highlights the governing equations plus solution methods needed for the remainder of the book. Figure 1.14 conveys that nowadays all transport phenomena with constant fluid properties are described by the Navier-Stokes system of equations, and it provides an overview of the fluidics topics and associated equations of interest.

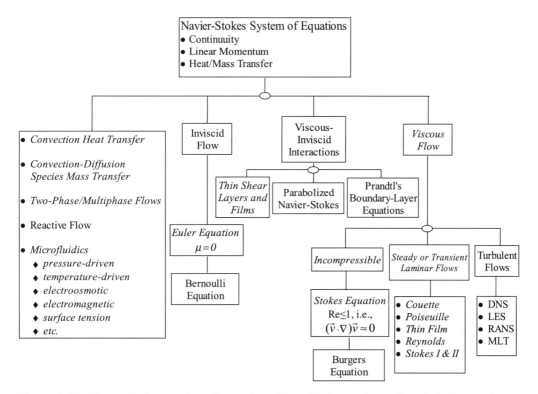

Figure 1.14 Navier-Stokes system of equations (<u>Note:</u> Topics and equations in italics are the main focus in this book)

Table 1.2   Solutions of Special Cases of the N-S Equations

Notes: • Laminar flow with constant fluid properties

Slip velocity plus temperature jump BCs may be needed in micro/nanofluidics (see Sect. 3.2.3)

   (i) Continuity Plus Linear Momentum:

$$\nabla \cdot \vec{v} = 0 \; ; \; \rho \frac{D\vec{v}}{Dt} = \sum \vec{f}_i = -\nabla p + \mu \nabla^2 \vec{v} + \rho \vec{g} + \vec{f}_{electric} + \vec{f}_{particle} + \ldots$$

and

   (ii) Scalar Transport Equations:

$$\frac{D\Phi}{Dt} \equiv \Gamma \nabla^2 \Phi \pm S_\Phi; \; \Phi = \begin{cases} T \\ c \end{cases}, \Gamma = \begin{cases} \alpha \\ \mathcal{D} \end{cases}, \; \alpha \equiv \frac{k}{\rho c_p}; \; \text{binary diffusivity } \mathcal{D}$$

(see Stokes-Einstein equation)

where the Stokes or material derivative is:

$$\frac{D}{Dt} \equiv \frac{\partial}{\partial t} + (\vec{v} \cdot \nabla) \; ; \text{e.g.,} \; \frac{D\vec{v}}{Dt} \equiv \vec{a}_{total} = \frac{\partial \vec{v}}{\partial t} + (\vec{v} \cdot \nabla) \vec{v}$$

Exact Solutions:	Approximate Solutions:

1-D: Simple geometry + fully developed flow
Couette, Poiseuille, thin-film
Direct integration $u'' = K$; $K \sim \Delta p / L$
$Q = \Delta p / R_{flow}$; hydraulic resistance
$R_{flow} \sim \mu$
2-D: Simple geometry
   Combined variable $\eta = \eta(x, t)$
Examples: • Stokes Problems I & II
       • Von Karman Problem
       • Blasius Problem
	• Prandtl's Boundary-Layer (BL) Theory:
	
	    $Re \gg 1, \dfrac{\delta}{l} \ll 1$
	
	• Euler's Equation $(\mu = 0)$
	
	• Stokes' Equation (Re <1.0)

Numerical Solutions

- Transient 3-D FE, FV, or FD solver
- Submodels for Exotic Fluids
- Submodels for Turbulence
- OpenFOAM: free access software (http://www.wikki.co.uk/)
- DNS (Direct Numerical Simulation)
- LBM (Lattice Boltzmann Method)
- MD (Molecular Dynamics) Simulation
- DSMC (Direct-Simulation Monte Carlo) Method

Example 1.10: Thermal Pipe Flow ($q_{wall} = \cancel{c}$)

Consider Poiseuille flow where a uniform heat flux, q_w, is applied to the wall of a pipe with radius r_0.

Set up the governing equations for the fluid temperature assuming thermally fully developed flow, i.e.,

$$\frac{T_w - T}{T_w - T_m} \equiv \Theta\left(\frac{r}{r_0}\right) \qquad \text{(E1.10-1)}$$

where $T_w(x)$ is the wall temperature, $T(r,x)$ is the fluid temperature, and $T_m(x)$ is the cross-sectionally averaged temperature, i.e.,

$$T_m = \frac{1}{\bar{u}A} \int_A uT\, dA \qquad \text{(E1.10-2)}$$

Note that $\Theta = \Theta(r)$ only, describing thermally fully developed flow.

Solve a reduced form of the heat transfer equation (1.75) and develop an expression for the Nusselt number, defined as:

$$\mathrm{Nu} = \frac{2r_0}{k}\frac{q_w}{T_w - T_m} := \frac{hD}{k} \qquad \text{(E1.10-3a,b)}$$

where k is the fluid conductivity and D is the pipe diameter.

Sketch:	Assumptions:	Concept:
	• As stated, i.e., $$u(r) = u_{max}\left(1 - \left(\frac{r}{r_0}\right)^2\right)$$ $\partial T / \partial t = 0$ <steady state> $$(\vec{v}\cdot\nabla)T \Rightarrow u\frac{\partial T}{\partial x}$$ $$\alpha\nabla T \Rightarrow \frac{\alpha}{r}\frac{\partial}{\partial r}\left(r\frac{\partial T}{\partial r}\right)$$	• Reduced heat transfer equation (Eq. (1.75)) based on assumptions

Solution:

With the reduced heat transfer equation in cylindrical coordinates from App. A (see also list of assumptions), we have:

$$\frac{u(r)}{\alpha}\frac{\partial T}{\partial x} = \frac{1}{r}\frac{\partial}{\partial r}\left(r\frac{\partial T}{\partial r}\right) \qquad \text{(E1.10-4)}$$

Employing the dimensionless temperature profile $\Theta(r/r_0) \equiv \Theta(\hat{r})$ given as Eq. (E1.10-1), we can rewrite Eq. (E1.10-4) as

$$-2\frac{hD}{k}\left(1-\hat{r}^2\right) = \frac{d^2\Theta}{d\hat{r}^2} + \frac{1}{\hat{r}}\frac{d\Theta}{d\hat{r}} \qquad \text{(E1.10-5)}$$

Specifically,

$$\text{for } q_w = \phi, \quad \frac{\partial T}{\partial x} = \frac{dT_w}{dx} = \frac{dT_m}{dx} = \frac{2}{r_0}\frac{q_w}{\rho c_p \bar{u}} = \phi \qquad \text{(E1.10-6a-c)}$$

as stated,

$$\frac{hD}{k} \equiv Nu_D := \frac{D}{k}\frac{q_w}{T_w - T_m} = \phi \qquad \text{(E1.10-7a, b)}$$

and with $d\Theta/d\hat{r}$ being finite at $\hat{r} = 0$, we obtain

$$T - T_w = -\left(T_w - T_m\right)Nu_D\left(\frac{3}{8} - \frac{\hat{r}^2}{2} + \frac{\hat{r}^4}{8}\right) \qquad \text{(E1.10-8)}$$

Now, by definition

$$T_w - T_m = \frac{2\pi}{\pi r_0^2 \bar{u}}\int_0^{r_0}\left(T_w - T\right)u(r)r dr \qquad \text{(E1.10-9)}$$

so that, when combining both equations and integrating, we have

$$1 = 4\mathrm{Nu}_D \int_0^1 \left(\frac{3}{8} - \frac{\hat{r}^2}{2} + \frac{\hat{r}^4}{8}\right)\left(1 - \hat{r}^2\right)\hat{r}\,d\hat{r} \tag{E1.10-10}$$

from which we finally obtain:

$$\mathrm{Nu}_D = \frac{48}{11} = 4.36 \tag{E1.10-11}$$

Comment:

It is interesting to note that for hydrodynamically and thermally fully developed flow in a tube, subject to a constant wall heat flux, the Nusselt number (or the heat transfer coefficient) is constant. The same holds for the isothermal wall condition; however, the Nu value is lower (see Kleinstreuer, 1997; or Bejan, 2002), as summarized in Table 1.3.

1.3.4 Entropy Generation Analysis

Background Information.   The second law of thermodynamics is the "increase of entropy" principle for any real process, i.e.,

$$\Delta S_{\mathrm{total}} = \Delta S_{\mathrm{system}} + \Delta S_{\mathrm{surrounding}} \equiv S_{\mathrm{gen}} > 0 \tag{1.97a-c}$$

An entropy value S [kJ/K] indicates the degree of molecular activity or *randomness* and the condition $S_{\mathrm{gen}} > 0$ is necessary for a process to proceed or a device to work. The source of entropy change is heat transferred from different sources. As Clausius stated, $dS > \delta Q/T$, implying all irreversibilities are contributing, e.g., due to heat exchange with internal and/or external sources as well as internal friction (or viscous effects) and net influx of entropy carried by fluid streams. The inequality (1.97c) can be recast as an entropy "balance" by recasting Eq. (1.97b) as:

$$\underbrace{\sum_{\mathrm{in}} m \cdot s - \sum_{\mathrm{out}} m \cdot s + \sum \frac{Q}{T_{\mathrm{ambient}}}}_{\Delta S_{\mathrm{surrounding}}} + \underbrace{S_{\mathrm{gen}}}_{\substack{\text{system}\\\text{irreversibilities}}} = \Delta S|_{\mathrm{C.V.}} \tag{1.98a}$$

where

$$\Delta S|_{\mathrm{C.V.}} = \underbrace{(ms)_{\mathrm{final}} - (ms)_{\mathrm{initial}}}_{\Delta S_{\mathrm{system}}} \tag{1.98b}$$

Table 1.3 Nusselt Numbers and Poiseuille Numbers for Fully Developed Flows in Different Conduits

Cross Section	b/a Ratios	$Nu = hD_h / k$		$Po = f\,Re_{D_h}$
		$q_{wall} = ¢$	$T_{wall} = ¢$	
⬤	—	4.36	3.66	64
a ▢ a	1.0	3.61	2.98	57
a ▢ b	1.43	3.73	3.08	59
a ▭ b	2.0	4.12	3.39	62
a ▭ b	3.0	4.79	3.96	69
a ▭ b	4.0	5.33	4.44	73
a ▭ b	8.0	6.49	5.60	82
▭	∞	8.23	7.54	96
Heated ▧ Insulated	∞	5.39	4386	96
△	—	3.11	2.49	53

Notes: (i) The Poiseuille number $Po = f\,Re_{D_h}$, where the Darcy friction factor $f = 8\tau_w / \rho v^2$, (ii) The Nusselt number $Nu = hD_h / k$.

Clearly, the larger S_{gen}, the more inefficient a process, device, or system is, i.e., S_{gen} is equivalent to "amount of waste generated." In convection heat transfer this "energy destruction" appears as viscous dissipation and random disorder due to heat input:

$$S_{gen} \sim \mu\Phi \text{ and } k(\nabla T)^2 \qquad (1.99a)$$

or

$$S_{gen}^{total} = S_{gen}^{friction} + S_{gen}^{thermal} \qquad (1.99b)$$

Entropy Generation Derivation.   For optimal system/device design it is important to find for a given objective the *best possible system geometry and operational conditions* so that S_{gen} is a minimum. Thus, within the framework of convection heat transfer with Newtonian fluids, it is of interest to derive an expression for

$$S_{gen} = S_{gen}(\text{thermal}) + S_{gen}(\text{friction}) \qquad (1.100)$$

Clearly, Eq. (1.100) encapsulates the irreversibilities due to heat transfer ($S_{gen,thermal}$) and viscous fluid flow ($S_{gen,friction}$).

Considering a point (x,y,z) in a fluid with convective heat transfer, the fluid element dx-dy-dz surrounding this point is part of a thermal flow system. Thus, the small element dx-dy-dz can be regarded as an open thermodynamic system, subject to mass fluxes, energy transfer, and entropy transfer interactions that penetrate the fixed control surface formed by the dx-dy-dz box of Figure 1.15. Hence, the local volumetric rate of entropy generation $\left(S_{gen} \text{ in } \left[\text{kW}/\left(\text{m}^3 \cdot \text{K}\right)\right]\right)$ is considered inside a viscous fluid with convective heat transfer without internal heat generation. The second law of thermodynamics for the dx-dy-dz box as an open system experiencing fluid flow and convective heat transfer then reads, based on the Clausius definition $dS = \delta Q/T\big|_{\text{reversible}}$ and Figure 1.15:

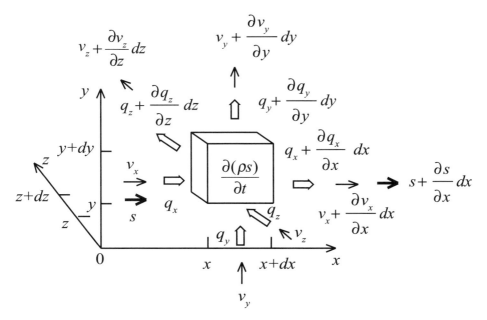

Figure 1.15 The local generation of entropy in a flow with a viscous fluid and conductive heat transfer

$$S_{gen} dxdydz = \frac{q_x + \frac{\partial q_x}{\partial x} dx}{T + \frac{\partial T}{\partial x} dx} dydz - \frac{q_x}{T} dydz + \frac{q_y + \frac{\partial q_y}{\partial y} dy}{T + \frac{\partial T}{\partial y} dy} dxdz - \frac{q_y}{T} dxdz$$

$$+ \frac{q_z + \frac{\partial q_z}{\partial z} dz}{T + \frac{\partial T}{\partial z} dz} dxdy - \frac{q_z}{T} dxdy + \frac{\partial(\rho s)}{\partial t} dxdydz$$

$$+ \left(s + \frac{\partial s}{\partial x} dx \right) \left(v_x + \frac{\partial v_x}{\partial x} dx \right) \left(\rho + \frac{\partial \rho}{\partial x} dx \right) dydz - sv_x \rho dydz$$

$$+ \left(s + \frac{\partial s}{\partial y} dy \right) \left(v_y + \frac{\partial v_y}{\partial y} dy \right) \left(\rho + \frac{\partial \rho}{\partial y} dy \right) dxdz - sv_y \rho dxdz$$

$$+ \left(s + \frac{\partial s}{\partial z} dz \right) \left(v_z + \frac{\partial v_z}{\partial z} dz \right) \left(\rho + \frac{\partial \rho}{\partial z} dz \right) dxdy - sv_z \rho dxdy \qquad (1.101a)$$

The first six terms on the right side of Eq. (1.101a) account for the entropy transfer associated with heat transfer. Combining terms 1 and 2, 3 and 4, and 5 and 6 and dividing by dx-dy-dz and taking the limit, the former six terms in Eq. (1.101a) can be reduced to:

$$\frac{T\frac{\partial q_x}{\partial x} - q_x \frac{\partial T}{\partial x}}{T\left(T + \frac{\partial T}{\partial x} dx\right)} + \frac{T\frac{\partial q_y}{\partial y} - q_y \frac{\partial T}{\partial y}}{T\left(T + \frac{\partial T}{\partial y} dy\right)} + \frac{T\frac{\partial q_z}{\partial z} - q_z \frac{\partial T}{\partial z}}{T\left(T + \frac{\partial T}{\partial z} dz\right)}$$

$$= \frac{1}{T}\left(\frac{\partial q_x}{\partial x} + \frac{\partial q_y}{\partial y} + \frac{\partial q_z}{\partial y} \right) - \frac{1}{T^2}\left(q_x \frac{\partial T}{\partial x} + q_y \frac{\partial T}{\partial y} + q_z \frac{\partial T}{\partial z} \right) \qquad (1.101b)$$

Terms 7 to 12 in Eq. (1.101a) represent the entropy convected into and out of the system, while the last term is the time rate of entropy accumulation in the dx-dy-dz control volume. Decomposing and combining the last seven terms as well as considering the limit, the last seven terms can be rearranged as:

$$\rho\left(\frac{\partial s}{\partial t} + v_x \frac{\partial s}{\partial x} + v_y \frac{\partial s}{\partial y} + v_z \frac{\partial s}{\partial z} \right)$$

$$+ s\left[\frac{\partial \rho}{\partial t} + v_x \frac{\partial \rho}{\partial x} + v_y \frac{\partial \rho}{\partial y} + v_z \frac{\partial \rho}{\partial z} + \rho\left(\frac{\partial v_x}{\partial x} + \frac{\partial v_y}{\partial y} + \frac{\partial v_z}{\partial z} \right) \right] \qquad (1.101c)$$

Combining Eq. (1.101b) and Eq. (1.101c), the local rate of entropy generation becomes:

$$
\begin{aligned}
S_{\text{gen}} &= \frac{1}{T}\left(\frac{\partial q_x}{\partial x}+\frac{\partial q_y}{\partial y}+\frac{\partial q_z}{\partial z}\right)-\frac{1}{T^2}\left(q_x\frac{\partial T}{\partial x}+q_y\frac{\partial T}{\partial y}+q_z\frac{\partial T}{\partial z}\right) \\
&\quad +\rho\left(\frac{\partial s}{\partial t}+v_x\frac{\partial s}{\partial x}+v_y\frac{\partial s}{\partial y}+v_z\frac{\partial s}{\partial z}\right) \\
&\quad +s\left[\frac{\partial \rho}{\partial t}+v_x\frac{\partial \rho}{\partial x}+v_y\frac{\partial \rho}{\partial y}+v_z\frac{\partial \rho}{\partial z}+\rho\left(\frac{\partial v_x}{\partial x}+\frac{\partial v_y}{\partial y}+\frac{\partial v_z}{\partial z}\right)\right]
\end{aligned}
\tag{1.102}
$$

Note that the last term of Eq. (1.102) (in square brackets) vanishes identically based on the mass conservation principle. Equation (1.102) can be recast, so that in vector notation the volume rate of entropy generation reads:

$$
S_{\text{gen}} = \frac{1}{T}\nabla\cdot\vec{q}-\frac{1}{T^2}\vec{q}\cdot\nabla T+\rho\frac{Ds}{Dt}
\tag{1.103}
$$

According to the first law of thermodynamics, the rate of change in internal energy per unit volume is equal to the net heat transfer rate by conduction, plus the work transfer rate due to compression, plus the work transfer rate per unit volume associated with viscous dissipation, i.e.,

$$
\rho\, D\tilde{u}/Dt = -\nabla\cdot\vec{q}-p\left(\nabla\cdot\vec{v}\right)+\mu\Phi
\tag{1.104}
$$

Employing the Gibbs relation in the form $d\tilde{u}=Tds-pd\left(1/r\right)$ and using the substantial derivative notation Eq. (1.104), we obtain:

$$
\rho\frac{Ds}{Dt} = \frac{\rho}{T}\frac{D\tilde{u}}{Dt}-\frac{p}{\rho T}\frac{D\rho}{Dt}
\tag{1.105}
$$

Combining Eq. (1.105) with $\rho\, Ds/Dt$ given by Eq. (1.103) and $\rho\, D\tilde{u}/Dt$ given by Eq. (1.104), the volumetric entropy generation rate can be expressed as:

$$
S_{\text{gen}} = -\frac{1}{T^2}\vec{q}\cdot\nabla T+\frac{\mu}{T}\Phi
\tag{1.106}
$$

If the Fourier law of heat conduction for an isotropic medium applies, i.e.,

$$
\vec{q} = -k\nabla T
\tag{1.107}
$$

the rate of volumetric entropy generation (S_{gen}) in three-dimensional Cartesian coordinates is then (see also Bejan, 1996):

$$S_{gen} \equiv S_G = \frac{k}{T^2}\left[\left(\frac{\partial T}{\partial x}\right)^2 + \left(\frac{\partial T}{\partial y}\right)^2 + \left(\frac{\partial T}{\partial z}\right)^2\right]$$

$$+\frac{\mu}{T}\left\{2\left[\left(\frac{\partial u}{\partial x}\right)^2 + \left(\frac{\partial v}{\partial y}\right)^2 + \left(\frac{\partial w}{\partial z}\right)^2\right] + \left(\frac{\partial u}{\partial y} + \frac{\partial v}{\partial x}\right)^2\right.$$

$$\left.+\left(\frac{\partial u}{\partial z} + \frac{\partial w}{\partial x}\right)^2 + \left(\frac{\partial v}{\partial z} + \frac{\partial w}{\partial y}\right)^2\right\} \qquad (1.108)$$

where u, v, and w are velocity vectors in the x-, y-, and z-directions, respectively; T is the temperature, k is the thermal conductivity, and μ is the dynamic viscosity.

Specifically, the dimensionless entropy generation rate induced by fluid friction and heat transfer can be defined as follows:

$$S_{G,F} = S_{gen}\,(\text{frictional}) \cdot kT_0^2 / q^2 \qquad (1.109)$$

where

$$S_{gen}\,(\text{frictional}) = \frac{\mu}{T}\left\{2\left[\left(\frac{\partial u}{\partial x}\right)^2 + \left(\frac{\partial v}{\partial y}\right)^2 + \left(\frac{\partial w}{\partial z}\right)^2\right]\right.$$

$$\left.+\left(\frac{\partial u}{\partial y} + \frac{\partial v}{\partial x}\right)^2 + \left(\frac{\partial u}{\partial z} + \frac{\partial w}{\partial x}\right)^2 + \left(\frac{\partial v}{\partial z} + \frac{\partial w}{\partial y}\right)^2\right\} \qquad (1.110)$$

while for the thermal entropy source,

$$S_{G,Y} = S_{gen}\,(\text{thermal}) \cdot \frac{kT_0^2}{q^2} \qquad (1.111)$$

where

$$S_{gen}\,(\text{thermal}) = \frac{k}{T^2}\left[\left(\frac{\partial T}{\partial x}\right)^2 + \left(\frac{\partial T}{\partial y}\right)^2 + \left(\frac{\partial T}{\partial z}\right)^2\right] \qquad (1.112)$$

Finally,

$$S_{G,\text{total}} = S_{\text{gen}} \frac{kT_0^2}{q^2} = S_{G,F} + S_{G,T} \qquad (1.113)$$

where T_0 is the fluid inlet temperature and q is the wall heat flux.

Example 1.11: Thermal Pipe Flow with Entropy Generation

Deriving the irreversibility profiles for Hagen-Poiseuille (H-P) flow through a smooth tube of radius r_0 with uniform wall heat flux $q\left[\text{W}/\text{m}^2\right]$ at the wall, the velocity and temperature for the fully developed regime are given by:

$$u = 2U\left[1 - \left(\frac{r}{r_0}\right)^2\right] \qquad (E1.11\text{-}1)$$

and

$$T - T_s = -\frac{qr_0}{k}\left[\frac{3}{4} - \left(\frac{r}{r_0}\right)^2 + \frac{1}{4}\left(\frac{r}{r_0}\right)^4\right] \qquad (E1.11\text{-}2)$$

Sketch:	*Assumptions:*	*Conceptions:*
	• Fully developed H-P flow • Constant properties and parameters	• Volumetric entropy generation rate Eq. (1.91) • Thermal entropy generation • Frictional entropy generation

Solution:

The wall temperature $T_s = T\left(r = r_0\right)$ can be obtained from the condition

$$\frac{\partial T}{\partial x} = \frac{dT_s}{dx} = \frac{2q}{\rho c_p U r_0} = \text{constant} \qquad (E1.11\text{-}3)$$

$$\frac{\partial T}{\partial r} = \frac{q}{k}\left[2\frac{r}{r_0} - \left(\frac{r}{r_0}\right)^3\right] \qquad (E1.11\text{-}4)$$

$$\frac{\partial u}{\partial r} = \frac{-4Ur}{r_0^2} \qquad (E1.11\text{-}5)$$

Hence, the dimensionless entropy generation for fully developed tubular H-P flow can be expressed as:

$$S_{gen}\frac{kT_0^{\,2}}{q^2}=\frac{4k^2}{\left(\rho c_p\bar{U}r_0\right)^2}\frac{T_0^{\,2}}{T^2}+\left(2R-R^3\right)^2\frac{T_0^{\,2}}{T^2}+\frac{16kT_0^{\,2}\mu\bar{U}^2}{q^2Tr_0^{\,2}}R^2$$

$$=\underbrace{\frac{16}{\mathrm{Pe}^2}\left(\frac{T_0}{T}\right)^2+\left(2R-R^3\right)^2\frac{T_0^{\,2}}{T^2}}_{\text{heat transfer}}+\underbrace{\varphi\frac{T_0}{T}R^2}_{\text{fluid friction}} \qquad \text{(E1.11-6)}$$

with

$$\mathrm{Pe}=\mathrm{Re}\cdot\mathrm{Pr}=\frac{2r_0\rho c_p U}{k} \qquad \text{(E1.11-7)}$$

and

$$R=\frac{r}{r_0} \text{ and } \varphi=\frac{16kT_0\mu U^2}{q^2r_0^{\,2}} \qquad \text{(E1.11-8a,b)}$$

Here, R is the dimensionless radius, T_0 is the inlet temperature which was selected as the reference temperature. On the right side of Eq. (E1.11-6), the first term represents the entropy generation by axial conduction, the second term is the entropy generated by heat transfer in the radial direction, and the last term is the fluid friction contribution. Parameter φ is the irreversibility distribution ratio $\left(S_{gen}\left(\text{fluid friction}\right)/S_{gen}\left(\text{heat transfer}\right)\right)$.

Graph:

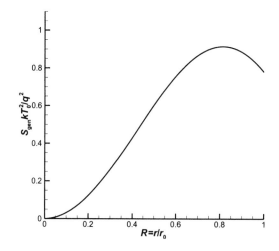

Comments:

As expected, according to Eq. (E1.11-6), at the center point, i.e., $R = 0$, only the first term on the right side contributed to the dimensionless entropy generation rate; however, for Pe $\gg 1$, the irreversibility due to axial conduction is negligible in the fully developed range. In contrast, in the wall region both thermal and frictional effects produce entropy with a maximum at $R \approx 0.8$ generated by dominant heat transfer induced entropy generation.

1.4 Homework Assignments

The suggested homework assignments either try to illuminate the text material via questions and tasks (Category I), probing basic understanding and "physical insight," or pose typical fluid dynamics problems whose solutions illustrate the chapter theory. These "text problems" (Category II) were accumulated selectively in revised and extended form over years of undergraduate teaching at North Carolina State University with books by Cengel & Cimbala, Fox et al., Potter & Wiggert, and White. Again, their contributions are gratefully acknowledged.

1.4.1 Physical Insight

1.1   Discuss additional implications of fluidic-device miniaturization down to the molecular level, i.e., provide examples of new (or unusual) material/fluid properties, fluid flow behavior, and device/system applications.

1.2   A force per unit area is labeled "pressure" or "stress"; however, one is a scalar and the other, in general, a tensor of rank 2: Explain the differences mathematically and physically.

1.3   Fluid flow through micro/nanochannels requires a driving force (e.g., in the form of a pressure drop). Contrast necessary "pumps" providing such a pressure differential for: (a) conventional (i.e., macroscale) systems and (b) micro/nanoscale systems.

1.4   Considering an ideal gas, derive expressions for pressure p, temperature T, and dynamic viscosity μ on the molecular level and explain.

1.5   Discuss the *continuum mechanics assumption* in light of "fluid flow" borderline cases, such as grain flow in a silo or jelly movement in food processing.

1.6   Is "open-channel flow" an example of *internal flow* or *external flow*?

1.7   Derive Eq. (1.10) and plot the case $\vec{v} = (0.5 + 0.8x)\hat{i} + (1.5 - 0.8y)\hat{j}$ in $[\text{m/s}]$.

1.8   Employing *scale analysis,* derive the following dimensionless groups: Peclet number, Euler number, Nusselt number, Weber number, and Sherwood number; also provide sample applications.

1.9   Considering Figure 1.3, set up a general energy balance (i.e., a first-order difference equation) for a transient open system, and show that any closed system is just a special case. What are the advantages/disadvantages of such a lumped-parameter approach vs. a differential approach with spatial resolution?

1.10 Recast Eqs. (1.15) to (1.17) so that the material properties all appear as (momentum, thermal, and mass) *diffusivities*, say, all in $[\text{m}^2\text{/s}]$.

1.11 Design an illustrative figure depicting the four "derivation approaches" mentioned in Sect. 1.3.1. Specifically, list math and physics aspects as well as possible approach interactions and applications.

1.12 Considering the extended cases of the RTT discussed in Sect. 1.3.2, expand Eq. (1.19) and derive two more general RTTs, i.e., one for a translating, accelerating, and deforming control volume and another one for a rotating control volume.

1.13 Concerning Example 1.3: (a) what are the limits on vertical coordinate z and (b) derive Eq. (E1.3-4), i.e., Torricelli's law.

1.14 Looking at Sect. 1.3.2.2, expand the D/Dt operator in the LHS of Eq. (1.28a) and express properly the RHS to arrive *directly* at the Navier-Stokes equation for linear momentum transfer (see Eq. (1.80)).

1.15 Provide some additional math and physics details to the "derivation" of Eq. (1.33), starting with the energy conservation law (1.31) in Sect. 1.3.2.3.

1.16 Starting with Eq. (1.37), derive a useful form of Eq. (1.40). Contrast that to the derivation approach when using the energy RTT.

1.17 What are the mathematical expressions for the three heat fluxes making up the total heat flux in Eq. (1.40)?

1.18 List and comment on the advantages of the conservation laws in *differential* form (see Sect. 1.3.3) over the integral form.

1.19 What is the mathematical reason and give a physical explanation for attaching the unit tensor to the pressure p in Eq. (1.52) and in other similar cases?

1.20 Explain the need and meaning of the transpose of the "velocity field gradient" in Eq. (1.54); also comment on the advantages/disadvantages of equations written/manipulated in "vector notation" versus "index notation."

1.21 Illustrate and contrast with an application each the following: the stress vector versus the stress tensor versus the total stress tensor.

1.22 Why is the (total) stress tensor symmetric?

1.23 Considering shear flow, illustrate Eq. (1.62) to explain Eqs. (1.63) and (1.64).

1.24 Prove Eq. (1.65) and illustrate/explain the "collapse" of the vorticity tensor to the vorticity vector

1.25 Equation (1.69) implies a host of boundary conditions. Discuss three basic cases.

1.26 Most solved book examples assume steady laminar fully developed flow, known as "Poiseuille-type" flow. Discuss the mathematical and physical implications and show three applications for which that flow field simplification would be *incorrect*.

1.27 In Sect. 1.3.3.3, four assumptions are listed for the transformation of the Cauchy equation to the Navier-Stokes equation: (a) justify these four assumptions; (b) derive Eq. (1.81) from Eq. (1.52); (c) give two examples when using the N-S equation would be inadequate.

1.28 Prandtl's boundary layer equations plus Euler's equation are still quite popular in aerodynamics and hardly used everywhere else. Why? In contrast, Stokes' equation (see Sect. 1.3.3.3) is often applicable in micro/nanofluidics. Why?

1.29 On a macroscale, enthalpy encapsulates internal energy plus flow energy and hence the essential contributions to "total energy" for many flow systems, devices, and processes. Hence, replacing e_{total} with h in Eq. (1.92), derive Eq. (1.93) and Eq. (1.95).

1.30 Starting with the "species mass" RTT, i.e., $B_s = m/\forall = c$, and assuming Fourier's law plus constant properties, derive Eq. (1.96). What are examples of S_c?

1.31 Show that, in general, hydrodynamic and thermal entrance lengths differ, i.e., provide math/physics/graphical explanations. Under what condition are they the same, if any?

1.32 Looking at Table 1.3, why is Nu (q_{wall} = const.) always greater than Nu (T_{wall} = const.)?

1.33 Derive Eq. (1.98) from the inequality stated in (1.97).

1.34 Explain/illustrate the two proportionalities for S_{gen} in Eq. (1.99a).

1.35 Derive Eqs. (1.104) and (1.105) and show that Eq. (1.106) is correct.

1.4.2 Text Problems

1.36 *Tailored continuity equations in differential form:* Set up the continuity equation for:

  (a) steady 1-D compressible flow in a tube;

  (b) compressible isothermal gas flow in terms of pressure p;

  (c) an incompressible fluid flowing just radially into: (i) a line sink or (ii) a point sink.

1.37 *Velocity components of incompressible flows:* Determine and plot:

  (a) $v(x,y)$ for planar flow with $u(x,y) = 10 + \dfrac{5x}{x^2 + y^2}$ if $v(x,0) = 0$

  (b) $v_r(r,\theta)$ for cylindrical flow with $v_\theta = -\left(10 + 4/r^2\right)\cos\theta$ if $v_r(2,\theta) = 0$

  (c) $v(x,y)$ for a near-surface 2-D flow with $u(x,y) = 10\left[2\left(\dfrac{y}{\delta}\right) - \left(\dfrac{y}{\delta}\right)^2\right]$, assuming $\delta(x) = Cx^{4/5}$, $v(x,0) = 0$, and $w = 0$

1.38 *Reduced system of Navier-Stokes equations:* Consider a large horizontal plate beneath a liquid to oscillate with $u_{wall} = u_0 \sin\omega t$. Propose the conservation laws for: (i) constant viscosity and (ii) temperature-dependent viscosity.

1.39 *Transient fully developed pipe flow:* Based on measurements, the following 1-D velocity profile has been constructed:

$$u(r,t) = u_m \left[1 - \left(\frac{r}{r_0} \right)^2 \right] \left(1 - e^{t/\tau} \right)$$

where $u_m = 2$ m/s, $r_0 = 0.02$ m, and $\tau = 10$ s. Find the fluid element velocity and acceleration for $t = 1$ s at $r = 0$ and $r = 0.25 r_0$. At what times t do the maximum values appear at these locations? Plot the parabolic profiles at different time levels during $0 \le t \le \infty$.

1.40 *Periodically changed temperature field:* In a steady flow field described by $u = u_m (1 - \hat{y}^2)$, the temperature varies as

$$T(\hat{y}, t) = T_m (1 - \hat{y}^2) \cos\left(\pi \frac{t}{\tau} \right) \text{ in } °C$$

where $\hat{y} \equiv y/d$, $u_m = 2$ m/s, $T_m = 20°C$, and $\tau = 100$ s. Find $T(\hat{y} = 0, t = 20 \text{ s})$, plot $T(\hat{y}, t)$, and comment!

1.41 *Is the pipe entrance length important?* Consider a horizontal 2 mm tube 3 m long connected to a constant-head reservoir. If 9 L/h of water at 15°C is collected, determine the ratio of $L_{entrance}/L_{tube}$ and comment.

1.42 *Show qualitatively that* $\left(\Delta p / \Delta x \right)_{entrance} > \left(\Delta p / \Delta x \right)_{fully\ developed}$. Now, consider an REV (radius R and thickness Δx) for steady laminar pipe flow, and obtain $\Delta p / \Delta x$ from a 1-D force balance. Based on physical insight conclude the stated results.

1.43 *Poiseuille flow in a slanted pipe:* Its radius is R and elevation change is h, with fluid properties μ and $\gamma = \rho g$. First show that the average velocity

$$v_{avg} = -\frac{R^2}{8\mu} \frac{d}{dx}(p + \gamma h)$$

Then compute the Reynolds number and wall shear stress for 20°C water flow in a pipe ($D = 4$ mm and $L = 10$ m) slanted upward by 10° and with a pressure rise of 6 kPa over the 10 m.

1.44 *Poiseuille flow in a horizontal pipe or in an annulus:*
   (a) Find the radial ratio r_x / R where $v(r_x) = v_{avg}$.
   (b) Find the radial ratio r_x / R where $\tau(r_x) = 0.5\tau_{wall}$.
   (c) Find $Q_{pipe} / Q_{annulus}$ where the pipe radius is R and the annulus is formed by $R_{inner} = R/2$ and $R_{out} = R$.

1.45 *Poiseuille flow in an annulus of radii R_1 and R_2, $R_2 > R_1$:*
   (a) Show that $Q = \frac{\pi}{8\mu} \left(\frac{dp}{dx} \right) \left[R_2^4 - R_1^4 - \frac{\left(R_2^2 - R_1^2 \right)}{\ln\left(R_2 / R_1 \right)} \right]$.
   (b) Determine the special case of $R_1 \to 0$ (pipe flow) and $R_1 \to R_2$ where $R_2 \gg 1$ (i.e., parallel plate flow).

1.46 *Run-off from a parking lot:* Consider a film of height $h = 10$ mm of water at 20°C flowing down a 100 m-wide plane which is sloped at 0.00015. Check the Reynolds number and determine the volumetric flow rate. Plot $Q(h)$ and comment.

1.47 *Torque on a shaft:* A cylinder of $D = 40$ cm rotates with 30 rad/s in a housing ($L = 80$ cm) and is lubricated with oil at 20°C, forming an 800-μm gap. Find the necessary torque, assuming first a linear velocity profile and then using the actual velocity distribution. Determine the induced error made. Plot the torque as a function of lubrication gap.

1.48 *Rotating cone:* A 90° cone of 10 cm side length rotates with 50 rad/s in a housing with a 2 mm lubrication film ($\mu = 0.01$ N·s/m^2). Find the necessary torque when assuming: (a) a linear velocity profile and (b) a more realistic velocity distribution. Comment on the error invoked.

1.49 In the process of tape coating, the tape (width w and thickness d) is pulled at constant velocity U through a housing of height H and length $L (H \ll L)$, containing the coating liquid of viscosity μ. Develop an expression for U_{max} not quite reaching the maximum tensile force, F_{pull}, the tape can withstand. Plot $U_{max}(\mu)$ and comment.

1.50 Pulling vertically a rigid sheet out of an oil-bath will leave a film on the surface. Given a constant U_{pull}, the sheet dimensions ($a \times b$), and the oil properties, develop a dimensionless equation for the velocity profile and estimate the oil thickness h and the necessary pulling force.

CHAPTER 2

Applications

By now it should be apparent that computing the *velocity field* is fundamental for the *detailed* description of the transport phenomena of a flow system, subject to appropriate initial and wall conditions as well as inlet and outlet profiles. Employing this "differential approach," once the velocity (and pressure) in space and time have been computed (or assumed), stresses, forces, and flow rates can be directly evaluated. In turn, determining the temperature and concentration is just a matter of solving (simplified) versions of the scalar heat and species mass transfer equations (see Table 1.2 and Figure 1.14).

2.1 Internal Fluid Flow

As implied in Sect. 1.3.3, the safe starting point is a reduced form of the Cauchy equation, i.e., the momentum equation in terms of stresses (see Sect. A.5.2). For example, when solving non-Newtonian fluid flow problems, Eq. (1.53) is needed because the apparent viscosity is dependent on the shear rate. If *constant fluid properties* can be assumed, the point of departure is the Navier-Stokes (N-S) equation (see Eq. (1.80) and Sect. A.5.3). A system of equations has to be considered when solving flow problems with temperature-dependent fluid properties, i.e., the coupled continuity, momentum, and heat transfer equations are solved iteratively.

For this section, the main goal is to find *analytic* problem solutions which can provide physical insight and are useful for validating more complex computer simulation models. Of course, a given problem has to be reduced to a simple partial differential equation (PDE) using separation of variables or a type of ordinary differential equation (ODE) suitable for direct integration or look-up in ODE solution books (e.g., Polyanin & Zaitsev, 1995). In order to set up a math problem description amiable to an analytic solution via direct integration, subject to suitable boundary conditions, the flow system's complexity has to be greatly reduced, while major transport phenomena are still preserved (see Table 2.1). Similar assumptions have to be invoked for directly solving heat transfer and species mass transfer problems.

A few basic fluid flow problems can be described with a second-order ODE (see Eq. (2.2c)), subject to two boundary conditions. Although being rather simple, they should illustrate the following:

- There is no mystery to setting up and finding exact (or useful approximate) solutions to *reduced forms* of the N-S equations, which are the conservation laws for fluid and species mass, as well as momentum and energy under the assumption of constant properties.
- These results (see Examples 2.2 to 2.6) provide some interesting insight to the physics of fluid flow. They form base-case solutions to a family of engineering applications, such as film coating, numerous internal flows, simple lubrication as well as non-Newtonian fluid flows, such as exotic oils, blood, paints, and polymeric liquids (see also Sect. 2.3.4).

For all practical purposes, the generic steps for *setting up flow problems for differential analysis* and solving them includes the following:

(i) Clever placement of the appropriate coordinate system into the system sketch is important to show the principal flow direction, which identifies the momentum component of interest, and to gain simple boundary conditions.
(ii) The velocity vector $\bar{v}$ and pressure gradient ∇p (also a vector) are the key unknowns; thus, based on the given flow system and with the help of Table 2.1, *postulates* for $\bar{v}$ and ∇p have to be provided first. The specific questions are which $\bar{v}$-component and ∇p-component is nonzero? What are their functional dependence based on the stated assumptions and boundary conditions? Is the continuity equation satisfied (see Example 2.1)?
(iii) Once functional postulates are determined, the nonzero component of the N-S equations (Sect. A.5.2) can be deduced and the resulting ODE integrated, subject to appropriate boundary conditions.

2.1.1 Problem-Solving Steps

Clearly, the background information in Table 2.1 has to be judiciously applied so that the flow problem can be reduced to a manageable ODE. Otherwise, a numerical solution using computer software (e.g., Matlab, Maple, COMSOL, CFX, Fluent, STAR CCM, etc.) is called for. The actual *solution procedure* can be divided into three essential steps (see Figure 2.1):

(i)   classification of the fluid flow system;
(ii) mathematical description of the system;
(iii) solution of the modeling equations and result graphing plus comments.

Table 2.1   Flow Assumptions and Their Impact

Flow Assumption	Explanation	Examples
Time dependence	$\partial/\partial t = 0$ implies steady state; but, flow is transient when $\vec{v} = \vec{v}(t)$	Transient flow: Pulsatile flow; fluid structure vibration; flow start-up/shut-down; etc.
Dimensionality	Required number of space coordinates	1-D: Couette flow; Poiseuille flow; thin-film flow 2-D: Boundary-layer flow; pipe entrance flow 3-D: Everything real …
Directionality	Required number of velocity components: $\vec{v} = (u,v,w)$, rectangular or $\vec{v} = (v_z,v_r,v_\theta)$, cylindrical or $\vec{v} = (v_r,v_\theta,v_\varphi)$, spherical	Usually same as "Dimensionality" with some exceptions: For example, for the rotating parallel disk (or viscous clutch) problem, $\vec{v} = (0,v_\theta,0)$ where $v_\theta = v_\theta(r,\theta)$, i.e., unidirectional; but the system is 2-D
Development phase	$\partial v/\partial s = 0$: fully developed flow ($s \hat{=}$ axial coordinate)	"Fully developed" implies no velocity profile changes in that direction; in this case, s
Symmetry	$\partial/\partial n = 0$: midplane or centerline ($n \hat{=}$ normal coordinate) $\partial/\partial\theta = 0$: axisymmetry	Self-explanatory

Flow regime: $\begin{cases} \text{Laminar} \rightarrow \text{Re}_{max} < \text{Re}_{critical} \\ \text{Turbulent} \rightarrow \text{Re} > \text{Re}_{critical} \end{cases}$ where

$\text{Re}_{critical} \approx \begin{cases} 2000 \text{ for pipe flow} \\ 5 \times 10^5 \text{ for flat-plate boundary-layer flow (see Fig. 1-11)} \end{cases}$

Now, by invoking suitable assumptions (see Table 2.1) the solution steps of Figure 2.1 are executed for a variety of basic flow problems. For the most part, they are *steady laminar incompressible 1-D (i.e., fully developed) flows* in tubes, between parallel plates and on inclines (i.e., thin films). In slightly more general terms, such flows are unidirectional (or parallel) where the velocity vector has only one component, e.g., $\vec{v} = (u,0,0)$. However, the axial (i.e., unidirectional) velocity u could be a function of t, y, and z as in transient duct flows (see Stokes equation (1.85)).

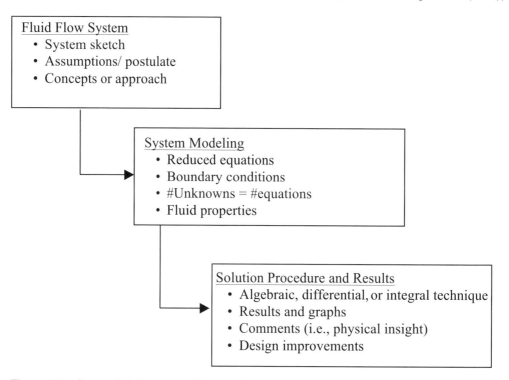

Figure 2.1   Sequential steps in problem solving

In unidirectional or *parallel flows,* the velocity vector has only one component, e.g., $\vec{v} = (u, 0, 0)$, where in general $u = u(y,z,t)$ and hence the Stokes equation (1.85) has to be solved. Unidirectional flow implies that the pressure gradient is constant, the flow is fully developed, and the resulting momentum equation expresses a dynamic equilibrium (or force balance) at all times between a 1-D driving force and frictional resistance:

$$\underbrace{\frac{\partial p}{\partial x} \text{ and/or } rg_x}_{\text{driving forces}} \sim \underbrace{\frac{\partial t_{xy}}{\partial y}}_{\text{resistance}} \tag{2.1a,b}$$

For steady 1-D incompressible flow $u = u(y)$, or $v_z = v_z(r)$ in cylindrical coordinates, as illustrated in Examples 1.9 and 2.1 to 2.3. Specifically, the reduced x-momentum of the Navier-Stokes equation based on the postulates $u = u(y)$ only and $\partial p / \partial x = ¢$ (¢ meaning constant) reads (see Sect. A.5.2):

$$0 = -\left(\frac{\partial p}{\partial x}\right) + \mu \frac{d^2 u}{dy^2} + \rho g_x \qquad (2.2a)$$

Rewriting Eq. (2.2a) with

$$-\frac{\partial p}{\partial x} = \frac{\Delta p}{\ell} = \frac{p_{in} - p_{out}}{\ell}$$

where ℓ is the system length, yields:

$$\frac{d^2 u}{dy^2} = \frac{1}{\mu}\left[-\left(\frac{\Delta p}{\ell}\right) - \rho g_x\right] = K = ¢ \qquad (2.2b)$$

It is apparent that steady laminar 1-D parallel flows can be described by a second-order ODE (see Sect. A.4) of the form:

$$u'' = K \qquad (2.2c)$$

Equation (2.2c) is subject to two boundary conditions, e.g., "no-slip" and symmetry.

2.1.2 Sample Solutions of the Reduced Navier-Stokes Equations

Steady Parallel or Near-Parallel Flows.   Knowledge of Sect. 2.1.1 material is now applied to solving problems which can be accurately described by reduced forms of the system of N-S equations, i.e., simplified versions of mass (continuity), momentum, and heat transfer. As mentioned, the upcoming examples serve a triple purpose: they provide physical insight, they constitute "role model" solutions important in both macrofluidics and microfluidics (Part B), and they are benchmark solutions for validation of complex computer simulation models. Thus, the student's goal is to be able to solve these problems independently.

Clearly, one of the key underlying assumptions for unidirectional flow is that the flow is *fully developed*. This is justifiable when the *entrance length* of conduits is short, i.e.,

$$L_{entrance} \ll L_{conduit} \qquad (2.3)$$

As outlined in Sect. 1.2, assuming steady laminar conditions, we have for:

- Macroscale flow

$$L_{\text{entrance}}^{\text{laminar}} = 0.05 \, \text{Re}_{D_h} \cdot D_h \; ; \; \text{Re}_{D_h} = \frac{vD_h}{v} \tag{2.4a,b}$$

$$L_{\text{entrance}}^{\text{turbulent}} = 10 \cdot D_h \; ; \; D_h = 4 \, A/P \tag{2.5a,b}$$

$$L_{\text{entrance}}^{\text{laminar}} \bigg|_{\text{thermal}} \approx 0.056 \, \text{Re}_{D_h} \cdot \text{Pr} \cdot D_h \; ; \; \text{Pr} = v/\alpha \tag{2.6a,b}$$

- Microscale flow

$$L_{\text{entrance}}^{\text{laminar}} = 0.5 \, \text{Re}_{D_h} \cdot D_h \tag{2.7}$$

===

Example 2.1: Poiseuille Flow

Consider steady laminar fully developed flow of an incompressible fluid in a pipe of length L and radius r_0, i.e., Poiseuille flow. Establish reduced forms of the N-S equations and solve for the axial velocity $u(r)$.

Sketch:	Assumptions:	Concept:
	• As stated, i.e., $\partial/\partial t = 0$ (steady-state) $\vec{v} \Rightarrow u$ (unidirectional) $\partial/\partial x = 0$ (fully developed) $\partial/\partial\theta = 0$ (axisymmetric)	• Reduced N-S equation based on the *postulates*: $\vec{v} = (u,0,0)$, where $u = u(r)$ only; and $-\nabla p \Rightarrow -\dfrac{\partial p}{\partial x} \triangleq \dfrac{\Delta p}{L}$ $= (p_1 - p_2)/L = \cancel{c}$

Solution:

Continuity, $\nabla \cdot \vec{v} = 0$, can be reduced to $\partial u/\partial x = 0$, which implies that the other two velocity components are zero. The N-S equation (1.80) reduces to:

$$0 = -\nabla p + \mu \nabla^2 \vec{v} \tag{E2.1-1}$$

or with the *postulates* and Sect. A.2:

$$0 = \left(\frac{\Delta p}{L} \right) + \mu \frac{1}{r} \frac{d}{dr} \left(r \frac{du}{dr} \right) \tag{E2.1-2}$$

subject to the boundary conditions $u(r = r_0) = 0$ (no slip) and at $r = 0$, $du/dr = 0$ (axisymmetry). Double integration and invoking the two boundary conditions yield:

$$u(r) = \frac{1}{4\mu} \left(\frac{\Delta p}{L} \right) (r_0^2 - r^2) = u_{max} \left[1 - \left(\frac{r}{r_0} \right)^2 \right] \tag{E2.1-3}$$

Comments:

The solution (E2.1-3) is the base case for steady laminar *internal* flows with numerous applications in terms of industrial pipe flows, such as pipe networks, heat exchangers, and fluid transport, as well as idealized tubular flows in biomedical, chemical, environmental, mechanical, and nuclear engineering. Equation (E2.1-3) not only provides the parabolic velocity profile $u(r)$ but can also be used to calculate the volumetric flow rate Q (Eq. (1.24)), wall shear stress τ_w (Eq. (1.68)), and pressure drop Δp (either via Eq. (1.24) or from the head loss h_f (cf. Eq. (1.91)). Example 2.2 further illustrates the use of the Poiseuille flow solution (E2.1-3).

Example 2.2: Steady Laminar Fully Developed Flow in a Pipe: Poiseuille Flow Revisited and Extended

Poiseuille flow is the base case of all laminar, fully developed internal flows. Thus, the following results will be frequently used in subsequent sections and chapters.

Sketch:	*Assumptions:*
	• Constant $-\dfrac{\partial p}{\partial x} \approx \dfrac{\Delta p}{\ell}$ Note: $\dfrac{\partial p}{\partial x} \approx \dfrac{p_1 - p_2}{x_1 - x_2}$ or $-\dfrac{\partial p}{\partial x} = \dfrac{p_{in} - p_{out}}{x_\ell - 0} = \dfrac{\Delta p}{\ell}$ • Constant fluid properties ρ and μ

Concepts:	Postulates:
• Reduced N-S equations in cylindrical coordinates $\dfrac{\partial}{\partial t} = 0$ <steady state> $\dfrac{\partial}{\partial \theta} = 0$ <axisymmetric> $\dfrac{\partial}{\partial x} = 0$ <fully developed>	• $\vec{v} = (u, 0, 0);\ u = u(r)$ only • $\nabla p \rightarrow \dfrac{\partial p}{\partial x} = \cancel{c}$

Solution:

Boundary Conditions: The obvious one is $u(r = r_0) = 0$ and the second one could be $u(r = 0) = u_{max}$; however, u_{max} is unknown. Thus, we use symmetry on the centerline, i.e., $du/dr\big|_{r=0} = 0$.

Continuity: $\dfrac{\partial u}{\partial x} + 0 + 0 = 0 \Rightarrow \dfrac{\partial u}{\partial x} = 0$ <fully developed flow confirmed>

x-Momentum:
$$0 = -\frac{\partial p}{\partial x} + \mu\left[\frac{1}{r}\frac{\partial}{\partial r}\left(r\frac{\partial u}{\partial r}\right)\right] - \rho g \sin\theta \tag{E2.2-1a}$$

or

$$\underbrace{\frac{\partial p}{\partial x} + rg\sin\theta}_{\text{driving forces}} = \underbrace{\mu\left[\frac{1}{r}\frac{\partial}{\partial r}\left(r\frac{\partial u}{\partial r}\right)\right]}_{\text{flow resistance}} \tag{E2.2-1b}$$

With $u = u(r)$ only and $-\partial p/\partial x = \Delta p/\ell$, we write:

$$\frac{1}{r}\left[\frac{d}{dr}\left(r\frac{du}{dr}\right)\right] = \frac{1}{m}\left[\left(\frac{-\Delta p}{\ell}\right) + rg\sin\theta\right] \equiv K = \cancel{c} \tag{E2.2-1c}$$

or after separation of variables

$$\int d\left(r\frac{du}{dr}\right) = K\int r\,dr + C_1 \tag{E2.2-2a}$$

so that

$$\frac{du}{dr} = \frac{K}{2} + \frac{C_1}{r} \qquad \text{(E2.2-2b)}$$

A second integration yields:

$$u = \frac{K}{4}r^2 + C_1 \ln r + C_2 \qquad \text{(E2.2-2c)}$$

From the first BC

$$0 = \frac{K}{4}r_0^2 + C_1 \ln r_0 + C_2 \qquad \text{(E2.2-3a)}$$

while the second BC [see Eq. (E2.2-2b)] yields

$$0 = 0 + \frac{C_1}{r}$$

which forces C_1 to be zero to avoid a physically impossible solution. From Eq. (E2.2-3a) we have

$$C_2 = -\frac{K}{4}r_0^2 \qquad \text{(E2.2-3b)}$$

so that with

$$K \equiv \frac{1}{\mu}\left(\frac{-\Delta p}{\ell} + rg \ \sin \ \theta\right)$$

we obtain:

$$u(r) = -\frac{Kr_0^2}{4}\left[1 - \left(\frac{r}{r_0}\right)^2\right] \qquad \text{(E2.2-4)}$$

For a horizontal pipe, $\theta = 0$ and hence (see (E2.1-3)):

$$u(r) = \frac{r_0^2}{4\mu}\left(\frac{\Delta p}{\ell}\right)\left[1 - \left(\frac{r}{r_0}\right)^2\right] = u_{max}\left[1 - \left(\frac{r}{r_0}\right)^2\right] \qquad \text{(E2.2-5a,b)}$$

Notes:

- The average velocity, $u_{av} \equiv \bar{u} = \frac{1}{A} \int u(r) dA$ where $dA = 2\pi r dr$ <cross-sectional ring>, so that

$$\bar{u} = \frac{1}{2\mu}\left(\frac{\Delta p}{\ell}\right)\int_0^{r_0}\left[1-\left(\frac{r}{r_0}\right)^2\right]r\,dr = \frac{r_0^2}{8\mu}\left(\frac{\Delta p}{\ell}\right) \qquad \text{(E2.2-6a,b)}$$

- The volumetric flow rate, $Q = u_{av}A$, where $A = \pi r_0^2$, can be used to calculate the necessary pressure drop to maintain the flow:

$$Q = \frac{pr_0^4}{8\mu}\left(\frac{\Delta p}{\ell}\right) \qquad \text{(E2.2-7a)}$$

so that with a given Q-value (or Reynolds number)

$$\Delta p = \frac{8\mu Q}{\pi r_0^4}\ell \qquad \text{(E2.2-7b)}$$

- The pressure drop $\Delta p = p_{in} - p_{out}$ is positive while the pressure gradient $\partial p/\partial x$ is negative.
- By inspection of Eq. (E2.2-5), $u(r=0) = u_{max}$, where (see also (E2.2-6b)):

$$u_{max} = \frac{r_0^2}{4\mu}\left(\frac{\Delta p}{\ell}\right) = 2u_{av} \qquad \text{(E2.2-8)}$$

- The wall shear stress, i.e., the frictional impact of viscous fluid flow on the inner pipe surface, is:

$$\tau_{wall} \equiv \tau_{rx}\big|_{r=r_0} = -\mu\frac{du}{dr}\bigg|_{r=r_0} \qquad \text{(E2.2-9a)}$$

where τ_{rx} was obtained from Sect. A.2, so that

$$\tau_w = \frac{r_0}{2}\left(\frac{\Delta p}{\ell}\right) \qquad \text{(E2.2-9b)}$$

- Thus, a second expression for the pressure drop (or gradient) is found:

$$\Delta p = \frac{2\tau_w}{r_0} \ell \qquad \text{(E2.2-10)}$$

Comment:

Equation (E2.2-10) will be most valuable in solving industrial pipe flow problems in conjunction with the extended Bernoulli equation (see Eq. (1.91)).

Example 2.3: Flow between Parallel Plates

Consider viscous fluid flow between horizontal or tilted plates a small constant distance h apart. For the test section $0 \le x \le \ell$ of interest, the flow is fully developed and could be driven by friction when the upper plate is moving at $u_0 = \text{\cent}$, or by a constant pressure gradient, dp/dx, and/or by gravity (see Couette flow Example 1.9)

Sketch:	*Assumptions:*	*Concepts/Postulates:*
	• Steady laminar unidirectional flow • Constant h, $\partial p / \partial x$, and fluid properties	• Reduced N-S equation where $\vec{v}=(u, 0, 0)$ and $-\partial p/\partial x \approx \Delta p/\ell = \text{\cent}$ • $u = u(y)$ only
	Note: If the constant pressure gradient is positive, i.e., adverse, back flow occurs.	

Solution:

Although this problem looks like a simple case of a moving plate on top of a flowing film, the situation is more interesting due to the additional pressure gradient, $\partial p/\partial x$ greater than, equal to, or less than 0 and the boundary condition $u(y=h) = u_0 = \text{\cent}$. With the postulates:

$$v = w = 0, \ u = u(y=h) \text{ only}, \ -\frac{\partial p}{\partial x} \approx \frac{\Delta p}{\ell} = \text{\cent}, \text{ and } f_{body} = \rho g \sin\phi$$

continuity is fulfilled and the x-momentum equation (see the equation sheet in Sect. A.27) reduces to:

$$0 = \frac{1}{\rho}\left(\frac{\Delta p}{l}\right) + v\frac{d^2 u}{dy^2} + g\sin\phi \qquad \text{(E2.3-1a)}$$

or

$$\frac{d^2 u}{dy^2} = \frac{-1}{\mu}\left[\left(\frac{\Delta p}{l}\right) + \rho g\sin\phi\right] = \cancel{c} \qquad \text{(E2.3-1b)}$$

subject to $u(y=0)=0$, and $u(y=h)=u_0$. Again, we have to solve an ODE of the form $u'' = K$. Introducing a dimensionless pressure gradient

$$P \equiv -\frac{h^2}{2\mu u_0}\left[\left(\frac{\Delta p}{l}\right) + \rho g\sin\phi\right] \qquad \text{(E2.3-2)}$$

we can write the solution $u(y)$, known as Couette flow, in a more compact form, i.e.,

$$\frac{u(y)}{u_0} = \frac{y}{h} + P\left(\frac{y}{h}\right)\left(1-\frac{y}{h}\right) \qquad \text{(E2.3-3)}$$

Graph ($\phi = 0$):

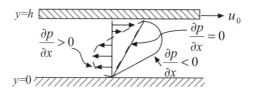

Comments:

- The pressure gradient, P or $\partial p/\partial x$, greatly influences $u(y)$. Clearly, $\partial p/\partial x < 0$ implies a "favorable" and $\partial p/\partial x > 0$ an "adverse" pressure gradient.
- For $u_0 = 0$, we have *flow between parallel plates*, i.e.,

$$u(y) = -\frac{h^2}{2\mu}\left[\left(\frac{\Delta p}{\ell}\right) - \rho g\sin\varphi\right]\left[\frac{y}{h} - \left(\frac{y}{h}\right)^2\right] \qquad \text{(E2.3-4)}$$

Examples 2.4A and 2.4B: Parallel Flows in Cylindrical Annulus

Case A: Consider flow through an annulus with tube radius R and inner concentric cylinder aR, where $a < 1$, creating a ring-like cross-sectional flow area.

Sketch:	*Assumptions:*	*Concepts:*
	• Steady laminar axisymmetric flow • Constant $\partial p/\partial z$ and constant ρ and μ	• Reduced Navier-Stokes equations in cylindrical coordinates

Postulates: $\vec{v} = (0, 0, v_z)$; $\nabla p \Rightarrow \partial p/\partial z \approx -\Delta p/\ell = \cancel{c}$; $v_z = v_z(r)$ only
$$\partial/\partial t = 0 < \text{steady state} >; \partial/\partial \varphi = 0 < \text{axisymmetric} >$$
Continuity Equation: $0 + 0 + \partial v_z/\partial z = 0$, i.e., fully developed flow
Boundary Conditions: $v_z(r = aR) = 0$ and $v_z(r = R) = 0$

Solution:

Of interest is the z-momentum equation, i.e., with the stated postulates (see the equation sheet in Sect. A.27):

$$0 = -\frac{\partial p}{\partial z} + \mu \left[\frac{1}{r} \frac{\partial}{\partial r} \left(r \frac{\partial v_z}{\partial r} \right) \right] \tag{E2.4-1a}$$

Thus, with $P \equiv \frac{1}{\mu} \left(\frac{-\Delta p}{\ell} \right)$

$$\frac{d}{dr} \left(r \frac{dv_z}{dr} \right) = P \cdot r \tag{E2.4-1b}$$

and hence after double integration,

$$v_z(r) = \frac{P}{4} r^2 + C_1 \ln r + C_2 \tag{E2.4-2}$$

Invoking the BCs,

$$0 = \frac{P}{4}(aR)^2 + C_1 \ln(aR) + C_2.$$

and

$$0 = \frac{P}{4}R^2 + C_1 \ln R + C_2$$

yields

$$v_z(r) = \frac{PR^2}{4}\left[1 - \left(\frac{r}{R}\right)^2 - \frac{1-a^2}{\ln(1/a)}\ln\left(\frac{R}{r}\right)\right] \qquad \text{(E2.4-3)}$$

and

$$\tau_{rz} = \mu\frac{dv_z}{dr} := \frac{\mu P}{2}R\left[\left(\frac{r}{R}\right) - \frac{1-a^2}{2\ln(1/a)}\left(\frac{R}{r}\right)\right] \qquad \text{(E2.4-4a,b)}$$

Notes:
For Poiseuille flow, i.e., no inner cylinder, the solution is (see Example 2.2):

$$v_z(r) = \frac{R^2}{4\mu}\left(\frac{\Delta p}{\ell}\right)\left[1 - \left(\frac{r}{R}\right)^2\right] \qquad \text{(E2.4-5)}$$

This solution is not recovered when letting $a \to 0$ because of the prevailing importance of the ln term near the inner wall.

The maximum annular velocity is not in the middle of the gap $aR \leq r \leq R$, but closer to the inner cylinder wall, where the velocity gradient is zero and hence

$$\tau_{rz}\big|_{r=bR} = 0$$

This equation can be solved for b so that $v_z(r = bR) = v_{max}$.

The average velocity is $v_{av} = \int v_z(r)dA$, where $dA = 2\pi r dr$ <cross-sectional ring of thickness dr>, so that

$$v_{av} = \frac{R^2}{8\mu}\left(\frac{\Delta p}{\ell}\right)\left[\frac{1-a^4}{1-a^2} - \frac{1-a^2}{\ln(1/a)}\right]$$

(E2.4-6)

and hence

$$Q = v_{av}\left[\pi R^2\left(1-a^2\right)\right] := \frac{\pi R^4}{8\mu}\left(\frac{\Delta p}{\ell}\right)\left[\left(1-a^4\right) - \frac{\left(1-a^2\right)^2}{\ln(1/a)}\right]$$

(E2.4-7)

The net force exerted by the fluid on the solid surfaces comes from two wall shear stress contributions:

$$F_s = \left(-\tau_{rz}\big|_{r=aR}\right)\left(2\pi aR\ell\right) + \left(\tau_{rz}\big|_{r=R}\right)\left(2\pi R\ell\right)$$

(E2.4-8a)

$$\therefore \ \ F_s = \pi R^2 \Delta p\left(1-a^2\right)$$

(E2.4-8b)

Case B: Consider Case A but now with $\partial p / \partial z \equiv 0$ and the inner cylinder rotating at angular velocity $\omega_0 = \cancel{c}$ <cylindrical Couette flow>; in general, the outer cylinder could rotate as well, say with $\omega_1 = \cancel{c}$.

Sketch:	Assumptions:	Concepts:
	• Steady laminar axisymmetric flow • Long cylinders, i.e., no end effects • Small ω's to avoid Taylor vortices	• Reduced N-S equations in cylindrical coordinates • Postulates: $\vec{v} = (0, v_\theta, 0)$, $\dfrac{\partial p}{\partial \theta} = \dfrac{\partial p}{\partial z} = 0$

Solutions:

With $v_r = v_z = 0, \partial/\partial t = \partial/\partial\theta = 0$, and $v_\theta = v_\theta(r)$ only (see the continuity equation and BCs), we reduce the θ-component of the Navier-Stokes equation (see App. A) to:

$$0 = 0 + \mu\left[\frac{\partial}{\partial r}\left(\frac{1}{r}\frac{\partial}{\partial r}(rv_\theta)\right)\right]$$

(E2.4-9)

subject to

$$v_\theta(r = aR) = \omega_0(aR) \text{ and } v_\theta(r = R) = \omega_1 R$$

Again, as in simple Couette flow after start-up, the moving-wall induced frictional effect propagates radially and the forced cylinder rotations balanced by the drag resistance generate an equilibrium velocity profile. Double integration yields:

$$v_\theta(r) = C_1 r + \frac{C_2}{r} \tag{E2.4-10a}$$

where

$$C_1 = \frac{\omega_1 R^2 - \omega_0(aR)^2}{R^2 - (aR)^2} \text{ and } C_2 = \frac{a^2 R^4(\omega_0 - \omega_1)}{R^2 - (aR)^2} \tag{E2.4-10b,c}$$

Notes:
The r-momentum equation reduces to:

$$-\frac{v_\theta^2}{r} = -\frac{1}{\rho}\frac{\partial p}{\partial r} \tag{E2.4-11}$$

Thus, with the solution for $v_\theta(r)$ known, Eq. (E2.4-11) can be used to find $\partial p/\partial r$ and ultimately the load-bearing capacity.

Applying this solution as a first-order approximation to a journal bearing where the outer tube (or sleeve) is fixed, i.e., $\omega_1 \equiv 0$, we have in dimensionless form:

$$\frac{v_\theta(r)}{\omega_0 R} = \frac{a^2}{1 - a^2}\left(\frac{R}{r} - \frac{r}{R}\right) \tag{E2.4-12}$$

The torque necessary to rotate the inner cylinder (or shaft) of length ℓ is

$$T = \int (aR)dF := (aR)\int_0^\ell \tau_{r\theta}|_{r=aR}\, dA \tag{E2.4-13}$$

where $dA = \pi(aR)dz$ and

$$\tau_{r\theta}|_{r=aR} = \mu\left[r\frac{d}{dr}\left(\frac{v_\theta}{r}\right)\right]_{r=aR}$$

Thus with:

$$\tau_{surface} \equiv \tau_{r\theta}\big|_{r=aR} = 2\mu \frac{\omega_0 R^2}{R^2 - (aR)^2} \tag{E2.4-14}$$

$$T = \tau_{surf} A_{surf}(aR) := 4\pi\mu(aR)^2 \ell \frac{\omega_0}{1-a^2} \tag{E2.4-15}$$

Graph:

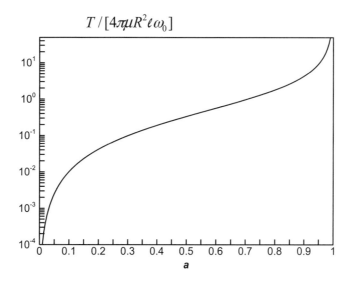

$$T / [4\pi\mu R^2 \ell \omega_0]$$

Comments:

- An electric motor may provide the necessary power, $P = T\omega_0$, which turns into thermal energy which has to be removed to avoid overheating.
- The graph depicts the nonlinear dependence of $T(a)$ for a given system. As the gap between rotor (or shaft) and stator reduces, the wall stress increases (see Eq. (E2.4-14)) as well as the surface area and hence the necessary torque.

Example 2.5: Film Coating

Consider "film coating," i.e., a liquid fed from a reservoir forms a film pulled down on an inclined plate by gravity. Obtain the velocity profile, flow rate, and wall shear force.

Sketch:
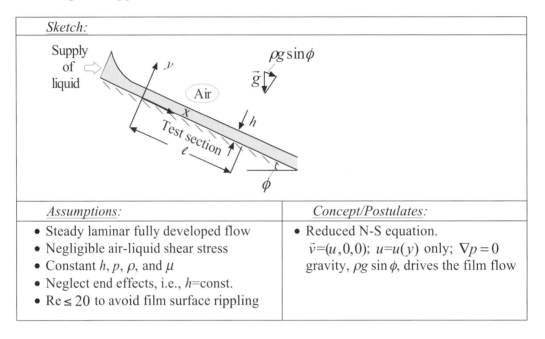

Assumptions:	Concept/Postulates:
• Steady laminar fully developed flow	• Reduced N-S equation.
• Negligible air-liquid shear stress	$\vec{v}=(u,0,0)$; $u=u(y)$ only; $\nabla p=0$
• Constant h, p, ρ, and μ	gravity, $\rho g \sin \phi$, drives the film flow
• Neglect end effects, i.e., h=const.	
• Re $\leq$ 20 to avoid film surface rippling	

Solution:

Concerning the Solution Steps (i) to (iii) just discussed, an inclined x-y coordinate system is attached to the plate, i.e., the x-momentum equation is of interest (note, the y- and z- momentum equations are both zero). Based on the sketch and assumptions, $\vec{v}=(u,0,0)$, $\nabla p \approx 0$ (thin-film condition, or no air pressure variation), and $\partial/\partial t = 0$ (steady state). With $v=w=0$, continuity $\nabla \cdot \vec{v}=0$ indicates that:

$$\frac{\partial u}{\partial x}+0+0=0 \qquad (E2.5\text{-}1)$$

which implies fully developed (or parallel) flow; thus, $u=u(y)$ only.

Consulting the equation sheet (Sect. A.27) and invoking the postulates $u=u(y)$, $v=w=0$, and $\partial p/\partial x = 0$, we have with the body force component $\rho g \sin \phi$

$$0=\mu \frac{d^2 u}{dy^2}+\rho g \sin \phi \qquad (E2.5\text{-}2)$$

subject to $u(y=0)=0$ <no slip> and $\tau_{\text{interface}}=\mu \, du/dy\big|_{y=h} \approx 0$, which implies $y=h \rightarrow du/dy=0$ <zero velocity gradient>.

Double integration of Eq. (E2.5-2) in the form $u'' = K$ (see App. A) yields after invoking the two BCs:

$$u(y) = -\frac{\rho g h^2 \sin\phi}{2\mu}\left[\left(\frac{y}{h}\right)^2 - 2y/h\right] \qquad \text{(E2.5-3a)}$$

Clearly,

$$u(y = h) = u_{max} = \frac{\rho g h^2}{2\mu}\sin\phi \qquad \text{(E2.5-3b)}$$

and the average velocity

$$u_{av} = \frac{1}{A}\int_A \vec{v}\cdot d\vec{A} := \frac{1}{h}\int_0^h u\,dy = \frac{\rho g h^2 \sin\phi}{3\mu} \qquad \text{(E2.5-4a)}$$

Hence,

$$u_{av} = \frac{2}{3}u_{max} \qquad \text{(E2.5-4b)}$$

In order to determine the film thickness, we first compute the volumetric flow rate Q, which is usually known:

$$Q = \int \vec{v}\cdot d\vec{A} := b\int_0^h u\,dy := u_{av}(hb) \qquad \text{(E2.5-5a)}$$

where b is the plate width. Thus,

$$Q = \frac{\rho\,g\,b\,\sin\phi}{3\mu}h^3 \qquad \text{(E2.5-5b)}$$

from which

$$h = \left(\frac{3\,\mu\,Q}{\rho\,g\,b\,\sin\phi}\right)^{1/3} \qquad \text{(E2.5-5c)}$$

The shear force (or drag) exerted by the liquid film onto the plate is

$$F_s = \int_A \tau_{yx}\, dA = b \int_0^\ell \left(-\mu \frac{du}{dy}\bigg|_{y=0} \right) dy \qquad\text{(E2.5-6a)}$$

Thus,

$$F_s = \rho g b h \ell \sin\phi \qquad\text{(E2.5-6b)}$$

which is the x-component of the weight of the whole liquid film along $0 \le x \le \ell$.

Graph:

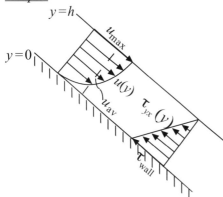

Comments:

$u_{average}$ can be used to compute the film Reynolds number, $\mathrm{Re}_h = u_{av} h / \nu$, and check if $\mathrm{Re}_h < \mathrm{Re}_{critical} \approx 20$.

Given Q (or $\dot{m} = \rho Q$), the coating thickness can be estimated.

Note:

The case of wire coating, i.e., falling film on a vertical cylinder, is assigned as a homework problem (see Sect. 2.6).

Example 2.6: Flow in a Slightly Converging Tube

Consider steady laminar flows in a slightly converging tube, i.e., Poiseuille flow in a mildly tapered tube where the radius changes as:

$$R(z) = R_1 - \frac{\Delta R}{L} z$$

Find the velocity solution.

Sketch:	Assumptions:	Concept:
	• Steady laminar axisymmetric flow • No-slip boundary condition	• Poiseuille flow

Solution:

With the underlying assumption of Poiseuille flow, the z-momentum equation reduces to (see A.5.3):

$$0 = -\frac{\partial p}{\partial z} + \mu \left[\frac{1}{r}\frac{\partial}{\partial r}\left(r\frac{\partial v_z}{\partial r} \right) \right] \tag{E2.6-1}$$

which implies that $v_r = 0$ (see also Sect.1.3.3). Now, as usual, the boundary condition at $r = 0$ is $dv_z / dr = 0$ because of symmetry; but the no-slip condition at the tube wall reads:

$$v_z \left[r = R(z) \right] = 0 \tag{E2.6-2}$$

which introduces the unique tube geometry and 2-D flow pattern. The solution is:

$$v_z = v_z \left(r, z \right) = v_{max} \left[1 - \left(\frac{r}{R(z)} \right)^2 \right] \tag{E2.6-3}$$

Checking the continuity equation for axisymmetric flow:

$$\frac{1}{r}\frac{\partial}{\partial r}\left(r\, v_r \right) + \frac{\partial v_z}{\partial z} = 0 \tag{E2.6-4}$$

we see that $v_r \neq 0$ because $v_z = v_z(z, r)$. In fact, Eq. (E2.6-4) can be employed to find an expression for $v_r \left(r, z \right)$ considering that $v_r \left(r = 0 \right) = 0$ or $v_r \left[r = R(z) \right] = 0$.

Comments:

Additional examples of nearly unidirectional flows include lubrication and stretching flows as given in Sect. 4.3; they are also discussed in Papanastasiou (1994), Kleinstreuer (1997), Middleman (1998), and Panton (2005), among others. ODE solutions to more difficult problems may be found in Polyanin & Zaitsev (1995).

Transient 1-D Sample Problem Solutions.   Often a system's transport phenomena can only be described in the form of a PDE because the dependent variable (i.e., v, T, or c) depends on two "coordinates," say, t and y or x and y. In some cases, based on the underlying physics, such PDEs can be transformed to one ODE (see Similarity Theory) or a pair of ODEs (see Separation-of-Variables Method), both techniques outlined in Greenberg (1998) among others. Both mathematical techniques are applied in a couple of sample problem solutions, as discussed next.

Specifically, industrial applications, i.e., sudden start-up of fluid flow, are given in Examples 2.7 and 2.8. One of the most famous examples of transient internal flow is pulsatile flow in a tube, e.g., blood flow in a straight artery, first solved analytically in 1955 by Womersley as discussed in Nichols & O'Rourke (1998). Example 2.9 then rounds out the section with steady fully developed flow in a square duct for which the axial flow $u = u(x, y)$, described by a PDE.

Example 2.7: Onset of Thin Shear Layer Flow (Stokes I)

Consider a horizontal plate or wall carrying a stagnant body of fluid (i.e., $u = 0$ when $t \leq 0$ for all y; see the sketch). Suddenly, the solid surface attains (at $y = 0$) a finite velocity, i.e., $u = U_0$ when $t > 0$. Recalling that $v_{wall} = v_{fluid}$ (no-slip condition), this plate motion sets up, within a growing layer, parallel flow of the viscous fluid, i.e., $u = u(y, t)$. The atmospheric pressure is constant everywhere. Find the region of moving-plate effect on the fluid body.

Sketch:	Assumptions:	Concepts:
	• Transient laminar unidirectional flow • No-slip boundary condition • Constant fluid properties	• Similar velocity profiles with time • Finite region/layer of influence
		Postulates:
		• Pressure $p = ¢$ everywhere • $\vec{v} = (u; 0; 0)$, where $u = u(y, t)$ only

Solution:

With the postulates

$$\vec{v} = u(y, t) \text{ only and } \nabla p = 0 \qquad \text{(E2.7-1a,b)}$$

we can reduce the *x*-momentum equation to (see Sect. A.5):

$$\frac{\partial u}{\partial t} = v \frac{\partial^2 u}{\partial y^2} \qquad \text{(E2.7-2)}$$

Equation (E2.7-2) is known as the transient one-dimensional *diffusion equation* [cf. Eqs. (1.95 and 1.96)]. In the present case, it describes "momentum diffusion" normal to the axial parallel flow induced by the wall motion. As implied, the associated initial/boundary conditions are:

$$u(t \le 0, y) = 0, \text{ but } u(t > 0; y = 0) = U_0; \text{ for } y \to \infty, u = 0 \qquad \text{(E2.7-3)}$$

Because the evolution of $u(y)$ with time shows similar profiles, the independent variables y and t can be combined in conjunction with the fluid viscosity v (see App. A). Thus, for the new *dimensionless* variable

$$\eta = \eta(y, t; v) \qquad \text{(E2.7-4a)}$$

we demand formally

$$[\eta] = y^a t^b v^c = [1] \qquad \text{(E2.7-4b)}$$

or by simple inspection with $a = 1$

$$\eta = \frac{y}{t^{0.5} v^{0.5}} \hat{=} [1] \qquad \text{(E2.7-4c)}$$

For convenience, the dimensionless independent variable can be written as:

$$\eta = \frac{y}{2\sqrt{vt}} \qquad \text{(E2.7-4d)}$$

It is apparent that

$$u(y, t) \sim f[\eta(y, t)] \tag{E2.7-5a}$$

where $f(\eta)$ is a dimensionless dependent variable. To turn the proportionality into an equation, we utilize the plate speed U_0, so that

$$u(y, t) = U_0 f(\eta) \tag{E2.7-5b}$$

Now, with Eqs. (E2.7-4d) and (E2.7-5b), the governing PDE (E2.7-2) can be transformed into a second-order ODE for $f(\eta)$, i.e.,

$$f'' + 2\eta f' = 0 \tag{E2.7-6}$$

subject to

$$f(\eta = 0) = 1 \text{ and } f(\eta \to \infty) \to 0 \tag{E2.7-7a,b}$$

The solution is $f(\eta) = 1 - \text{erf}(\eta)$, where $\text{erf}(\eta)$ is the error function (Sect. A3.3), so that

$$\frac{u}{U_0} = 1 - \frac{2}{\sqrt{\pi}} \int_0^\eta \exp(-\eta^2) d\eta \tag{E2.7-8}$$

When plotting Eq. (E2.7-8), it turns out that for $\eta = 2.0$ the moving-plate effect on the fluid body peters out, i.e., $f(h = 2) \equiv u/U_0 \approx 0.01$. This implies that the region of frictional influence, i.e., $0 \le y \le \delta$, can be estimated from $y(\eta = 2) = \delta$ as

$$\delta \approx 4\sqrt{vt} \tag{E2.7-9a}$$

Replacing t in terms of the plate travel time, i.e., $t = x/U_0$, Eq. (E2.7-9a) can be written as:

$$\delta \approx 4\sqrt{v\frac{x}{U_0}} \tag{E2.7-9b}$$

which can also be expressed as:

$$\frac{\delta}{x} \approx \frac{4}{\sqrt{\text{Re}_x}} \tag{E2.7-9c}$$

where $\mathrm{Re}_x = U_0 x / \nu$ is the local Reynolds number and δ is the extent of the shear layer in which $u = u(y)$. Outside the shear layer, i.e., $y \geq \delta$, $u = 0$ in this case.

Notes:
- $\delta(x)$ is fundamental to laminar thin-shear-layer (TSL), or boundary-layer (B-L) theory (see Sect. 1.3.3.3; Eq. (1.84)).
- Akin to the Stokes first problem discussed here is his second problem solution, that of an oscillating flat plate.
- Even more complicated is laminar flow generated by start-up of a rotating disk in a reservoir of a viscous fluid. It is three-dimensional because fluid exits radially the finite disk (v_r-component) because of the centrifugal force; this vanishing fluid is constantly replaced by swirling, incoming fluid (v_θ- and v_z-components).
- Setting up these and other transient flow problems is part of the homework assignments in Section 2.6.

Example 2.8: Start-Up Flow in a Tube

Consider a viscous fluid at rest in a horizontal tube when suddenly a constant pressure gradient, $\Delta p / L$, is applied. For example, a valve connecting a pipe to a reservoir is suddenly opened. Find an expression for the resulting $u(r, t)$.

Sketch:	Assumptions:	Concepts:
	• Pressure gradient $-\partial p / \partial x \approx \Delta p / L = \cancel{c}$ at all times • Transient laminar 1-D flow • Constant fluid properties	• Reduced N-S equations • Superposition of steady and transient contributions

Solution:

Postulate that the actual velocity $u(r, t)$ can be decomposed into a steady-state part and a transient part, i.e.,

$$u(r, t) = u(r)\big|_{ss} + u(r, t)\big|_{tr} \tag{E2.8-1}$$

With $\vec{v} = \left[u(r, t); 0; 0 \right]$ and $-\nabla p \rightarrow \Delta p / L = \not{c}$ the governing momentum equation is

$$\frac{\partial u}{\partial t} = \frac{1}{\rho}\left(\frac{\Delta p}{L}\right) + \frac{v}{r}\frac{\partial}{\partial r}\left(r\frac{\partial u}{\partial r}\right) \qquad \text{(E2.8-2)}$$

subject to

$$u(r, t = 0) = 0, \ u(r = r_0, t) = 0, \text{ and } \left.\frac{\partial u}{\partial r}\right|_{r=0} = 0 \qquad \text{(E2.8-3a-c)}$$

Clearly, the steady-state part, u_{ss}, is the Poiseuille flow solution, i.e.,

$$u_{ss}(r) = u_{max}\left[1 - \left(\frac{r}{r_0}\right)^2\right] \qquad \text{(E2.8-4a)}$$

where

$$u_{max} = -\frac{1}{4\mu}\left(\frac{\Delta p}{l}\right)r_0^2 \qquad \text{(E2.8-4b)}$$

Knowing $u_{ss}(r)$ and employing the dimensionless variables

$$\hat{u} = \frac{u_{tr}}{u_{max}}, \ \hat{r} = \frac{r}{r_0}, \text{ and } \hat{t} = \frac{vt}{r_0} \qquad \text{(E2.8-5a-c)}$$

Equation (E2.8-2) can be transformed to the well-known form:

$$\frac{\partial \hat{u}}{\partial \hat{t}} = \frac{1}{\hat{r}}\frac{\partial}{\partial \hat{r}}\left(\hat{r}\frac{\partial \hat{u}}{\partial \hat{r}}\right) \qquad \text{(E2.8-6)}$$

subject to

$$\hat{u}(\hat{t} = 0) = 1 - \hat{r}^2, \ \hat{u}(\hat{r} = 1) = 0, \text{ and } \left.\frac{\partial \hat{u}}{\partial \hat{r}}\right|_{\hat{r}=0} = 0 \qquad \text{(E2.8-7a-c)}$$

The solution is an infinite series in $\hat{r}$ times a decaying exponential function in $\hat{t}$, i.e.,

$$\hat{u} \sim \sum \text{fct}(\hat{r}) \cdot e^{-i} \qquad (E2.8\text{-}8)$$

Graph:

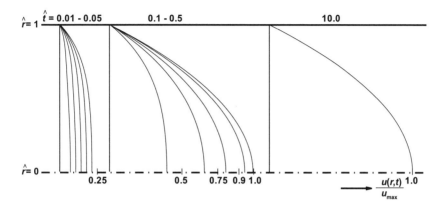

Comments:

The final solution for the axial tubular velocity $u(\hat{r}, \hat{t})$ is graphed for $0 \le \hat{r} \le 1$ and $0 \le \hat{t} \le 10.0$. When at $\hat{t} \approx 10.0$, $u_{tr} \to 0$ and $u = u_{ss}$, i.e., Poiseuille flow has been established.

It is interesting to note that the suddenly elevated tube inlet pressure starts the core fluid off almost uniformly (for $0.01 \le \hat{t} < 0.05$) and then, after $\hat{t} = 0.2$, in conjunction with the no-slip condition a parabolic velocity profile forms.

Example 2.9: Steady Fully Developed Flow in a Square Duct

Consider a steady one-dimensional fully developed flow in a square duct under a constant pressure gradient $\partial p / \partial z = -\Delta p / L = \cancel{c}$ in the z-direction. The square duct is of side $2a$ and length L. The following equation has been proposed for the approximate velocity profile (see Wilkes, 2006):

$$v_z = v_{max}\left[1 - \left(\frac{x}{a}\right)^2\right]\left[1 - \left(\frac{y}{a}\right)^2\right] = \frac{1}{2\mu}\left(-\frac{\partial p}{\partial z}\right)a^2\left[1 - \left(\frac{x}{a}\right)^2\right]\left[1 - \left(\frac{y}{a}\right)^2\right]$$

Use the function to determine the average velocity $v_{ave} = Q/A$ in the duct as a ratio of v_{max}.

Sketch:	Assumptions:	Concepts:
	• Fully developed flow • Pressure gradient $\partial p/\partial z \approx -\Delta p/L = \cancel{c}$ • Constant fluid properties	• Poiseuille flow • Reduced N-S equation based on the postulates $v_z = \mathrm{fct}(x, y)$ only and $v_x = v_y = 0$

Solution:

$$Q = \int v_z \, dA = \iint_A v_z \, dx \, dy \tag{E2.9-1}$$

while

$$v_z = v_{max}\left[1-\left(\frac{x}{a}\right)^2\right]\left[1-\left(\frac{y}{a}\right)^2\right] = \frac{1}{2\mu}\frac{\Delta p}{L}a^2\left[1-\left(\frac{x}{a}\right)^2\right]\left[1-\left(\frac{y}{a}\right)^2\right] \tag{E2.9-2}$$

From Eqs. (E2.9-1) and (E2.9-2), we have:

$$Q = \frac{1}{\mu}\frac{\Delta p}{L}\frac{8}{9}a^4 \tag{E2.9-3}$$

and

$$v_{ave} = Q/A = \frac{\dfrac{1}{2\mu}\dfrac{\Delta p}{L}\dfrac{16}{9}a^4}{4a^2} = \frac{1}{\mu}\frac{\Delta p}{L}\frac{2}{9}a^2 \tag{E2.9-4}$$

Thus,

$$v_{ave} = \frac{4}{9}v_{max} \tag{E2.9-5}$$

Calculating the wall shear stress and hence the friction factor f, the Poiseuille number Po $= f$ Re for laminar, fully developed square-duct flow can be determined, i.e.,

$Po = 57$. For tubular flow, $Po = 64$. The Po numbers for specific rectangular and triangular ducts are tabulated as follows:

Rectangular		Isosceles triangle	
b/a	Po	θ, deg	Po
0.0	96.00	0	48.0
0.05	89.91	10	51.6
0.1	84.68	20	52.9
0.125	82.34	30	53.3
0.167	78.81	40	52.9
0.25	72.93	50	52.0
0.4	65.47	60	51.1
0.5	62.19	70	49.5
0.75	57.89	80	48.3
1.0	56.91	90	48.0

Note:

The Poiseuille number is defined as $Po = f \, Re_{D_h}$, based on the Darcy friction factor $f = 8\tau_w / \rho v^2$

2.2 Porous Medium Flow

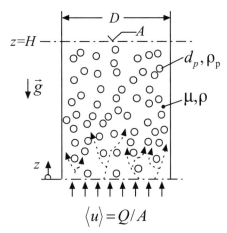

$$\langle u \rangle = Q/A$$

Figure 2.2   Porous medium column

In numerous natural and industrial processes, a fluid flows (or just migrates) through a *fully saturated* porous medium. Examples include blood flow via capillaries through tissue, groundwater flow through geologic media, oil or steam dispersion through sand and porous rock, fluid flow through a container packed with spheres, pellets, or granular material (with upward flow known as fluidized beds), moisture migration through a porous composite, mixture flow across membranes or filters, coolant flow through an array of microchannels, etc. In any case, as indicated in Figure 2.2, the local velocity field through pores, capillaries, fissures, microchannels, and/or around packed spheres, pellets, cylinders, fibers, cells, or granular material is very complicated. For that reason, a volume-averaged, i.e., *superficial velocity*

$$\langle u \rangle = \frac{1}{\forall} \int \vec{u} d\forall = \frac{Q}{A} \qquad (2.8\text{a,b})$$

is introduced, where the length scale of the control volume, $\forall^{1/3}$, is smaller than the characteristic length of the system, say, channel height H or pipe diameter D; but, also, $\forall^{1/3} >> d_p$, i.e., the pore (or pellet) diameter, so that we have $d_p << \forall^{1/3} << \ell_{\text{system}}$. Note, the volume integral extends over the volume space occupied by the fluid. Clearly, $\langle \vec{u} \rangle$, the velocity vector, is a function of the driving forces (e.g., pressure gradient and/or gravity), fluid properties (i.e., viscosity and density), as well as the porous medium structure and material. As with all complicated fluid flow problems, possible solutions start with dimensional analysis and experimental investigations. Specifically, based on Darcy's observations in 1856, the uniform superficial 1-D velocity of a viscous fluid through a homogeneous porous medium is (Figure 2.2):

$$\langle u \rangle = -\frac{\text{æ}}{\mu}\left(\frac{dp}{dz} + \rho g_z \right) \qquad (2.9\text{a})$$

where $\text{æ} \left[L^2 \right]$ is the permeability, i.e., $\sqrt{\text{æ}} \; [L]$ is a length-scale representative of the effective pore diameter. Note, $\text{æ}/\mu \equiv K$ is the "hydraulic conductivity." Not surprising, æ has been correlated to the porosity e, for example by Ergun (1952) as:

$$æ = \frac{\varepsilon^3 d_p^2}{C(1-\varepsilon)^2} \tag{2.9b}$$

where $C = 150 - 180$.

Assuming negligible gravitational effects, Eq. (2.9a) can be used to determine the key driving force, i.e., the pressure gradient as:

$$\frac{dp}{dz} = \frac{C\mu \langle u \rangle}{d_p^2} \frac{(1-\varepsilon)^2}{\varepsilon^3} \tag{2.10}$$

The porosity, or void fraction, is a volume ratio, i.e., $\varepsilon = \forall_{void} / \forall_{total} \approx 1 - \forall_{particles} / \forall_{total}$. It can be employed to construct a particle Reynolds number; for example, for a uniformly packed bed of spheres,

$$\mathrm{Re}_p = \frac{\rho \langle u \rangle d_p}{(1-\varepsilon)\mu} \tag{2.11a}$$

Alternatively, the Reynolds number is based on $\sqrt{æ}$, i.e.,

$$\mathrm{Re} = \frac{\langle u \rangle \sqrt{æ}}{\nu} \tag{2.11b}$$

In any case, Eq. (2.9a) holds strictly for $\mathrm{Re} \leq 1$, but it is often used up to $\mathrm{Re} \approx 10$. When Eq. (2.9a) without the gravitational term is compared with Eq. (E2.2-6b) for Poiseuille flow in a horizontal tube of radius R, i.e.,

$$v_{av} = \frac{R^2}{8\mu}\left(-\frac{dp}{dx}\right) \tag{2.12}$$

the similarity is obvious. Thus, it is conducive that a homogeneous porous medium could be modeled as an assemblage of tiny straight tubes (i.e., pores) with laminar fully developed flow of a viscous fluid.

Several additions to Eq. (2.9a) have been proposed because of the obvious shortcomings of Eq. (2.9a), e.g., $\mathrm{Re} \leq O(1)$, 1-D flow only, the no-slip condition cannot be enforced, and local variations in the flow field and porous material structure are averaged out. The two most famous ones are the Brinkman (1947) and Forchheimer (1901) extensions to Darcy's law, where in 3-D:

$$\nabla p = -\frac{\mu}{\text{æ}}\vec{u} - C_F \underbrace{\frac{\rho\vec{u}|\vec{u}|}{\sqrt{\text{æ}}}}_{\substack{\text{Forchheimer}\\\text{term}}} + \underbrace{\hat{\mu}\nabla^2\vec{u}}_{\substack{\text{Brinkman}\\\text{term}}} \tag{2.13}$$

Here, $\vec{u} \triangleq \langle\vec{u}\rangle$ and C_F is a dimensionless form-drag constant which depends on the characteristics of the porous medium and the bounding walls, if any; typically, $C_F \approx 0.55$. The effective viscosity $\hat{\mu} \approx \mu$; however, the ratio $\hat{\mu}/\mu$ can depend again on the nature of the porous medium.

When comparing Eq. (2.13) with the Navier-Stokes equation (1.82), it is apparent that the nonlinear Forchheimer term relates to the inertia term. It is a necessary addition for porous media flow with $Re > 10$, i.e., when the pore flow is still laminar but the form drag posed by the porous material structure becomes important. The second-order Brinkman term relates to the viscous drag term. It is necessary to enforce no-slip boundary conditions, i.e., the term is significant in thin wall shear layers, typically of thickness $\sqrt{\text{æ}} \ll \ell_{\text{system}}$.

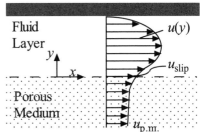

Figure 2.3 Unidirectional channel flow with porous medium flow

Quite frequently, channel flow interacts with (saturated) porous medium flow (see Figure 2.3), where at the interface between fluid layer and medium the magnitude and gradient of the velocity have to match. As a first approximation, Beavers & Joseph (1967) suggested:

$$\frac{\partial u}{\partial y}\bigg|_{y=0} = \frac{\alpha}{\sqrt{\text{æ}}}\left(u_{\text{slip}} - u_{\text{p.m.}}\right) \tag{2.14}$$

where $u_{\text{slip}} = u(y=0)$ and $u_{\text{p.m.}}$ is the averaged superficial velocity well below the interface. More realistic interface conditions were given by Ochoa-Tapia & Whitaker (1995):

$$u = v \tag{2.15a}$$

and

$$\frac{1}{\varepsilon}\frac{du}{dy} - \frac{dv}{dy} = \frac{\beta}{\sqrt{\text{æ}}}u \tag{2.15b}$$

where $u \triangleq u_{\text{slip}}$, $v \triangleq u_{\text{p.m.}}$, and α, β are measured coefficients.

Example 2.10: Darcy's Experiment

The given sketch depicts the basic set-up of Darcy's experiment. In terms of the form factor $\varphi \equiv A/\ell$ of the porous slab, find the flow rate for the given data, i.e., $A = 10$ m^2, $\ell = 1.5$ m, $h = 1.0$ m, and the Darcy coefficient $k \equiv \gamma K = 0.5$ cm/s; $\gamma \equiv \rho g$. Derive also Darcy's law.

Sketch:	*Assumptions:*
Feed Q_{in} Sand Filter (area A; height ℓ) ℓ h $\downarrow \vec{g}$ Q_{out} Permeability K z	• Steady laminar flow with Re ≤ 1.0 • Height h kept constant • Constant properties • Homogeneous filter without wall effects

Solution:

Based on dimensional analysis we can state:

$$Q = Q\left(\underbrace{æ, \mu;}_{\sim k \text{ or } K} v; A; \underbrace{h, \ell}_{\sim \Delta p} \right) \tag{E2.10-1}$$

From laboratory observations:

$$\frac{Q}{A} \sim \frac{h}{\ell} \quad \text{or} \quad \frac{Q}{A} \sim \frac{\Delta p}{\rho g \ell} \quad \text{after Bernoulli}$$

Darcy employed a coefficient of proportionality k, so that

$$Q = \left(k \frac{h}{\ell} \right) A = vA \tag{E2.10-2}$$

That coefficient depends on both the porous matrix as well as fluid properties, i.e.,

$$k = æ \rho g / \mu \equiv \gamma K \tag{E2.10-3a,b}$$

where æ is the permeability, $\gamma = \rho g$ the specific weight, and K the hydraulic permeability.

From these findings the previous equations can be formulated, i.e.,

$$v \equiv \langle u \rangle = -K \frac{\partial p}{\partial z} \qquad (E2.10\text{-}4)$$

and as a 3-D extension:

$$\nabla p = -\frac{\mu}{\text{æ}} \vec{v} \qquad (E2.10\text{-}5)$$

For the present problem, $\varphi \equiv A/\ell := 6.6$ m and $Q = kh\varphi := 0.033$ m^3/s.

Note:

Typical Darcy coefficients, i.e., $[k] \hat{=}$ length/time, and average grain sizes are given below for different soil types.

Soil Type	Clean Gravel	Fine Gravel	Coarse Sand	Fine Sand	Clay
Grain size [mm]	4–7	2–4	0.5	0.1	0.002
k-Range [cm/s]	2.5–4.0	1.0–3.5	0.01–1.0	0.001–0.05	10^{-9}–10^{-6}

Example 2.11: Evaluation of the Hydraulic Conductivity

As discussed, a possible porous medium model is a structure with n parallel tubes of diameter d, representing straight pores or capillaries. Assuming steady laminar fully developed flow in horizontal tubes, find an expression for $K = \text{æ}/\mu$.

Sketch:	*Concept:*
	• Equate volumetric flow rate obtained from Poiseuille flow with Darcy's law, where $dp/dx \approx \cent$ and $$\langle u \rangle = \frac{Q}{A} = -K \frac{dp}{dx}$$

Solution:

Using Eq. (E2.2-7a) to obtain the flow rate per pore or tube, we have:

$$\frac{Q}{n} = v_{av} A_{tube} = -\frac{\pi d^4}{128\mu}\left(\frac{dp}{dx}\right) \tag{E2.11-1a}$$

As indicated in Figure 2.2, $\langle u \rangle = Q/A := \varepsilon v_{av}$, so that $A = \frac{n}{\varepsilon} A_{tube}$. Hence,

$$\frac{Q}{n} = \langle u \rangle \frac{1}{\varepsilon} A_{tube} = -K\left(\frac{dp}{dx}\right)\frac{d^2\pi}{4\varepsilon} \tag{E2.11-1b}$$

so that by inspection

$$K = \frac{d^2\varepsilon}{32\mu} \tag{E2.11-2}$$

where $0 < \varepsilon < 1.0$.

Comment:

Once the average pore diameter has been estimated, setting $\varepsilon = 0.5$ and the fluid viscosity known, K and æ can be calculated and hence suitable porous medium flow analyses can be carried out.

═══════════════════════════════════

═══════════════════════════════════

Example 2.12: Creeping Flow in a Channel Filled with a Porous Medium

A slab of homogeneous porous medium of thickness h, bound by impermeable channel walls, represents a good base case for axial flow through porous insulation, a filter, a catalytic converter, tissue, i.e., extracellular matrix, etc.

Sketch:	*Assumptions:*	*Concepts:*
	• Creeping flow, i.e., $\langle u \rangle \equiv u$, where $u(y)$ only $\partial p/\partial x = ¢$ • Constant properties • Symmetry	• Use of the Darcy-Brinkman equation to invoke no slip at channel walls

Solution:

The 1-D form of Eq. (2.13) without the "high-speed" Forchheimer term reads:

$$\frac{dp}{dx} = -\frac{\mu}{\text{æ}}u + \mu\frac{d^2u}{dy^2} = \cancel{0} \qquad \text{(E2.12-1a)}$$

or

$$\frac{d^2u}{dy^2} - \frac{1}{\text{æ}}u = \frac{1}{\mu}\left(\frac{dp}{dx}\right) = \cancel{0} \qquad \text{(E2.12-1b)}$$

subject to $u(y = h/2) = 0$ and $du/dy = 0$ at $y = 0$.

The homogeneous solution to Eq. (E2.12-1b) can be found on page 130 of Polyanin & Zaitsev (1995) or other math handbooks:

$$u(y) = -\frac{\text{æ}h}{\mu}\left(\frac{dp}{dx}\right)\left[1 - \frac{\cosh(y/\sqrt{\text{æ}})}{\cosh(h/\sqrt{\text{æ}})}\right] \qquad \text{(E2.12-2)}$$

An *effective* hydraulic conductivity can be estimated, given the channel flow rate per unit depth, i.e.,

$$\hat{Q} = 2\int_0^{h/2} u\,dy := -\frac{\text{æ}h}{\mu}\left(\frac{dp}{dx}\right)\left[1 - \frac{2\sqrt{\text{æ}}}{h}\tanh\left(\frac{h}{2\sqrt{\text{æ}}}\right)\right] \qquad \text{(E2.12-3a)}$$

Defining K_{eff} as $\bar{u}/(dp/dx) = -\dfrac{\hat{Q}/h}{dp/dx}$ we obtain:

$$\frac{K_{\text{eff}}}{\text{æ}/\mu} \equiv \frac{K_{\text{eff}}}{K} = 1 - \frac{2\sqrt{\text{æ}}}{h}\tanh\left(\frac{h}{2\sqrt{\text{æ}}}\right) \qquad \text{(E2.12-3b)}$$

Comments (see Graphs):

The first graph depicts a $u(y)$-profile determined by the $1 - \cosh y$ function (i.e., compare Eq. (E2.12-1b) with Eq. (E2.2-5) for Poiseuille flow). The second graph shows the expected nonlinear increase of the nondimensionalized K_{eff} with channel height h.

Graphs:

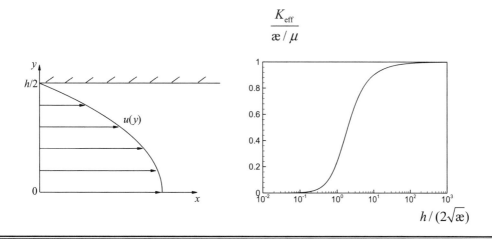

Example 2.13: Radial Flow through a Porous-Walled Tube

Consider pressure-driven flow in the radial direction through a porous tubular wall. Applications include radial flow in a porous pipe, ultrafiltration, tubular filter or membrane, seepage into a lymph vessel, etc. For the given system (see sketch), find the radial $p(r)$ distribution and $v(r)$ profile as well as the added mass flow rate over the tube length L.

Sketch:	*Assumptions:*	*Concepts:*
p_0 $v_0=Q/A_{surf}$ $r=R_2$ $r=R_1$ p_i $\Rightarrow \dot{m}$ $r=0$ $z=0$ $z=L$	• Steady radial seepage in $R_1 \leq r \leq R_2$ • Constant pressures and properties • No gravity effect	• Darcy's law: $$\vec{v} = -\frac{\kappa}{\mu}\nabla p$$ where $\vec{v}=(0,v,0)$ and $v=v(r)$ • Mass balance

Solution:

Writing Eq. (2.8) in its basic form (see assumptions and concepts) as:

$$\vec{v} = -\frac{\text{æ}}{\mu}\nabla p \tag{E2.13-1}$$

and taking the divergence of Eq. (E.2.13-1) we have with $\nabla \cdot \vec{v} = 0$ (continuity equation for incompressible fluids)

$$0 = -\frac{\text{æ}}{\mu}\nabla^2 p \tag{E2.13-2}$$

where ∇^2 is the Laplace operator (App. A). For our 1-D case in cylindrical coordinates, Eq. (2.13-2) is simply

$$\frac{1}{r}\frac{d}{dr}\left(r\frac{dp}{dr}\right) = 0 \tag{E2.13-3}$$

subject to $p(r = R_1) = p_i$ and $p(r = R_2) = p_0$. Double integration leads to

$$p(r) = C_1 \ln r + C_2$$

And finally for $p_0 > p_i$:

$$\frac{p - p_i}{p_0 - p_i} = \frac{\ln(r/R_1)}{\ln(R_2/R_1)} \tag{E2.13-4}$$

Now, with Darcy's law in the r-direction, i.e.,

$$v = -\frac{\text{æ}}{\mu}\left(\frac{dp}{dr}\right) \tag{E2.13-5a}$$

we obtain with the given $p(r)$ and $\Delta p \equiv p_0 - p_i$

$$v(r) = -\frac{\text{æ}}{\mu}\frac{\Delta p}{\ln(R_2/R_1)}\frac{1}{r} \tag{E2.13-5b}$$

A radial mass balance provides the added mass flow rate

$$\Delta \dot{m} = -\rho v\big|_{r=R_1} A_{\text{surface}} \tag{E2.13-6a}$$

where $A_{\text{surface}} = 2\pi R_1 L$, so that

$$\Delta \dot{m} = \frac{2\pi \ae \rho \Delta p}{\mu \ln\left(R_2 / R_1\right)} L \qquad\qquad \text{(E2.13-6b)}$$

Graph:

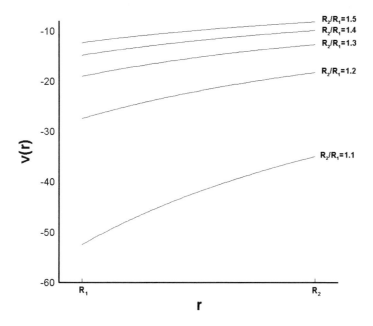

Comments:

The $v(r)$-function (E2.13-5b) is hyperbolic, i.e., for a given Δp, $\ae$, μ, and geometry, v decreases inversely with r. As expected radial mass influx increases with wall permeability, tube length, and pressure difference.

2.3 Mixture Flows

2.3.1 Introduction

Natural and industrial mixture flows, e.g., two-phase flows, such as particle suspensions, are all around us ranging from dust storms in arid regions to bubbly flows in pipes or air fuel injection in ICEs and nanofluid flow in microchannels. Traditionally, two-phase flow theory and application were the domain of applied mathematicians, as well as chemical, environmental, and nuclear engineers. However, for industrial pipe network design, pump sizing, and applied force evaluation, mechanical engineers have to know basic two-phase flow modeling techniques. Furthermore, biomedical engineers encounter fluid-particle dynamics problems in both the cardiovascular and pulmonary systems.

By definition, two-phase flow is the interactive flow of two distinct phases with common interfaces in, say, a conduit. Each phase, representing a volume fraction (or mass fraction) of solid, liquid, or gaseous matter, has its own properties, velocity, and temperature. Typical dilute (or dense) particle suspension flows include droplets in gas flow, liquid-vapor, i.e., bubbly, flow, as well as liquid or gas flow with solid particles. In addition to predicting the flow phases, it is also important to know the flow regimes, i.e., characteristic flow patterns based on the interfaces formed between the phases. Two-phase systems can be grouped into flows of separated phases, mixed phases, and dispersed phases. Examples of *separated* flows include liquid layers on a wall in gas flow, e.g., the mucus layer in lung airways, and liquid jets in gas flow (or vice versa). *Mixed-phase* flows are encountered in phase-change processes, such as boiling nuclear reactor channels and steam pipes with vapor core and annular liquid wall film as well as heat pipes with large vapor bubbles and evaporating liquid layers on heated surfaces. Most frequently, two-phase flows appear as *dispersed* phases, such as dilute particle suspensions in gas or liquid flows, droplets in gas flow (e.g., sprays), or bubbles in liquid flows (e.g., chemical reactors). Clearly, alternative two-phase flow classifications exist. For example, solid particles, droplets, or bubbles form the *dispersed (or particle) phase* while the carrier fluid is the *continuous (or fluid) phase*. The degree of phase coupling, i.e., from one-way for very dilute suspensions to four-way in dense suspensions. In the latter case, not only fluid flow affects particle motion and vice versa but particle-particle interactions due to collision are expected and particle-induced flow fields affect other particles, as in drafting.

Critical heat transfer may change the thermodynamic state of a phase or may generate two-phase flow in the first place, as discussed by Naterer (2002) and Faghri & Zhang (2006). Other recent two-phase or multiphase books include the texts by Crowe et al. (1998) and Kleinstreuer (2003) and the handbook edited by Crowe (2006).

2.3.2 Modeling Approaches

It should be evident from reading the Introduction (Sect. 2.3.1) that a very dilute suspension of uniformly distributed (micron- or nano-) particles, which only slightly affect the mixture properties, constitute the simplest two-phase flow example and hence the easiest case to solve. Such well-mixed suspensions, which are actually *pseudo* two-phase flows, can be described with homogeneous fluid mechanics equations and fall into the category of *flow mixture models* (see Figure 2.4). In contrast, when a distinct particle phase interacts with the continuous phase, more complex *separate flow models* are needed to describe the two-phase dynamics.

Definitions.   Focusing on dispersed flows, the particle phase is characterized by the particle volume fraction, mass concentration, and loading. In light of the continuum assumption, all quantities are defined as volume ratios with limits from the mixture volume sample $\delta\forall'$ which excludes the molecular range. It should be noted that *subscripts c, f, or 1* indicates the continuous phase or carrier fluid, while *subscripts p, d, or 2* refers to the particle or dispersed phase. For example, the volume (or void) fraction of the dispersed phase is:

$$\alpha_d \equiv \alpha = \lim_{\delta\forall \to \delta\forall'} \frac{\delta\forall_d}{\delta\forall} := \frac{\forall_{particles}}{\forall_{mixture}} \tag{2.16}$$

Then, the volume fraction of the continuous fluid phase is:

$$\alpha_c = \lim_{\delta\forall \to \delta\forall'} \frac{\delta\forall_c}{\delta\forall} := \frac{\forall_{fluid}}{\forall_{mixture}} \tag{2.17}$$

so that

$$\alpha_d + \alpha_c = 1 \tag{2.18}$$

With the two-phase volume fractions defined, the mixture (or effective) density is:

$$\rho_m = \alpha_c \rho_c + \alpha_d \rho_d := \bar{\rho}_c + \bar{\rho}_d \tag{2.19a,b}$$

where $\bar{\rho}_d = nm_p$ with n being the number of particles per unit volume and m_p being the particle's mass. The dispersed-phase concentration is given as:

$$c = \bar{\rho}_d / \bar{\rho}_c \tag{2.20}$$

(a) Hierarchy of multiphase flow models

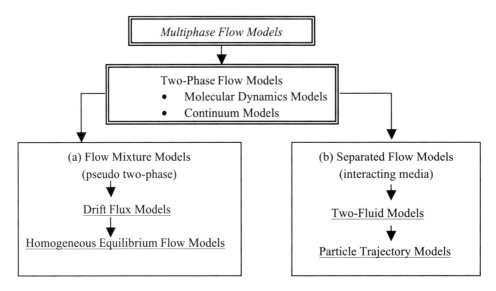

(b) Two-phase flow model applications

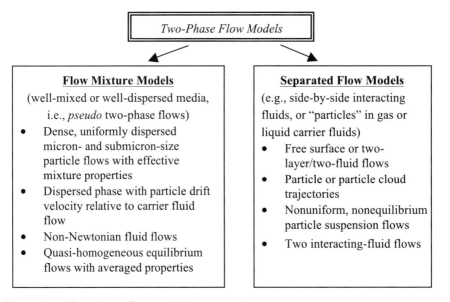

Figure 2.4   Two-phase flow modeling categories

The local loading is the mass flux ratio:

$$r = \bar{\rho}_d \bar{v}_p / (\bar{\rho}_c \bar{v}_c) \tag{2.21a}$$

while the total loading is:

$$\kappa = \dot{m}_d / \dot{m}_c \approx \bar{\rho}_d / \bar{\rho}_c = c \tag{2.21b}$$

implying that κ is approximately the concentration c.

Clearly, the mixture mass flow rate, important for internal flows, is:

$$\dot{m}_m = \dot{m}_c + \dot{m}_d = (\rho Q)_c + (\rho Q)_d \tag{2.22a,b}$$

which leads to the "quality" (i.e., particle mass concentration):

$$x = \dot{m}_d / \dot{m}_m \tag{2.23}$$

while the mixture mass flux is:

$$G_m = \dot{m}_m / A = G_c + G_d \tag{2.24a,b}$$

and the volume flux is:

$$j_m = \frac{G_m}{\rho_m} = \frac{Q_c + Q_d}{A} = j_c + j_d = v_m \tag{2.25a-d}$$

where v_m is the superficial velocity of the mixture, which is composed of the phase superficial velocities:

$$v_{c,s} = \alpha_c v_c \text{ and } v_{d,s} = \alpha_d v_d \tag{2.26a,b}$$

while the actual phase velocities are:

$$v_c = j_c / \alpha_c \text{ and } v_d = j_d / \alpha_d \tag{2.27a,b}$$

Finally, the relative (or slip) velocity between the two phases is:

$$v_r = v_c - v_d \tag{2.28}$$

The drift velocity, which indicates deviatory motion of the particle phase from the mixture flow, is:

$$v^{\text{drift}} = v_d - v_m \tag{2.29}$$

With these basic phase definitions established, we can now consider mixture properties, such as density ρ_m and dynamic viscosity μ_m (see Figure 2.5). Specifically,

$$\rho_m = \alpha\rho_2 + (1-\alpha)\rho_1 \tag{2.30a}$$

or

$$\frac{1}{\rho_m} = \frac{x}{\rho_2} + \frac{1-x}{\rho_1} \tag{2.30b}$$

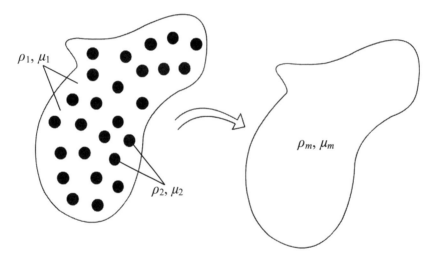

ρ_1, μ_1

ρ_2, μ_2

ρ_m, μ_m

Figure 2.5 Approximation of two-phase (dispersed) flow to a uniform mixture flow

where $\alpha \equiv \alpha_2 = \forall_2/\forall = \forall_d/\forall_m$ and $x = \dot{m}_2/\dot{m}$. At low volume fractions of spherical particles (say, $\alpha \le 0.05$) the mixture viscosity is (Soo, 1990; Zapryanov & Tabakova, 1999):

$$\mu_m = \mu_1\left(1 + 2.5\alpha\frac{\mu_2 + 0.4\mu_1}{\mu_2 + \mu_1}\right) \tag{2.31a}$$

which for $\mu_2 \gg \mu_1$ reduces to:

$$\mu_m\big|_{\text{solid spheres}} = \mu_1(1+0.4\alpha) \tag{2.31b}$$

or

$$\mu_m\big|_{\text{gas bubbles}} = \mu_1(1+\alpha) \tag{2.31c}$$

Similar to Eq. (2.31b), for well-dispersed gas-liquid flows, e.g., droplets in air or bubbles in liquid flow, we have

$$\frac{1}{\mu_m} = \frac{x}{\mu_2} + \frac{1-x}{\mu_1} \tag{2.32a}$$

or

$$\mu_m = x\mu_2 + (1-x)\mu_1 \tag{2.32b}$$

One of the important dilute particle suspensions are *nanofluids*, i.e., nanoparticles in liquids at low volume fractions (see Part C). Some basic viscosity expressions applicable to certain nanofluids are listed in Table 2.2.

Table 2.2   Conventional Viscosity Models for Nanofluids

Model	Expression	Comments
Einstein (1906)	$\mu_{nf} = \mu_{bf}(1+2.5\varphi)$	Spherical particles and low volume fraction, i.e., $\varphi < 2\%$
Brinkman (1952)	$\mu_{nf} = \dfrac{\mu_{bf}}{(1-\varphi)^{2.5}}$	Extended Einstein expression
Batchelor (1977)	$\mu_{nf} = \mu_{bf}(1+2.5\varphi+6.5\varphi^2)$	Extended Einstein equation by considering the effect of Brownian motion on the bulk stress
Graham (1981)	$\mu_{nf} = \mu_{bf}\left(1+2.5\varphi+4.5\left[\dfrac{1}{\left(\dfrac{c}{d_p}\right)\left(2+\dfrac{c}{d_p}\right)\left(1+\dfrac{c}{d_p}\right)^2}\right]\right)$	d_p is the particle diameter and c is the interparticle spacing

Note: $\varphi \triangleq$ volume fraction and subscripts nf $\triangleq$ nanofluid and bf $\triangleq$ base fluid.

Effective Dynamic Viscosity of Nanofluids.   Most of the reported data for nanofluid viscosities have been discussed in terms of formulations proposed by Einstein (1906), Brinkman (1952), Batchelor (1977), and Graham (1981), to name a few. The conventional viscosity models of nanofluids are summarized in Table 2.2. It turns out that none of the models mentioned can predict the viscosity of nanofluids for a wide range of nanoparticle volume fraction very well.

Example 2.14: Poiseuille-Type Mixture Flow

Consider steady laminar fully developed pipe flow (radius R, length L) where the void fraction of solid particles in air ranges from $\alpha = 0$ to $\alpha = 0.05$. Find $u(r)$ and plot $u/u_{\max}$ versus r/R.

Sketch:	*Assumptions*:	*Concepts*:
	• Poiseuille flow • One-way coupling • Quasi-homogeneous equilibrium flow • Constant mixture properties	• Reduced N-S equations • Constant $-dp/dx = \Delta p / L$ • Mixture viscosity $\mu_m = \mu_1 (1 - 0.4\alpha)$ • $\alpha \equiv \forall_{particles} / \forall_{mixture}$

Solution:

Based on the assumptions, continuity is fulfilled and the x-momentum equation reduces to (see Example 2.1):

$$0 = -\frac{dp}{dx} + \frac{\mu}{r}\frac{d}{dr}\left(r\frac{du}{dr}\right)$$

(E2.14-1)

subject to $u(r = R) = 0$ and $du/dr\big|_{r=0} = 0$. Hence,

$$u(r) = \underbrace{\frac{R^2}{4\mu_m}\left(\frac{\Delta p}{L}\right)}_{u_{\max}}\left[1 - \left(\frac{r}{R}\right)^2\right]$$

(E2.14-2)

where according to Eq. (2.31b):

$$\mu_m = (1 + 0.4\alpha)\mu_{\text{fluid}} \quad \text{for } 0 \le \alpha \le 0.05 \tag{E2.14-3}$$

Graph:

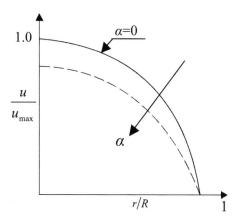

Comments:

Clearly, for $\alpha = 0$ the original Poiseuille flow is recovered. When $\alpha > 0$, the flow rate decreases for the same (given) pressure drop because of the higher resistance, and hence the velocity profile is flatter.

Example 2.15: Bubbly Pipe Flow

Of interest are the total pressure drop and void fraction of steady turbulent bubbly flow in a vertical pipe ($D = 25$ mm, $L = 45$ cm). The upward flow is a mixture of water ($m_\ell = 0.42$ kg/s, $\rho_\ell = 10^3$ kg/m^3) and air bubbles ($m_g = 0.01$ kg/s, $\rho_g = 1.177$ kg/m^3).

Sketch:	*Assumptions*:	*Concepts*:
	• Steady fully developed turbulent flow with uniform velocities • Thermodynamic equilibrium, i.e., $v_g = v_\ell$ and $T = \cancel{c}$	• Constant mixture properties ρ_m and μ_m • Extended Bernoulli equations • Blasius friction factor for smooth pipes $f = 0.316\,\text{Re}_D^{-1/4}$ • Void fraction $\alpha = Q_g / Q$

Solution:

Mass conservation: $\dot{m} = \dot{m}_\ell + \dot{m}_g$ and hence $v = v_\ell = v_g = \dot{m}/(\rho_m A)$, while quality $x = \dot{m}_g/\dot{m}$.

Mixture properties (see Eq. (2.30b)):

$$\rho_m = \left(\frac{x}{\rho_g} + \frac{1-x}{\rho_\ell}\right)^{-1} \text{ and } \mu_m = \left(\frac{x}{\mu_g} + \frac{1-x}{\mu_\ell}\right)^{-1} \tag{E2.15-1a,b}$$

Thus, the numerical property values are then with $x = 0.023$ and $v = 17.65$ m/s as follows: $\rho_m \equiv \rho = 49.64$ kg/m^3, $\mu_m \equiv \mu = 4.435 \times 10^{-4}$ kg/(m·s), and Re$_D = \rho v D / \mu = 49,388$.

After Blasius (see Kleinstreuer, 1997) the pipe friction factor is:

$$f = \frac{0.316}{\text{Re}_D^{0.25}} = 0.0212 \tag{E2.15-2}$$

The extended Bernoulli equation reads:

$$\frac{p_1}{\rho g} + \frac{v_1^2}{2g} + z_1 = \frac{p_2}{\rho g} + \frac{v_2^2}{2g} + z_2 + h_f \tag{E2.15-3a}$$

where

$$h_f = f\left(\frac{L}{D}\right)\frac{v^2}{2g} \tag{E2.15-3b}$$

The numerical results are:

Pressure Drop:

$$\Delta p = \Delta z + h_f := 31.7 \text{ kN/m}^2 \tag{E2.15-4}$$

Volumetric Flow Rates:

$$Q_i = \left(\frac{\dot{m}}{\rho}\right)_i := \begin{cases} 0.00833 \text{ m}^3/\text{s} & \text{<bubbles>} \\ 0.00042 \text{ m}^3/\text{s} & \text{<water>} \end{cases} \tag{E2.15-5}$$

Void Fraction:

$$\alpha = \frac{\forall_g}{\forall} = \frac{Q_g}{Q} := 0.96 \qquad \text{(E2.15-6)}$$

Comments:

Although the quality $(x = 0.023)$ is rather low, the almost 1000-fold density difference between the carrier fluid and the air bubbles generates 20x the volumetric flow rate for air when compared to water flow. As a result, $\alpha = 0.96$, i.e., the pipe is mainly filled with air (bubbles) under the assumption of homogeneous equilibrium flow.

2.3.3 Homogeneous Flow Equations

As indicated, certain two-phase flows can be treated as a single-phase flow if both phases are in quasi-thermodynamic equilibrium, i.e., spatially their properties, velocities, and temperatures do not deviate significantly. The flow regimes for gas-liquid flows are typically bubbly or misty flows. For solid-particle suspension flows the void fraction is very low, say, less than 5%, and the relative velocity $v_{\text{fluid}} - v_{\text{particle}}$ is minor. In order to yield one-dimensional transport equations, both phases are area averaged and mixture properties are used (see Eqs. (2.30 and 2.31)). Specifically, for a conduit of area A and inclined angle θ we obtain the following (see Sect. 2.6):

Continuity:

$$A\frac{\partial \rho}{\partial t} + \frac{\partial}{\partial x}(\rho v A) = 0 \qquad (2.33)$$

Momentum:

$$A\frac{\partial(\rho v)}{\partial t} + \frac{\partial}{\partial x}(\rho v v A) = -\frac{\partial(pA)}{\partial x} - \tau_w P - \rho g A \sin\theta \qquad (2.34)$$

Energy:

$$\frac{\partial(\rho e)}{\partial t} + \frac{1}{A}\frac{\partial}{\partial x}(v A \rho e) = \frac{P}{A}q_w + q_{\text{gen}} + \frac{\partial p}{\partial t} \qquad (2.35)$$

Here, τ_w is the wall shear stress; P is the conduit perimeter; the specific energy is $e = h + v^2/2 + gx\sin\theta$, with $h = u + p/\rho = h_c + x(h_\alpha - h_c)$ being the enthalpy; q_w is the wall heat flux; and q_{gen} is the internal heat generation, say, due to viscous effects or chemical reaction.

For steady homogeneous mixture flow in a rigid pipe of diameter D, the momentum equation (2.34) reduces to:

$$-\frac{dp}{dx} = \frac{\Delta p}{L} = \frac{\partial}{\partial x}(\rho v v) + \frac{4\tau_w}{D} + \rho g \sin \theta \qquad (2.36)$$

while the energy equation (2.35) becomes with $dh = c_p dT$ the heat transfer equation:

$$\rho c_p \frac{dT}{dx} = -\frac{d}{dx}\left(\frac{\rho}{2}v^2\right) + \frac{4q_w}{vD} + \frac{q_{gen}}{v} - \rho g \sin \theta \qquad (2.37)$$

Knowing the quality $x = \dot{m}_d / \dot{m}$, we can express the mixture density as:

$$\rho = \frac{\rho_d \rho_c}{\rho_c x + \rho_d (1-x)} \qquad (2.38)$$

and the void fraction as:

$$\alpha = \frac{x}{x + (1-x)\rho_d / \rho_c} \qquad (2.39)$$

while the momentum flux is:

$$\rho v = (\dot{m}_c + \dot{m}_d)/A = \dot{m}/A \qquad (2.40)$$

For the wall shear stress, we recall $\tau_w \sim \Delta p / L$ which can be expressed as the frictional loss $h_f = h_f(f, L/D, v^2)$ as discussed in undergraduate fluids text and assigned in Sect. 2.6. Specifically,

$$-\frac{dp}{dx}\bigg|_{friction} = \frac{4\tau_w}{D} = \frac{2f\rho v^2}{D} \qquad (2.41a)$$

where Beattie & Whalley (1982) suggested for the annular and bubbly regimes:

$$f^{-1/2} = 3.48 - 4\log_{10}\left[2\left(\frac{\varepsilon}{D}\right) + \frac{9.35}{\mathrm{Re}\sqrt{f}}\right] \qquad (2.41b)$$

The Reynolds number $\mathrm{Re}_D = \rho v D / \mu$ requires the evaluation of the dynamic viscosity, e.g.,

$$\mu = \left(\frac{x}{\mu_d} + \frac{1-x}{\mu_c} \right) \tag{2.42}$$

Summary.   The two Examples 2.20 and 2.21 as well as Eqs. (2.33) to (2.41) show that solving quasi-homogeneous flow problems is rather straightforward. It is required that the two phases form a uniform mixture with effective properties (see Figure 2.5) and that the phases are in thermodynamic equilibrium, i.e., no velocity slip and equal temperatures. As indicated in Figure 2.4, flow of non-Newtonian fluids (see Sect. 2.3.4) as well as flows with drift flux (Kleinstreuer, 2003) fall also into the category of flow mixture models. More complicated are *separated flow* models (see Figure 2.4). Examples include two-layer fluids flowing with a smooth interface and (dilute) spherical particle suspension flows (Sect. 2.3.5). More realistic aspects of two-phase flows can be described with two-way coupled *two-fluid models* where the phases interact, i.e., they influence each other.

2.3.4 Non-Newtonian Fluid Flow

We recall that gases and small-molecule liquids (e.g., water and basic oils) are Newtonian. The reason is that the random thermally driven molecular motions (i.e., spin, vibration, collision) within such fluids are sufficiently vigorous that they completely overcome any tendency of the fluid flow forces to produce a molecular configuration state (i.e., local molecular restructuring) that differs significantly from the isotropic, homogeneous state of statistical equilibrium. Clearly, subject to a shear stress fluid mass displaces, i.e., it flows continuously without changing the fluid configuration on the molecular level. For such Newtonian fluids, e.g. air, water, and basic oils, Stokes' hypothesis of a *linear* relationship between shear stress and shear rate holds (see Figure 2.6). In contrast, some fluids, such as polymeric liquids, exotic lubricants, latex paints, food stuff, paste, and certain particle suspensions, exhibit nonlinear viscous effects. Shear rate dependence and/or memory of the viscosity of *non-Newtonian fluids* is due to their component make-up and/or molecular structure (Tanner & Walters, 1998; Macosko, 1994; Bird et al., 1987). Assuming steady incompressible isothermal fluid flow, only shear-rate- (or shear-stress-) dependent liquids are considered, i.e.,

$$\tau_{ij} = \mu \dot{\gamma}_{ij} \text{ (Newtonian) is now replaced by } \tau_{ij} = \eta(\dot{\gamma}, \tau) \dot{\gamma}_{ij} \tag{2.43a,b}$$

where η is the non-Newtonian (or apparent) viscosity.

Thus, non-Newtonian fluid flow phenomena such as rotating-rod climbing and jet swelling after extrusion or the viscoelastic effect of fluid recoil, stress relaxation, and overshoot are not discussed.

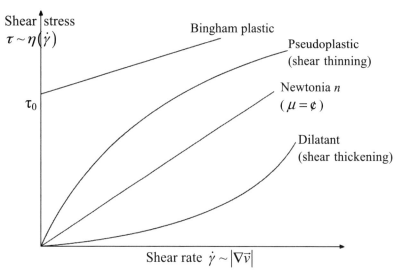

Figure 2.6   Stress/shear rate behavior of fluids

Generalized Newtonian Liquids.   As indicated with Eqs. (2.43a,b), viscous inelastic liquids are also labeled generalized Newtonian liquids because of their similar constitutive equations. The simplest empiricism for $\eta(\dot{\gamma})$ is the two-parameter *power law* expression:

$$\eta = m\dot{\gamma}^{n-1} \tag{2.44}$$

where the constants m and n characterize the fluid. Clearly, when $n = 1$, $m = \mu$ and Eq. (2.43a) is recovered. If $n < 1$, the fluid exhibits shear thinning (i.e., pseudoplastic) behavior, while for $n > 1$ the fluid is called dilatants (or shear thickening) as shown in Figure 2.6. Although the power law is being widely used, it cannot describe the viscosity at very low shear rates (e.g., blood) and the parameters m and n are *not* actual fluid properties.

Power law modeling. When using Eq. (2.44) to model viscous inelastic liquid flows, the point of departure is the reduced *equation of motion* (i.e., the Cauchy equation of App. A), not the Navier-Stokes equation which implies constant fluid properties. For example, considering steady unidirectional flow, the simplified forms of Eq. (1.53) read:

$$0 = -\frac{dp}{dx} + \frac{d\tau_{yx}}{dy} \qquad \text{for planar Couette flows} \qquad (2.45)$$

$$0 = \frac{d\tau_{yx}}{dy} + \rho g \sin \theta \qquad \text{for thin-film flow} \qquad (2.46)$$

$$0 = -\frac{dp}{dx} + \frac{1}{r}\frac{d}{dr}(r\tau_{rx}) \qquad \text{for tubular Poiseuille flow} \qquad (2.47)$$

Examples 2.16 to 2.18 illustrate the use of Eqs. (2.45) to (2.47).

Bingham fluid. For thick suspensions and pastes (e.g., ketchup and toothpaste), no flow occurs until a certain critical stress, called the *yield stress* τ_0, is reached as the result of an applied force. Then, the mixture flows like a Newtonian fluid (see Figure 2.6). Thus, the two-parameter Bingham model:

$$\eta = \begin{cases} \infty & \text{for} \quad \tau < \tau_0 \\ \mu_0 + \dfrac{\tau_0}{\dot{\gamma}} & \text{for} \quad \tau \geq \tau_0 \end{cases} \qquad (2.48\text{a,b})$$

is an illustration of a constitutive equation for a "viscoplastic" material.

More accurate but also more complex non-Newtonian fluid models for specific applications are discussed in Bird et al. (1987), Macosko (1994), and Kleinstreuer (2006), among other texts. Convection heat transfer results in terms of Nusselt number correlations for pipe and slit flows with power law fluids are summarized in Bird et al. (2002).

Example 2.16: Power Law Fluid Flow in a Slightly Tapered Tube

Sketch:	*Assumptions/Postulates:*	*Concepts:*
	• Steady laminar unidirectional flow • Power law fluid • $\vec{v} = [v_z(r,z); 0; 0]$ • Constant pressure gradient $-\partial p / \partial z = \Delta p / L = \cancel{c}$	• Eq. (2.42) with $v_z(z)$ dependence via no-slip condition

Solution:

• Slightly tapered tube:

$$R(z) = R_0 - \frac{R_0 - R_L}{L} z \tag{E2.16-1}$$

• Shear stress and power law:

$$\tau_{rz} = \eta \frac{dv_z}{dr} \text{ and } \eta = m\dot{\gamma}^{n-1} = m\left(\frac{dv_z}{dr}\right)^{n-1} \tag{E2.16-2}$$

so that

$$\tau_{rz} = m\left(-dv_z/dr\right)^n \tag{E2.16-3}$$

where the negative sign assures that $\dot{\gamma}$ stays a positive quantity.

• From a 1-D force balance for fully developed flow:

$$\tau_{rz} = -\frac{\Delta p}{2L} r = \tau_w \frac{r}{R} \tag{E2.16-4a,b}$$

which holds for turbulent flow as well.

• z-Momentum equation (2.47):

$$\frac{1}{r} \frac{d}{dr}\left(r\tau_{rz}\right) = -\frac{p_0 - p_L}{L} \tag{E2.16-5a}$$

or after integration

$$\tau_{rz} = -\frac{\Delta p}{2L} r + \frac{C_1}{r} \tag{E2.16-5b}$$

Clearly, $C_1 \equiv 0$ because at the centerline $r = 0$ but τ_{rz} is finite, i.e., $\tau_{rz}(r = 0) = 0$ (see Eq. (E2.16-4)). Combining (E2.16-5b) with (E2.17-4) and (E2.17-3) yields:

$$m\left(-\frac{dv_z}{dr}\right)^n = \tau_w \frac{r}{R} \tag{E2.16-6}$$

Taking the *n*th root of both sides and integrating result in

$$v_z = -\left(\frac{\tau_w}{mR}\right)^{1/n} \frac{nr^{\frac{n+1}{n}}}{n+1} + C_2 \tag{E2.16-7}$$

subject to

$$v_z\left[r = R(z) = R_0 - \frac{R_0 - R_L}{L} z\right] = 0$$

Thus,

$$v_z(r,\, z) = \underbrace{\left[\left(\frac{\tau_w}{mR(z)}\right)^{1/n} \frac{nR(z)}{n+1}\right]}_{v_{max}}\left[1 - \left(\frac{r}{R(z)}\right)^{\frac{n+1}{n}}\right] \tag{E2.16-8}$$

where $\tau_w = -\left(\Delta p/2L\right)R$ from Eq. (E2.16-4).

• Pressure drop based on volumetric flow rate: given a *Q*-value and with $Q = \int_A \vec{v} \cdot d\vec{A} = 2\pi \int_0^R v_z(r,\, z)r\, dr$, we obtain

$$\Delta p = \frac{2mL}{3n}\left[\frac{Q}{n\pi}(3n+1)\right]^n \left(\frac{R_L^{-3n} - R_0^{-3n}}{R_0 - R_L}\right) \tag{E2.16-9}$$

Graph:

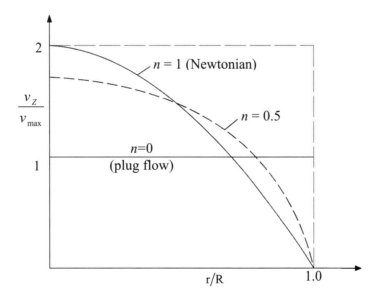

Comment:

With $n < 1.0$, the axial velocity profile becomes flatter because of the shear-thinning effect.

Note: Compare these profiles to the results of Example 2.14 for dilute homogeneous particle suspension flows.

Example 2.17: Film Thickness of a Flowing Polymer

Consider a steady laminar thin layer $(h = \not\subset)$ of a power law fluid (m, n) moving down an incline (angle θ, width w) with a volumetric flow rate Q. Find the film thickness h.

Sketch:	*Assumptions*:	*Concepts*:
	• Steady laminar unidirectional flow • No surface ripples, i.e., $h = $ constant	• Equation (2.46) subject to no-slip and zero interface stress conditions

Solution:

Integration of Eq. (2.46) subject to $\tau_{yz}(y = h) = 0$ yields:

$$\tau_{yx} = \rho g h \left(1 - \frac{y}{h} \right) \sin\theta \qquad \text{(E2.17-1)}$$

where $\tau_{yx}(y = 0) = \tau_{wall} = \rho g h \sin\theta$. Now, with $\tau_{yx} = m(-du/dy)^n$ and after taking the nth root plus integration of

$$-\frac{du}{dy} = \left[\frac{\rho g h \sin\theta}{m} \left(1 - \frac{y}{h} \right) \right]^{1/n} \qquad \text{(E2.17-2a)}$$

subject to $u(y = 0) = 0$ we obtain:

$$u(y) = C\left[1 - \left(1 - \frac{y}{h}\right)^{\frac{n+1}{n}}\right] \qquad \text{(E2.17-2b)}$$

where

$$C \equiv \left(\frac{\rho g h \, \sin \, \theta}{m}\right)^{1/n} \frac{nh}{n+1} \qquad \text{(E2.17-2c)}$$

The volumetric flow rate is:

$$Q = w\int_0^h u(y)\,dy = whC\left(1 - \frac{n}{(2n+1)}\right) \qquad \text{(E2.17-3a)}$$

$$Q = \frac{wh^2 n}{2n+1}\left(\frac{\rho g h \sin\theta}{m}\right)^{1/n} \qquad \text{(E2.17-3b)}$$

Given Q, we can solve for the film thickness, i.e.,

$$h = \left(\frac{m}{\rho g \sin\theta}\right)^{1/(2n+1)}\left[\frac{(2n+1)Q}{nw}\right]^{n/(2n+1)} \qquad \text{(E2.17-4)}$$

Graph:

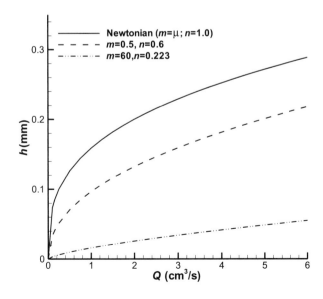

Comments:

- As expected, $h(Q)$ is a nonlinearly increasing function.
- Recalling that power law fluids with $n < 1.0$ are shear-thinning, blunter velocity profiles are generated (see Example 2.16), and the film thicknesses decrease with lower n-values.

Example 2.18: Cylindrical Couette Flow with a Bingham Plastic

Consider two *concentric* cylinders (length L, $R_{inner} = \kappa R$ and $R_{outer} = R$) where the outer one rotates at angular velocity ω_0 due to torque T, setting a Bingham fluid into steady laminar motion. Develop a relationship for $T = T(\omega_0; R, \kappa, L; \tau_0, \mu_0)$.

Sketch:	*Assumptions*:	*Concepts*:
$\kappa R \le r \le R$	• As stated • $\vec{v} = [0, v_\theta(r), 0]$ • $\nabla p = 0$ • Constant torque on outer cylinder and slow rotation	• Reduced θ-momentum equation with only $\tau_{r\theta}$ being nonzero • Bingham fluid model $\eta(\tau)$

Solution:

While continuity is preserved, the θ-momentum equation in cylindrical coordinates (see App. A) reduces to:

$$0 = \frac{1}{r^2}\frac{d}{dr}\left(r^2\tau_{r\theta}\right)$$

(E2.18-1)

Integration yields:

$$\tau_{r\theta} = \frac{C}{r^2}$$

(E2.18-2)

where

$$\tau_{r\theta}(r=R) = \tau_{wall} = \frac{T}{2\pi LR^2} = \frac{C}{R^2} \tag{E2.18-3a-c}$$

and hence

$$\tau_{r\theta} = \frac{T}{2\pi L}r^{-2} \tag{E2.18-4}$$

As indicated with Eqs. (2.48a,b) there is a radial location r_0 where $\tau_{r\theta} = \tau_0$, the yield stress. Clearly, from Eq. (E2.18-4):

$$\tau_0 = \left(\frac{T}{2\pi L r_0}\right)^{1/2} \tag{E2.18-5}$$

where r_0 has to be between κR and R to observe some form of fluid flow (see Eq. (2.48a)). Specifically, for $\kappa R < r < r_0$ there will be viscous flow and for $r_0 \le r \le R$ there will be uniform (or plug) flow.

With shear rate $\dot{\gamma}_{r\theta} = r\,d(v_\theta/r)/dr$ (see App. A), we rewrite Eq. (2.43b) as:

$$\tau_{r\theta} = \eta\left[r\frac{d}{dr}\left(\frac{v_\theta}{r}\right)\right] \tag{E2.18-6}$$

and with Eq. (2.48b) we have:

$$\eta = \mu_0 + \frac{\tau_0}{\dot{\gamma}} = \mu_0 + \frac{\tau_0}{r\dfrac{d}{dr}\left(\dfrac{v_\theta}{r}\right)} \tag{E2.18-7a,b}$$

so that

$$\tau_{r\theta} = \tau_0 + \mu_0 r\frac{d}{dr}\left(\frac{v_\theta}{r}\right) \tag{E2.18-8}$$

Combining (E2.18-6) and (E2.19-4) to solve for $v_\theta(r)$, where $\kappa R < r \le R$, we obtain:

$$\frac{d}{dr}\left(\frac{v_\theta}{r}\right) = \frac{T}{2\pi L\mu_0}r^{-3} - \frac{\tau_0}{\mu_0}r^{-1} \tag{E2.18-9}$$

Integration and invoking the BC $v_\theta\left(r=r_0\right)=r_0\omega_0$ yields for $\kappa R \le r \le r_0$:

$$v_\theta(r)=\omega_0 r+\frac{T}{4\pi L\mu_0 r_0}\left(\frac{r}{r_0}\right)\left[1-\left(\frac{r_0}{r}\right)^2\right]-\frac{\tau_0 r}{\mu_0}\ln\frac{r}{r_0} \qquad \text{(E2.18-10a)}$$

while for $r_0 \le r \le R$:

$$v_\theta=\omega_0 r_0 \qquad \text{(E2.18-10b)}$$

The first graph shows schematically, in terms of the $\tau_{r\theta}(r)$-function (E2.18-4), the impact regions expressed in Eqs. (2.48a,b).

Graph I:

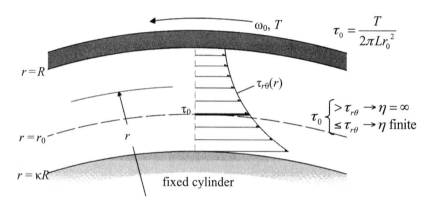

The second graph depicts $v_\theta(r)$ when r_0 is between κR and R as shown in Graph I.
Graph II:

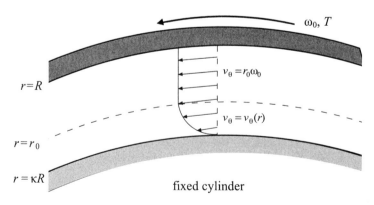

Now, if the yield stress τ_0 is exceeded in the entire gap, i.e., if $r_0 \geq R$, Eq. (E2.18-10a) yields with the BC $v_\theta(r=R)=\omega_0 R$:

$$v_\theta(r)=\omega_0 r+\frac{T}{4\pi L\mu_0 R}\left(\frac{r}{R}\right)\left[1-\left(\frac{R}{r}\right)^2\right]-\frac{\tau_0 r}{\mu_0}\ln\frac{r}{R} \qquad \text{(E2.18-11)}$$

Invoking the no-slip condition $v_\theta(r=\kappa R)=0$ yields the desired expression for the torque:

$$T=\frac{4\pi L\mu_0(\kappa R)^2}{1-\kappa^2}\left[\omega_0-\frac{\tau_0}{\mu_0}\ln\kappa\right] \qquad \text{(E2.18-12)}$$

Comments:

Equation (E2.18-12) is known as the (80-year-old) Reiner-Rivlin equation. With the geometry of the "concentric-cylinder viscometer" given and T and ω_0 measured, the Bingham plastic parameters μ_0 and τ_0 can be determined.

2.3.5 Particle Transport

There are typically two size-dependent categories when modeling spherical noninteracting particles. For microspheres the Euler-Lagrange approach is employed and nanoparticles in the range of $1<d_p<100$ nm are modeled in the Euler-Euler frame. Here, "Euler" implies continuum solution of the conservation laws and "Lagrange" implies particle tracking, i.e., the solution of Newton's second law. Transport simulation of lager nanoparticles may require inclusion of inertia effects. Much more challenging is the dynamics modeling of interacting particles as well as nonspherical particles.

Microsphere Trajectory Models.   As discussed in Sect. 2.3.1 and indicated in Figure 2.4, suspensions of distinct particles with effective diameters, typically greater than 1 μm, fall into the category of *separated flows*. Consequently, two separate sets of equations are needed. One equation describes the particle dynamics and the other one describes the fluid flow; both may contain coupling terms which reflect possible two-phase interactions (see Michaelides, 1997). Quite frequently the dispersed phase, i.e., solid particles, droplets, or bubbles, is *uncoupled* from the continuous phase (or carrier fluid). Specifically, when considering solid, nonrotating spheres with $d_p>1$ μm and a high particle-to-fluid-density ratio in laminar flow, the trajectory equation can be significantly simplified. In any case, the combination of continuous fluid flow and discrete particle transport modeling is known as the Euler-Lagrange approach.

Considering relatively small quasi-spherical micron particles, as well as small particle and shear Reynolds numbers, i.e., $\mathrm{Re}_p = d_p|v - v_p|/v \ll 1$ and $\mathrm{Re}_s = vd_p^2/(vL) \ll 1$, respectively, Newton's second law of motion is applicable in the form (see Crowe et al., 1998; Kleinstreuer, 2003; among others):

$$m_p \frac{d\vec{v}_p}{dt} = \vec{F}_{\mathrm{drag}} + \vec{F}_{\mathrm{pressure}} + \vec{F}_{\mathrm{interactive}} + \vec{F}_{\mathrm{lift}} + \vec{F}_{\mathrm{Basset}} + \underbrace{\vec{F}_{\mathrm{gravity}} + \vec{F}_{\mathrm{virtual\ mass}}}_{\Sigma \vec{F}_{\mathrm{body}}} \qquad (2.49)$$

Here, m_p is the particle mass and $\vec{v}_p = d\vec{x}/dt$ is the particle velocity vector, while all external forces are *point forces* acting on the particle. For laminar flow with negligible particle lift $(\omega_p \approx 0)$, one-way coupling prevails, i.e., the particle presence does not influence the fluid flow and a high density ratio, i.e., $\rho_p/\rho_c \gg 1$, assures that only $\vec{F}_{\mathrm{drag}}$ and perhaps $\vec{F}_{\mathrm{gravity}}$ are important (see Crowe et al., 1998; Buchanan et al., 2000; among others). Hence,

$$m_p \frac{d\vec{v}_p}{dt} = \vec{F}_D + \vec{F}_G \qquad (2.50a)$$

where

$$\vec{F}_D = \frac{\pi}{8} \rho d_p^2 C_{Dp} (\vec{v}_p - \vec{v})|\vec{v}_p - \vec{v}| \qquad (2.50b)$$

which always keeps F_{drag} opposite to the flow direction. Furthermore,

$$\vec{F}_G = m_p \vec{g} \text{ and } m_p = \rho_p \pi d_p^3 / 6 \qquad (2.50c,d)$$

and

$$C_{D_p} = C_D / C_{\mathrm{slip}} ; \ C_D = \frac{24}{\mathrm{Re}_p}(1 + 0.15\,\mathrm{Re}_p^{0.687}) \qquad (2.50e,f)$$

As mentioned, the particle Reynolds number $\mathrm{Re}_p = \rho d_p|v - v_p|/\mu$ is small, say, $\mathrm{Re}_p \le 10$, and C_{slip} is the slip correction factor $\vartheta(1)$ after Clift et al. (1978). Knowing the flow field $\vec{v}(\vec{x}, t)$, the individual particle velocities $\vec{v}_p(t)$ can be obtained, subject to given initial conditions. A second integration, $\vec{x} = \int \vec{v}_p \, dt$, provides then the particle locations, i.e., trajectories. It is typically assumed that a particle has deposited on a surface when it approaches within one radius, i.e., the particle touches the wall. For spherical, noninteracting *droplets*, alternative C_D correlations apply (see Clift et al., 1978) because of the friction-induced internal circulation.

For $Re_p \ll 1$, i.e., Stokes flow, $C_D = 24/Re_p$ and Eq. (2.50) reduces for microsphere flow in a horizontal tube of diameter D to:

$$\frac{D}{U} St \frac{d\vec{v}_p}{dt} = \left(\vec{v} - \vec{v}_p\right) + v_{settling}\hat{g} \tag{2.51a}$$

where U is the mean fluid velocity, e.g., $0.5u_{max}$ in Poiseuille flow, D is the tube diameter, St is the Stokes number, and $v_{settling}$ is the terminal velocity of a sphere. Specifically, for $C_{slip} = 1.0$:

$$St = U\rho_p d_p^2 /\left(18\mu D\right) \tag{2.51b}$$

and after Stokes:

$$v_{settling} = \rho_p g d_p^2 /\left(18\mu\right) \tag{2.51c}$$

An application of Eq. (2.51) for micron-particle suspension flow in a horizontal pipe ($D = 0.2$ cm and $L = 1$ cm) is shown in Figure 2.7 for $U = 10$ and 20 m/s. The deposition efficiency (DE) is defined as the number ratio of particles deposited in a specific region to particles which have entered this region (Kleinstreuer et al., 2007).

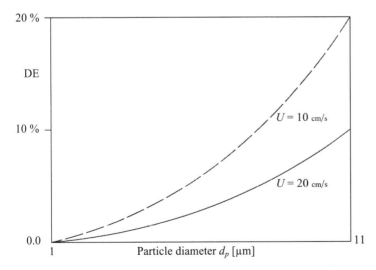

Figure.2.7   Gravitational deposition of micron particles in a horizontal pipe

Example 2.19: Particle Being Accelerated from Rest by a Steady Uniform Airstream

Sketch:	*Assumptions*:	*Concepts*:
$u_\infty \equiv U$ a_p, u_p m_p, d_p	• Spherical particle • Steady unidirectional (air) flow with $\rho/\rho_p \ll 1$ • Drag is the dominant point force	• Carrier fluid $\vec{v} = (u \equiv U, 0, 0)$ • Particle trajectory: $m_p \, du_p/dt = F_D$ $F_D \sim C_D \, A_{\text{projected}}$

Solution:

Using Eq. (2.50a) with $\vec{F}_G = 0$, the 1-D form reads:

$$m_p \frac{du_p}{dt} = \frac{\rho}{2} A C_D (u - u_p)|u - u_p| \qquad (\text{E.2.19-1})$$

where $u - u_p = u_{\text{relative}} \equiv u_r$, C_D is given with Eq. (2.50f), $A_{\text{proj}} = d_p^2 \pi/4$ and $m_p = \rho_p \pi d_p^3/6$. With $u \equiv U = \cancel{c}$, $du_r/dt = -du_p/dt$ and hence Eq. (E2.19-1) can be rewritten as:

$$-\frac{du_r}{dt} = \frac{18\mu}{\rho_p d_p^2}\left[1 + 0.15\left(\frac{\rho d_p}{\mu}\right)^{0.687} u_r^{0.687}\right]u_r \qquad (\text{E2.19-2a})$$

or

$$\frac{du_r}{dt} = A\left[1 + Bu_r^{0.687}\right]u_r \qquad (\text{E2.19-2b})$$

where $A = -18\mu/(\rho_p d_p^2)$ and $B = 0.15(\rho d_p/\mu)^{0.687}$; while Eq. (E2.19-2b) is subject to $u_p(t=0) = 0$, i.e., $u_r(t=0) = u \equiv U$. Separation of variables and integration yield:

$$u_r = U - u_p = \left[(B + U^{-0.687}) \cdot \exp(-0.687At) - B\right]^{-1/0.687} \qquad (\text{E2.19-3})$$

Graph:

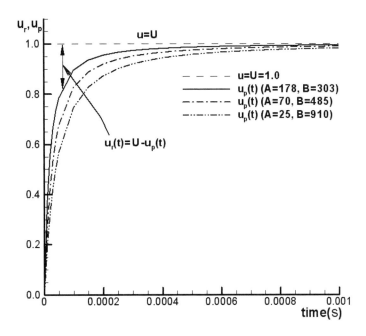

Comment:

Pushed by the free stream, the particle accelerates from rest and reaches exponentially the fluid velocity at a point in time which depends on $A = 18\mu/(\rho d^2)_p$, i.e., $t_e = \ln 1/(0.687A)$.

Nanoparticle Transport. For submicron particles, Brownian motion becomes effective, i.e., random particle motion due to molecular bombardment by the surrounding fluid. That results in enhanced particle diffusion with decreasing nanoparticle diameter. Thus, rather than employing the Euler-Lagrange modeling approach (i.e., solving the fluid flow equations and then the particle trajectory equation), nanomaterial transport is best described in the Eulerian-Eulerian frame. Again, assuming $d_p \leq 100$ nm and laminar flow with low nanoparticle loading, say, below 6% by volume to assure only one-way coupling, the momentum equation is solved first and then the mass transfer equation. Specifically, the equation for nanoparticle convection and diffusion reads:

$$\frac{\partial Y}{\partial t} + \frac{\partial}{\partial x_i}(u_i Y) = \frac{\partial}{\partial x_i}\left[\mathcal{D}_{nano}\frac{\partial Y}{\partial x_i}\right] \pm S_y \qquad (2.52)$$

where $Y \equiv c/c_0$ is the nanoparticle mass fraction, u_i is the fluid velocity vector, $\mathcal{D}_{nano}$ is the nanomaterial diffusion coefficient, and S_y is a possible nanoparticle sink or source. Specifically, according to Stokes-Einstein:

$$\mathcal{D}_{nano} = \frac{k_B T C_{slip}}{3\pi\mu d_p} \qquad (2.53)$$

where $k_B = 1.38 \times 10^{-23} \, \text{JK}^{-1}$ is the Boltzmann constant, T is the temperature in Kelvin, and $C_{slip} = \vartheta(1)$ is the Cunningham slip correction factor (Clift et al., 1978). Based on Fick's law, the regional deposition efficiency can be computed as:

$$DE = \frac{\left(\mathcal{D} \left. \frac{\partial Y}{\partial n} \right|_{n=0} A \right)}{\left(Q_{in} Y_{in} \right)} \qquad (2.54)$$

where n is the surface normal, A is the surface area, Q_{in} is the inlet volumetric flow rate, and Y_{in} is the inlet mass fraction.

Applications of nanoparticle transport and deposition related to biomedical engineering is discussed in Kleinstreuer (2006), while nanofluid flow in microchannels is given in Sect. 6.3.

Example 2.20: Nanoparticle Convection, Diffusion, and Uptake from a Planar Source

Consider steady 1-D flow of a liquid through a porous plug which releases a low constant concentration of nanoparticles which disperse and dissolve/vanish according to a first-order reaction.

Sketch:	*Assumptions:*	*Concepts:*
	• Steady 1-D isothermal plug flow • Constant properties • Low-volume nanoparticle release • Idealized nanoparticle sink	• Uniform flow $v = Q/A = \cancel{c}$ • Reduced form of Eq. (2.52) • Axial diffusion only

Solution:

In light of the assumptions, Eq. (2.52) can be reduced, with $Y \to c, u \to v$, and $S_c \to -kc$, to:

$$v \frac{dc}{dx} = \mathcal{D} \frac{d^2c}{dx^2} - kc \tag{E2.20-1}$$

where $v = Q/A$, while the binary diffusion coefficient $\mathcal{D}$ is given with Eq. (2.53), and k is a constant reaction coefficient.
    Equation (E2.20-1) can be recast as:

$$c'' - \frac{v}{\mathcal{D}} c' - \frac{k}{\mathcal{D}} c = 0 \tag{E2.20-2}$$

subject to the BCs:

$$c(x = 0) = c_0 \text{ and } c(x \to \infty) \to 0 \tag{E2.20-3a,b}$$

The trial solution $c(x) = e^{ax}$ satisfies Eq. (E2.20-2) where

$$a = \frac{v}{2\mathcal{D}} \left[1 - \sqrt{1 + \frac{4k\mathcal{D}}{v^2}} \right]$$

to match the BCs. Hence,

$$c(x) = c_0 \exp \left[-\frac{vx}{2\mathcal{D}} \left(\sqrt{1 + \frac{4k\mathcal{D}}{v^2}} - 1 \right) \right] \tag{E2.20-4}$$

Parameter Values:
    The nanoparticle diffusion coefficient for $T = 300$ K, $C_{slip} \approx 1.0$, $dp = 10$ nm, and $\mu_{water} = 0.9 \times 10^{-3}$ kg/(m·s) is:

$$\mathcal{D} = 4.88 \times 10^{-10} \text{ m}^2/\text{s}$$

Assuming the ratio $4k\mathcal{D}/v^2$ in Eq. (E2.20-4) to be in the range of

$$0.1 \le \frac{4k\mathcal{D}}{v^2} \le 10$$

we can now graph a family of curves:

$$\frac{c(x)}{c_0} \text{ versus } \frac{v}{2\mathcal{D}}x$$

Graph:

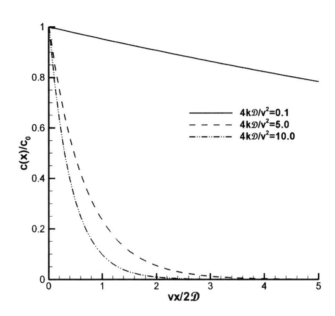

Comment:

As expected, $c(x)$ decays swiftly for $4k\mathcal{D}/v^2 > 1.0$, i.e., due to the first-order uptake.

Modeling Aspects in Fluid-Particle Dynamics. Clearly, so far Sect. 2.3.5 has provided only a basic introduction to spherical micron/nanoparticle transport modeling and simulation. Additional aspects are the particle characteristics and behavior as well as the type of mixture flow. Thus, for fluid-particle system identification and the related solution approach one has to consider the following.

(A) Type of Particle Considerations:

- **Shape:** Spherical versus nonspherical (e.g., cube-like, cylinder/disk, ellipsoid, arbitrary shape) employing direct numerical simulation (DNS) or sphere-equivalent approaches
- **State:** Solid, liquid, or vapor/molecules

- **Size:** Vapor molecules ($d_p \leq 5$ nm) and nanoparticles (see Euler-Euler approach with Stokes-Einstein diffusivity for $d_p \leq 100$ nm) versus micron particles (see Lagrangian tracking, i.e., Newton's second law)
- **Surface:** Neutral versus electrically charged versus chemically coated
- **Behavior:** Dissolved species (see mass transfer equation) or noninteracting monodisperse versus interacting (i.e., influencing the flow field, collisions, aggregation) microspheres, or shape change of droplets (growth, shrinking)

Concerning *nonspherical particles*, as all naturally occurring and most man-made solid particles are, two geometric parameters, Φ and SF, capture deviations from spheres:

$$\underline{\textit{Sphericity:}} \qquad \Phi = \frac{\text{Surface area of volume-equivalent sphere}}{\text{Surface area of considered particle}} \qquad (2.55)$$

and

$$\underline{\textit{Shape factor:}} \qquad SF = \frac{3}{\Phi} \qquad (2.56)$$

Specifically,

$$\Phi \equiv \frac{A_s}{A_p} = \frac{\pi^{1/3}\left(6\forall_p\right)^{2/3}}{A_p} \qquad (2.57)$$

where $\forall_p$ is the actual particle volume with an equivalent sphere diameter of

$$D_{\text{equiv.}} = \left(\frac{6\forall_p}{\pi}\right)^{1/3} \qquad (2.58)$$

Clearly, if the actual particle is spherical, then $\Phi = 1$ and $SF = 3$.

Now, for uniform laminar flow in the range $10^{-1} \leq Re \leq 10^4$, where $Re = UD_{\text{equiv.}}/\nu$, the drag force is

$$F_D = 3\pi\mu\, f U D_{\text{equiv.}} \qquad (2.59)$$

Equation (2.59) is based on the Stokes drag expression, assuming $Re \leq 1.0$.

The effective drag factor $f = f(\Phi, \mathrm{Re})$, as given by Crowe et al. (1998), is shown in Figure 2.8.

It should be noted that the $f = f(\Phi, \mathrm{Re})$ correlations hold for near-spherical particles, e.g., not for cylinders (see Sect. B.2), fibers, or rods.

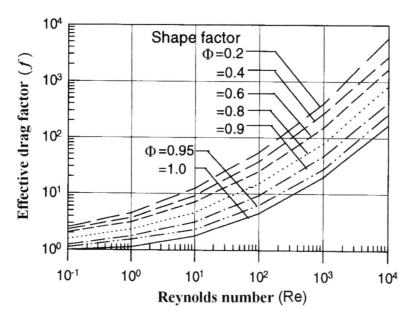

Figure 2.8   Effective drag factor as a function of Reynolds number and sphericity

---

Example 2.21: Drag Force on a Cubicle Particle

Consider a 1-mm cube exposed to an airstream at 25°C ($\rho = 1.184$ kg/m³; $\mu = 1.849 \times 10^{-5}$ kg/(m·s)) with $U = 1$ m/s. Find the diameter and drag for the equivalent sphere.

Sketch:	_Assumptions_:	_Concepts_:
⟨cube⟩ $\hat{=}$ ⟨sphere $D_{eq.}$⟩	• Steady laminar uniform flow • Constant properties	• $F_D = 3\pi\mu f U D_{equiv.}$ • $f = f(\Phi, \mathrm{Re})$ • $\mathrm{Re} = U D_{equiv.} / \nu$

Solution:

- Calculate sphericity Φ and Re to obtain $f = f(\Phi, \mathrm{Re})$ from Figure 2.8.
- With $\forall_p = 10^{-9}$ m^3 and $A_p = 6 \times 10^{-6}$ m^2, $\Phi = \pi^{1/3}\left(6\forall_p\right)^{2/3}\Big/A_p := 0.8$ and $D_{\mathrm{equiv.}} = \left(6\forall_p/\pi\right)^{1/3} := 1.24 \times 10^{-3}$ m
- $\mathrm{Re} = \rho U D_{\mathrm{equiv.}}/\mu := 794$
- Hence, $f = f(\Phi, \mathrm{Re}) \xrightarrow{\text{Fig.2-8}} \approx 70$ and so $F_D = 3\pi\mu\, f U D_{\mathrm{equiv.}} := 1.513 \times 10^{-5}$ N
- Note, $C_D = F_D \Big/ \left[(\rho/2)U^2 A\right] \xrightarrow{\text{App.B.1}} \approx 0.8$ for a sphere of $D = D_{\mathrm{equiv.}}$ and with $\mathrm{Re}_D = 794$. Finally, $F_D = 0.572 \times 10^{-6}$ N

Comments:

- Figure 2.8 is not valid to estimate the drag forces for cylinders, rods, or fibers.
- The calculated drag force is miniscule.

(B) *Type of Mixture Flow Considerations:*

- Coupling: Dilute vs. dense particle suspensions
- Flow Regime: Laminar or turbulent fluid-particle dynamics
- Pairing: Bubbles in liquids, droplets in gas, solid particles in liquid or gas

In micro-/nanofluidics *laminar* fluid-particle flow with solid particles in liquids is most frequently encountered. Simulation of accurate trajectories of *nonspherical* particles requires significant computer resources (see Shenoy and Kleinstreuer, 2010; among others). Considering the interactions of multiple particles with DNS can be prohibitive. Thus, even for a particle suspension with *spheres* it is advantageous when one-way coupling, i.e., assuming a dilute mixture, can be assumed. There are three particle suspension parameters which describe the characteristics of particle suspension flows.

(i) Stokes number:

$$\mathrm{St} = \frac{\rho_p d_p^2 U}{18\mu D} \sim \frac{F_p^{\mathrm{inertia}}}{F_f^{\mathrm{inertia}}} \Rightarrow \begin{cases} \to 0, \ i.e., \ d_p \ll 1; \ \text{perfect mixture} \ \left(v_p = v_f\right) \\ \to \infty, \ i.e., \ d_p \gg 1; \ \text{flow past large fixed spheres} \ \left(v_p \approx 0\right) \end{cases}$$

(ii) Particle loading: $\kappa = \dfrac{(\rho v)_p}{(\rho v)_f} \approx \dfrac{\rho_p}{\rho_f} \to \begin{cases} <6\% & \text{dilute} \\ >10\% & \text{dense} \end{cases}$

(iii) Momentum coupling: $K_M = \dfrac{\kappa}{1 + \mathrm{St}} \Rightarrow \begin{cases} \ll 1 & \text{one-way coupling} \\ >1 & \text{two-way coupling} \end{cases}$

In case the particle suspension is "dense," two modeling approaches should be considered (see Figure 2.4), i.e., well-mixed suspensions or the more complex separated flows.

(C) *Well-Mixed Flow Considerations:*

- Quasi-homogeneous (i.e., single-phase) flows with adjusted mixture (or non-Newtonian fluid) properties
- Mixture flow with particle drift, i.e., due to inertial effects the particle velocity vector is somewhat different from the velocity vector of the carrier fluid
- Uniform dense mixture flow with new mixture properties; especially, μ_m, ρ_m, and k_m

(D) *Separated Flow Considerations:*

- Individual (micro- or macrosize) particle tracking using Newton's second law of motion:

$$m_p \frac{d\vec{v}_p}{dt} = \sum \vec{F}; \; F = F\left(d_p, \vec{v}_{\text{fluid}}, \vec{v}_p, \text{Re}, \text{St, etc.}\right); \; \vec{v}_p = d\vec{x}_p/dt$$

- Momentum equation with particle-fluid interaction:

$$\frac{D\vec{v}}{Dt} = \sum \vec{f}; \; f = f\left(\nabla p, \tau; d_p, v_p, \text{etc.}\right)$$

- Dispersed nanoparticle transport using the Euler-Euler solution approach for $d_p \leq 100$ nm:

$$\frac{Dc}{Dt} = \nabla \cdot \left(\mathcal{D}\nabla c\right) \pm S_c; \; \mathcal{D} = \frac{k_B T C_{\text{slip}}}{3\pi\mu d_p} \quad \text{(Stokes-Einstein equation)}$$

- Two interacting "fluids," i.e., one continuum phase (Fluid I) and the other particle phase (Fluid II)

2.4 Heat Transfer

While undergraduate engineering students are generally acquainted with heat transfer via introductory courses, solving species mass transfer problems is traditionally the domain of chemical, biomedical, and environmental engineers. As already indicated in Sect. 1.3.3.4, both transport phenomena are described by similar scalar equations. Thus, after reviewing forced convection heat transfer, the extension to species mass transfer is not that difficult and is very important in numerous micro/nanofluidics applications.

2.4.1 Forced Convection Heat Transfer

Figure 2.9 summarizes the interactions between fluid mechanics and heat transfer.

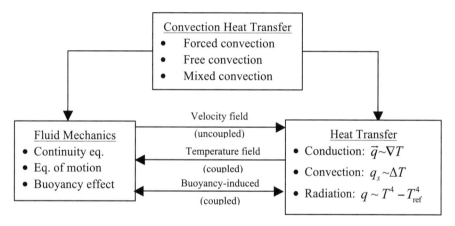

Figure 2.9   Convection heat transfer components

Typically, fluid properties are assumed to be constant which decouples the momentum equation from the heat transfer equation. In contrast (two-way) coupling occurs when, say, the viscosity and/or density change greatly with temperature. Specifically, in Sect. 1.2.4 the heat flux vector, i.e., Fourier's law, was introduced as:

$$\vec{q} = -k\nabla T \tag{2.60}$$

where k is the (isotropic) thermal conductivity. Then, in Sect. 1.3.3.4 the heat transfer equation:

$$\frac{\partial T}{\partial t} + (\vec{v} \cdot \nabla)T = \alpha \nabla^2 T + \frac{\mu}{\rho c_p}\Phi \pm S_{\text{heat}} \tag{2.61}$$

was derived, where $\alpha \equiv k/(\rho c_p)$ is the thermal diffusivity, $\mu \Phi = \tau_{ij} \partial v_i/\partial x_j$ is the viscous dissipation function, and S_{heat} is a possible source or sink of energy, say, due to chemical reaction. In rectangular coordinates we have (see Sect. A.2):

$$\Phi = 2\left[\left(\frac{\partial u}{\partial x}\right)^2 + \left(\frac{\partial v}{\partial y}\right)^2 + \left(\frac{\partial w}{\partial z}\right)^2\right]$$
$$+ \left[\left(\frac{\partial u}{\partial y} + \frac{\partial v}{\partial x}\right)^2 + \left(\frac{\partial v}{\partial z} + \frac{\partial w}{\partial y}\right)^2 + \left(\frac{\partial w}{\partial x} + \frac{\partial u}{\partial z}\right)^2\right]$$
$$- \frac{2}{3}\left(\frac{\partial u}{\partial x} + \frac{\partial v}{\partial y} + \frac{\partial w}{\partial z}\right)^2 \tag{2.62}$$

where obviously the $(\partial u/\partial y)^2$ term in Eq. (2.62) is most significant. Obviously, flow field areas with steep velocity gradients and fluids of high viscosity may generate measurable temperature increases. Equation (2.60) was used for Eq. (2.61), i.e., the net heat conduction term is for constant fluid properties:

$$-\nabla \cdot \vec{q} = k\nabla^2 T \tag{2.63}$$

While in Eq. (2.61) heat conduction, $\alpha \nabla^2 T$, is a diffusional transport phenomenon, heat transfer by convection, $(\vec{v} \cdot \nabla)T$, occurs typically much more rapidly, at least in macrochannels. Now, in order to avoid temperature gradients and hence simplify things, the surface heat flux, q_s, from a hot surface of temperature T_s into a moving stream with reference temperature T_{ref} can be based on the temperature difference $\Delta T = T_s - T_{\text{ref}}$. Thus, rather than employing Eq. (2.60), Newton's "law of cooling" is used (see Sect. 2.4.2):

$$q_s = h(T_s - T_{\text{ref}}) \tag{2.64}$$

Here, h is the convective heat transfer coefficient, $T_{\text{reference}}$ is either T_{mean}, the cross-sectionally averaged fluid temperature, i.e., $T_m(x)$, or T_∞ as in thermal boundary-layer theory, or the fluid bulk temperature $T_b = (T_{\text{in}} + T_{\text{out}})/2$, with T_∞ and T_b being constant. As always, the heat flow rate is then:

$$\dot{Q} = A_s q_s \tag{2.65}$$

2.4.2 Convection Heat Transfer Coefficient

The heat transfer coefficient h encapsulates all possible system parameters, such as temperature difference, thermal boundary-layer thickness, Reynolds number, fluid Prandtl number, and wall geometry. Clearly, h is not a property such as k; but it is a convenient artifact to calculate q_s or T_s and ultimately $\dot{Q}$, which also greatly depends on the heat transfer area A_s. For example, for boundary-layer flow:

$$h = \frac{q_s}{T_s - T_\infty} = \frac{-k}{T_s - T_\infty} \frac{\partial T}{\partial y}\bigg|_{y=0} \qquad (2.66a,b)$$

Typical h-values range from 20 to 300 for gases and from 5000 to 50000 W/(m² · °C) for liquid metals. Figure 2.10 visualizes Eq. (2.66).

Newton's law of cooling:

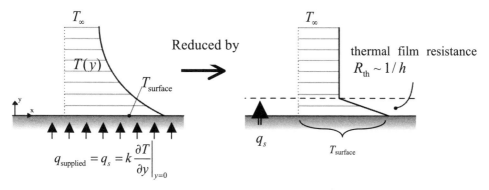

Thermal boundary layers:

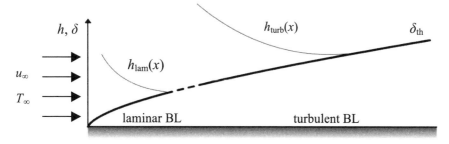

Figure 2.10   Dependence of heat transfer coefficient on flow regime

The Nusselt Number.   Equation (2.66a) can be nondimensionalized by inspection, using the axial coordinate x as a length scale, i.e.,

$$\frac{hx}{k} \equiv \mathrm{Nu}_x = \frac{q_s x}{k(T_s - T_\infty)} \qquad (2.67\mathrm{a,b})$$

where Nu_x is known as the local Nusselt number. Similarly, the average Nusselt number based on system length L, where L could be a plate length or pipe diameter, is:

$$\overline{\mathrm{Nu}}_L = \frac{\bar{h}L}{k} \qquad (2.67\mathrm{c})$$

where

$$\bar{h} = \frac{1}{L}\int_0^L h(x)\,dx \qquad (2.68)$$

For forced convection, neglecting buoyancy and viscous dissipation,

$$\mathrm{Nu}_x = \mathrm{Nu}_x\left(\frac{x}{L};\ \mathrm{Re},\ \mathrm{Pr}\right) \qquad (2.69)$$

In summary, the main objective for a differential analysis is to find the temperature field $T = T(x, y, z, t)$ and then obtain h (or Nu_x) in order to calculate the surface heat flux or temperature. The following solution steps are for constant-property fluids, i.e., one-way coupled problems:

(i)   Solve, subject to appropriate boundary conditions, a reduced form of Eq. (2.61) after securing a computed (or measured) velocity function (see, for example, Sect. 2.1).
(ii) Calculate the wall temperature gradient and obtain, via Eq. (2.66), $h(x)$ and $\mathrm{Nu}(x)$.

The Reynolds-Colburn Analogy.   Note that as an *alternative approach* Reynolds and Colburn (R-C) established an analogy between heat and momentum transfer. It is based on the similarity between dimensionless temperature and velocity profiles in boundary layers (see Sect. 2.6):

$$\frac{1}{2}C_f(x) = \text{St}_x \, \text{Pr}^{2/3} \quad \text{for } 0.6 < \text{Pr} < 60 \qquad (2.70)$$

where the skin friction coefficient and Stanton number are:

$$C_f = \frac{2\tau_{\text{wall}}}{\rho u_\infty^2} \text{ and St}_x = \frac{\text{Nu}_x}{(\text{Re}_x\text{Pr})} = \frac{h(x)}{(\rho c_p u_\infty)} \qquad (2.71\text{a-c})$$

Clearly, once the wall shear stress of a thermal boundary-layer problem is known, $\text{Nu}(x)$ or $h(x)$ can be directly obtained.

Example 2.22: Simple Couette Flow with Viscous Dissipation

As an example of planar lubrication with significant heat generation due to oil-film friction, consider simple thermal Couette flow with adiabatic wall and constant temperature of the moving plate.

Sketch:	*Assumptions:*	*Approach:*
$y=d$ u_0 $T_s=T_o$ $u(y)$ $\left(\frac{\partial p}{\partial x}=0\right)$ y $q_s=0$ x	• Steady laminar 1-D flow $\nabla p = 0$; u_0 and d are constant	• Reduced N-S equations and HT equation. • Constant thermal wall condition.

Solution:

Based on the postulates $\vec{v} = [u(y), 0, 0]$ and $\nabla p \equiv 0$, the Navier-Stokes equations reduce to:

$$0 = 0 \quad \text{<continuity>}$$

and

$$0 = \frac{d^2 u}{dy^2} \quad \text{<x-momentum>} \qquad (\text{E2.22-1a})$$

subject to $u(y=0) = 0$ and $u(y=d) = u_0$. Thus,

$$u(y) = u_0\, y/d \qquad\qquad \text{(E2.22-1b)}$$

The heat transfer equation (2.61) with $\Phi = \left(\partial u/\partial y\right)^2$ from Eq. (2.62) reduces to:

$$k\frac{d^2 T}{dy^2} = -\mu\left(\frac{u_0}{d}\right)^2 \qquad\qquad \text{(E2.22-2)}$$

subject to $dT/dy\big|_{y=0} = 0$ and $T\left(y=d\right) = T_0$. Double integration yields:

$$T(y) = T_0 + \frac{\mu u_0^2}{2k}\left[1 - \left(\frac{y}{d}\right)^2\right] \qquad\qquad \text{(E2.22-3)}$$

At the plate surface $q_s = q\left(y=d\right) = -k\left(\partial T/\partial y\right)\big|_{y=d}$ we have:

$$q_s = \mu u_0^2/d \qquad\qquad \text{(E2.22-4)}$$

Comments:

    Clearly, as μ and u_0 increase and the spacing decreases, q_s shoots up. For simple Couette flow du/dy evaluated at $y=d$ is equal to u_0/d so that $q_s = u_0\tau_{\text{wall}}$ here, which is a simple example of the heat transfer and momentum transfer relation (see Reynolds-Colburn analogy). Of interest would be the evaluation of the mean fluid temperature, $T_m = \left(1/\dot{m}\right)\int_A \rho u T\, dA$, to estimate h from $q_s = h\left(T_0 - T_m\right)$.

Example 2.23: Reynolds-Colburn Analogy Applied to Laminar Boundary-Layer Flow

Consider a heated plate of length L and constant wall temperature T_w, subject to a cooling airstream $\left(u_\infty, T_\infty\right)$. Find a functional dependence for $q_w\left(x\right)$.

Sketch:	*Assumptions:*	*Approach:*
	• Thermal Blasius flow (see Sect. 1.3.3.3) • Constant properties	• Reynolds-Colburn analogy

Solution:

Rewriting Eq. (2.70) with Eq. (2.71c) and Eq. (2.68a) we have:

$$\frac{q_w(x)}{\rho c_p u_\infty (T_w - T_\infty)} = C_f \, \mathrm{Pr}^{-2/3} \qquad \text{(E2.23-1)}$$

we can deduce that

$$C_f \sim \mathrm{Re}_x^{-1/2} \qquad \text{(E2.23-2)}$$

where actually $C_f = 0.664/\sqrt{\mathrm{Re}_x}$ as shown by a homework assignment in Sect. 2.6. Now, with everything else being constant, the wall heat flux distribution is:

$$q_w(x) \sim \frac{K}{\sqrt{x}}, \quad K = \cancel{c} \qquad \text{(E2.23-3a, b)}$$

Graph:

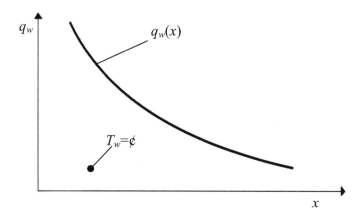

Comment:

The wall heat flux from the plate surface decreases nonlinearly with plate distance because of the increasing $\mathrm{Re}(x)$ or, better, the larger thermal boundary-layer thickness, $\delta_{th}(x)$, and hence milder wall temperature gradients (see Eq. (2.60)).

Example 2.24: Lubricated Shaft Rotation with Heat Generation

Consider "cylindrical" Couette flow where, somewhat similar to Example 2.22, the rotating shaft (R_i, ω_i) is adiabatic and the stationary housing (R_0, T_0) is isothermal. In light of viscous dissipation, find $T(r)$ as well as T_{max} at $r = R_i$ and $Q_{wall}(r = R_0)$.

Sketch:	Assumptions:	Approach:
	• Steady laminar 1-D axisymmetrical flow • No gravity or end effects • Constant properties	• Reduced Θ-momentum equation • Postulate $v_\theta = v_\theta(r)$ only

Solution:

Based on the postulate and assumptions, the continuity equation is satisfied and the Θ-momentum equation in cylindrical coordinates reduces to (see the equation sheet in Sect. A.27):

$$\frac{d}{dr}\left(\frac{1}{r}\frac{d}{dr}(rv_\theta)\right) = 0 \qquad \text{(E2.24-1a)}$$

subject to

$$v_\theta(r = R_i) = (\omega R)_i \text{ and } v_\theta(r = R_0) = 0 \qquad \text{(E2.24-1b,c)}$$

Double integration and invoking the BCs yield:

$$v_\theta(r) = \frac{\omega_i R_i (R_0/R_i)^2}{\left(\dfrac{R_0}{R_i}\right)^2 - 1}\left[\frac{R_i}{r} - \frac{r}{R_i}\right] \qquad \text{(E2.24-2)}$$

The heat transfer equation (see the equation sheet in Sect. A.27) reduces to:

$$0 = -\frac{k}{r}\frac{d}{dr}\left(r\frac{dT}{dr}\right) - \mu\Phi \qquad \text{(E2.24-3a)}$$

where

$$\Phi = \left(\frac{dv_\theta}{dr} - \frac{v_\theta}{r}\right)^2 \qquad \text{(E2.24-3b)}$$

and as stated:

$$\left.\frac{dT}{dr}\right|_{r=R_i} = 0; \; T(r = R_0) = T_0 \qquad \text{(E2.24-3c,d)}$$

With $v_\theta(r)$ given, Eq. (E2.24-3b) can be determined and hence Eq. (E2.24-3a) can be integrated subject to Eqs. (E2.24-3c,d). Thus,

$$T(r) = T_0 + \frac{\mu}{4k}\left[\frac{2\omega_i R_i}{1-(R_i/R_0)^2}\right]^2\left[\left(\frac{R_i}{R_0}\right)^2 - \left(\frac{R_i}{r}\right)^2 + 2\,\ln\left(\frac{R_0}{r}\right)\right] \qquad \text{(E2.24-4)}$$

Now, either by inspection of (E2.24-4) or setting dT/dr to zero, $T_{\max}$ occurs at $r = R_i$. In dimensionless form,

$$\frac{T_{\max} - T_0}{\dfrac{\mu}{4k}\left[\dfrac{2\omega_i R_i}{1-(R_i/R_0)^2}\right]^2} = \left[\left(\frac{R_i}{R_0}\right)^2 - 1 + 2\ln\left(\frac{R_0}{R_i}\right)\right] \qquad \text{(E2.24-5)}$$

The wall heat transfer rate per unit length is:

$$\hat{Q}_{\text{wall}} = (2\pi R_0)q(r = R_0)$$

where

$$q(r = R_0) = -k\left.\frac{dT}{dr}\right|_{r=R_0} \qquad \text{(E2.24-6a,b)}$$

Hence,

$$\hat{Q}_w = \frac{4\pi\mu(\omega R)_i^2}{1-(R_i/R_0)^2}$$

(E2.24-6c)

Graphs:
   (a) Velocity profiles

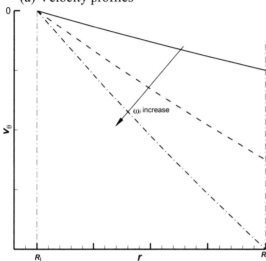

   (b) Temperature profiles

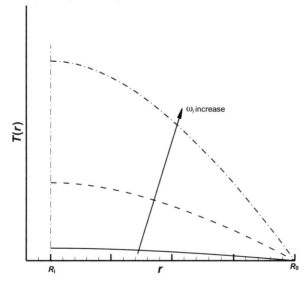

(c) Wall heat flow rate

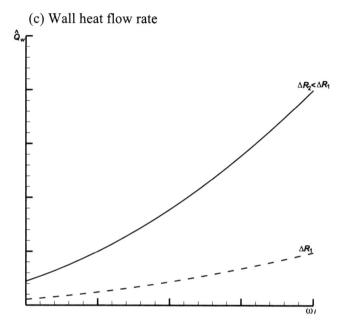

Comments:

- For small gaps, i.e., $R_0 - R_i \ll 1$ the velocity profiles are almost linear, despite the hyperbolic term in Eq. (E2.24-2). Clearly, with $\Delta R \ll 1$, $v_\theta(r)$ is "linearized."
- This is not the case for $T(r)$ due to the strong viscous heating effect (see graph b).
- As expected, $\hat{Q}_w(\omega_i)$ decreases with a strong nonlinear influence of the gap size (see graph c)

2.5 Convection-Diffusion Mass Transfer

Transport of chemical compounds or molecular and nanosize particles in carrier fluids (typically water or air) via convection and/or diffusion is a challenging problem area with numerous applications in environmental, chemical, and biomedical engineering. Part III of the seminal book by Bird et al. (2002) is a good source to study and refresh the basics in mass transfer. Here we ignore some finer points of convection-diffusion mass transfer and simply review the basic modeling and solution approaches. In addition, we provide some examples in *compartmental modeling* as needed in some microfluidics and nanofluidics applications.

2.5.1 Modeling Approaches

Diffusion is the process by which molecules, ions, or other small particles mix. A dual interpretation of this process is possible. Either species are moving from regions of high to low concentrations controlled by a diffusion coefficient, or a concentration difference drives species mass transfer, controlled by a mass transfer coefficient (see Eq. (2.74) versus Eq. (2.75)). Being akin to heat transfer, the mass transfer equation describing diffusion and convection of species (e.g., dissolved chemical compounds or tiny nanoparticles) was already introduced in Sect. 1.3. In mathematical form:

$$\frac{Dc}{Dt} = \underbrace{\frac{\partial c}{\partial t}}_{\substack{\text{Temporal} \\ \text{changes in } c}} + \underbrace{\vec{v} \cdot \nabla c}_{\substack{\text{Convection} \\ \text{of } c}} = \underbrace{\nabla \cdot \vec{j}}_{\substack{\text{Diffusion} \\ \text{of } c}} \pm \underbrace{S_c}_{\substack{\text{Sinks or} \\ \text{sources of } c}} \tag{2.72a}$$

where Fick's law states:

$$\vec{j} = -\boldsymbol{\mathcal{D}}\nabla c \tag{2.72b}$$

Compared to convection, diffusion in the same direction is negligible in macrofluidics application. For example, for predominantly axial mixture flow we can assume:

$$u\frac{\partial c}{\partial x} \gg \frac{\partial j_c}{\partial x} = \boldsymbol{\mathcal{D}}\frac{\partial^2 c}{\partial x^2} \tag{2.73a,b}$$

In microchannels, e.g., being part of LoCs (labs-on-chips) or bioMEMS, diffusion processes are often dominant because of the very small distances and high diffusivities but very low velocities.

For simplicity, considering total one-dimensional species diffusion flux, Eq. (2.72b) can be written as:

$$J = Aj_x = A\mathcal{D}\frac{\partial c}{\partial x} \quad \left[\frac{kg}{s}\right] \tag{2.74}$$

where A is the area across which diffusion occurs. An alternative description of mass transfer is appropriate when two well-mixed solutions, naturally separated by an interface, have different species concentrations.

Then,

$$J = Ak(c_1 - c_2) \quad \left[\frac{kg}{s}\right] \tag{2.75}$$

where J is the rate of mass transferred, i.e., $j = J/A$ (see Eq. (2.72b)) in comparison, A is the interfacial area, k is the mass transfer coefficient, and $\Delta c = c_1 - c_2$ is the concentration difference. Clearly, Eq. (2.75) resembles Newton's law of cooling (see Eq. (2.65)).

A mass balance for a well-mixed compartment of volume $\forall$ and time-dependent concentration c yields:

$$\frac{d(\forall c)}{dt} = -J = -kA\Delta c \tag{2.76}$$

subject to some initial conditions $c(t=0) = c_0$ (see Example 2.24). Clearly, when comparing a 1-D version of Eq. (2.72a) with the simpler Eq. (2.76), it is apparent that the transient term is preserved in Eq. (2.76); but the convection-diffusion terms are gone. Thus, the RHS term of Eq. (2.76) can be interpreted as a source term.

========

Example 2.25: Air Humidification in a Tank

Consider a tank $(\forall = 20\ L, A = 150\ cm^2)$ with 1 L of water at 25°C. Water is evaporating into the initially dry air, where a measured mass transfer coefficient is $k = 5 \times 10^{-2}$ cm/s. How long will it take to reach 90% saturation (see Cussler, 2009)?

Sketch:	*Assumptions:*	*Concepts:*
Air $c(t)$ Interface Water	• Instantaneously well-mixed compartments • Constant air volume • Constant mass transfer coefficient	• Lumped-parameter approach • Mass conservation • Use Eq. (2.76)

Solution:

Equation (2.76) can be rewritten as a balance of H_2O vapor accumulation in the (dry) air and the H_2O evaporation rate:

$$\forall \frac{dc}{dt} = -kA(c_{sat} - c) \qquad \text{(E2.25-1)}$$

where A is the interface (or tank) area and $c_{sat} = 100\%$; $c(t=0)=0$ while $c(t=t_{final}) = c_{final} = 90\%$.
    Separation of variables and integration yield:

$$\frac{dc}{c_{sat} - c} = -\frac{kA}{\forall} \text{ and } \frac{c}{c_{sat}} = 1 - \exp\left(-\frac{t}{\tau}\right) \qquad \text{(E2.25-2)}$$

where $\tau \equiv \forall/kA$ is the system rate constant. Alternatively,

$$t = -\tau \ln\left(1 - \frac{c}{c_{sat}}\right) \qquad \text{(E2.25-3)}$$

which yields $t = 1.71\,\text{h}$ for $c/c_{sat} = 0.9$.

Graph:

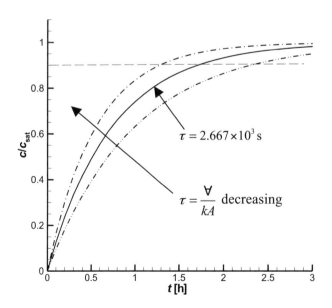

Comment:

As the system rate constant τ decreases, a longer time is required to reach 90% saturation.

Examples 2.26a and 2.26b: Liquid-in-Liquid and Gas-in-Liquid Mass Transfer

(a) Consider droplets ($\forall$, A) of bromine, a heavy reddish-brown nonmetallic liquid, starting at $t = 0$ to be dissolved in water. Find the mass transfer coefficient, k, based on the measurement that the bromine concentration is about half saturated in 3 min, i.e., $c/c_{sat}(t = 3 \text{ min}) = 0.5$.

Sketch:	*Assumptions:*	*Concepts:*
Bromine droplets · Water · c(t)	• Uniform (1-D) diffusion • Homogenous mixture • Constant k, $\forall$, and A	• Eq. (2.76) • $æ = \dfrac{\text{surface area}}{\text{droplet volume}}$ • $c_{interface} = c_{sat}$ • $c_{bulk} = c$

Solution:

Rewriting Eq. (2.76) we have:

$$\frac{d}{dt}(c\forall) = Ak(c_{sat} - c) \qquad \text{(E2.26a-1)}$$

With $æ = A/\forall$,

$$\frac{dc}{dt} = kæ(c_{sat} - c) \qquad \text{(E2.26a-2)}$$

subject to $c(t = 0) = 0$. Integration yields:

$$\frac{c}{c_{\text{sat}}} = 1 - \exp(-k\text{æ}t)$$ (E2.26a-3)

Solving for æk we obtain:

$$\text{æ}k = -\frac{1}{t}\ln\left(1 - \frac{c}{c_{\text{sat}}}\right)$$ (E2.26a-4)

With $t = 3$ min and $c/c_{\text{sat}} = 0.5$:

$$\text{æ}k = 3.9 \times 10^{-3} \text{ s}^{-1}$$ (E2.26a-5)

Say, for $\text{æ} = 3.9 \times 10^{5}$ m^{-1}, $k = 1.0 \times 10^{-8}$ m/s.

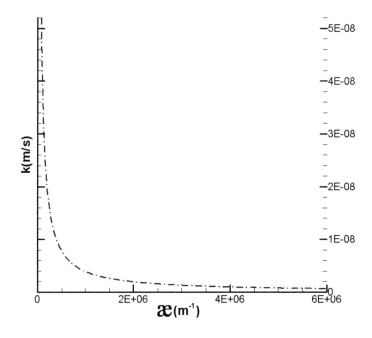

(b) Consider an oxygen bubble, with shrinking diameter $D(t=0) = 10^{-3}$ m and $D(t = 10 \text{ min}) = 10^{-4}$ m dissolving into well-mixed water. Recall that at standard conditions the O_2 concentration inside the bubble is $c_0 = 1 \text{ mol}/22.4$ L, while $c_{\text{sat}} = 1.5 \times 10^{-3}$ mol/L which is the O_2 concentration at saturation in water. Find the mass transfer coefficient k.

Sketch:	Assumptions:	Concepts:
	• Uniform radial diffusion • Constant c_0, c_{sat}, and k • $c_{water}(t=0)=0$	• Form of Eq. (2.76) • $\forall_{sphere}=4\pi r^3/3,\ A_{surf}=4\pi r^2$ • $c_{interface}=c_{sat}$ • $\Delta c = c_{sat}-c_{water}=c_{sat}-0$

Solution:

$$\frac{d}{dt}(c\forall)=-kA\Delta c \rightarrow \frac{d}{dt}\left(c_0\frac{4}{3}\pi r^3\right)=-k4\pi r^2 c_{sat} \qquad \text{(E2.26b-1)}$$

or

$$\frac{c_0}{3}\frac{dr^3}{dt}=-r^2 kc_{sat},\ \text{where}\ \frac{dr^3}{dt}=\frac{dr^3}{dr}\frac{dr}{dt}=3r^2\frac{dr}{dt} \qquad \text{(E2.26b-2)}$$

Hence,

$$\frac{dr}{dt}=-k\frac{c_{sat}}{c_0}=0.0336k \qquad \text{(E2.26b-3)}$$

Finally, subject to $r(t=0)=0.5\times10^{-3}$ m

$$r(t)=0.05\ \text{cm}-0.0336kt \qquad \text{(E2.26b-4)}$$

Inserting the measured data point, i.e., $D(t=10\ \text{min})=10^{-4}$ m, yields:

$$0.005\ \text{cm}=0.05\ \text{cm}-0.0336k(600\ \text{s}) \qquad \text{(E2.26b-5)}$$

Thus,

$$k=2.23\times10^{-3}\ \text{cm/s} \qquad \text{(E2.26b-6)}$$

Comment:

The k-value is only accurate initially, i.e., at $t=0$ when $c_{water}=0$.

2.5.2 Compartmental Modeling

Most industrial transport phenomena, including convective-diffusive mass transfer, are rather complex. Hence, traditionally such problems have been analyzed with a lumped-parameter (or "black-box") approach, relying on the RTT (see Sect. 1.3) from which coupled first-order rate equations for species mass transfer can be derived. The study of drug, toxin, and/or metabolite kinetics in the human body is a major application, known as pharmacokinetics modeling.

In general, a compartment is a "well-mixed box," e.g., a homogeneous isotropic region, an entity or device of constant volume which has uniform inlet and outlet streams of different velocities, enthalpies, and/or concentrations. Of interest are integral (or global) quantities, such as flow rates and exerted forces, or the time rate of change of species concentrations inside the compartment (and hence its outlet), as well as species material transport, deposition, and conversion. Examples include:

(i)   Flow problems which can be solved with the mass/momentum RTT, e.g., fluid-mass conservation in a pipe network or drag forces on submerged bodies (see Examples 1.4 and 1.7).

(ii) Energy flow exchange, work done, and heat transfer for closed and open systems, employing the first law of thermodynamics via a "lumped-parameter" approach, based on the energy RTT.

(iii) Species mass transfer in complex systems (e.g., the human body, part of the environment, or a petroleum refinery) where the system is compartmentalized and transport phenomena described by *mass balances*, typically a coupled set of convective first-order rate equations.

While examples for Groups (i) and (ii) have been provided, compartmental analysis of complex species mass transfer systems requires further discussion. Clearly, Group (iii) applications have been traditionally the domain of biomedical and chemical engineers, rather than mechanical engineers. To begin with, a representative system is decomposed into a network of perfectly mixed, constant-volume compartments which are connected by ducts with negligible impact, i.e., no volume, dispersion, and losses in the conduits.

Because of their many simplifications, compartment models have their limitations. In many applications the material in the compartment is not homogeneous. Furthermore, not all the inflowing/outflowing material plus material conversion due to chemical reaction can be fully accounted for. Examples include varying species concentrations, such as vertical nutrient concentrations in a lake, or an incomplete water balance of an estuary.

For a system with n compartments, the first-order mass balance equation reads (Godfrey, 1983; among others):

$$\frac{dc_i}{dt} = f_{io} + \sum_{\substack{j=1 \\ j\neq i}}^{n}\left(f_{ij} - f_{ji}\right) - f_{oi}, \quad i = 1, 2, \ldots, n \tag{2.77a}$$

where c_i is the amount of material in compartment i, f_{ij} is the mass flow rate to compartment i from compartment j, and subscript o indicates the surrounding.

The flows f_{ij} are related to the state variables c_1 to c_n as:

$$f_{ij} = k_{ij} f_j\left(c_i\right), \quad i = 0, 1, \ldots, n; j = 0, 1, \ldots, n; i \neq j \tag{2.77b}$$

Focusing on linear, time-invariant compartmental models for which $f_{ij} \approx k_{ij}c_i$, where k_{ij} are the rate constants, Eq. (2.77) can be rewritten as (Figure 2.11):

$$\frac{dc_i}{dt} = \sum_{\substack{j=1 \\ j\neq i}}^{n} k_{ij}c_j - \sum_{\substack{j=1 \\ j\neq i}}^{n} k_{ji}c_i - k_{oi}c_i + u_i\left(t\right), i = 1, 2, \ldots, n \tag{2.78}$$

It should be noted that the term $-k_{oi}c_i$ (discharge of material to the surrounding) could also be a material sink, while input $u_i\left(t\right)$ could also be a species source due to chemical reaction, etc.

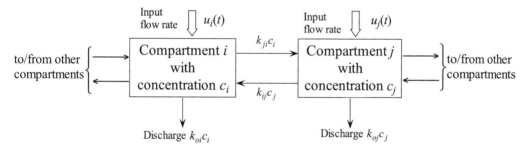

Figure 2.11   Two compartments of a generalized "lumped-parameter" model

Alternatively to using the RTT to derive Eq. (2.78), the transient 1-D mass transfer equation, i.e., Eq. (2.72a) could be simplified:

$$\frac{\partial c}{\partial t} + u\frac{\partial c}{\partial x} = \mathcal{D}\frac{\partial^2 c}{\partial y^2} \pm S_c \tag{2.79}$$

That is accomplished by considering n compartments, neglecting diffusion, and approximating the convective term as:

$$\frac{\Delta(uc)}{\Delta x} \approx f_{ji} - f_{ij}; \; f_{ij} = k_{ij}c_i \qquad (2.80a,b)$$

i.e., net rate of material transferred between compartments i and j. In a more compact form Eq. (2.78) then reads:

$$\frac{dc_i}{dt} = \sum_{j=1}^{n}\left(k_{ij}c_j - k_{ji}c_i\right) + S_i(t) \qquad (2.81)$$

Its homogeneous solution is of the form

$$c_i(t) = \sum_{k=1}^{n} A_k e^{-\alpha_k t} \qquad (2.82)$$

where the coefficients A_k are proportional to k_{ij}, $c_i(t=0)$ and $S_i(t)$, while the eigen-frequencies $\alpha_k \sim k_{ij}$. The particular solution for $c_i(t)$ depends on the information for k_{ij}, $c_i(t=0)$, and $S_i(t)$. Linearity of Eq. (2.81) allows one to compute $c_i(t)$ in steps and then sum up all the (independent) solutions (see Example 2.27 as well as Hoffman, 2001; among others).

─────────────────────────────

Example 2.27: *Two-Compartment System in Series*

Consider a two-compartment system (see Truskey et al., 2004; among others). For example, drug injection (species mass M) into a main compartment I of volume $\forall_1$ and initial concentration $c_1(t=0) = c_0 = M/\forall_1$, and with drug clearance, i.e., discharge or inactivation. Compartment I intersects via two-way species mass transfer (i.e., rate constants k_1 and k_2) with compartment II, where $c_2(t=0) = 0$

(a) Based on suitable assumptions, develop mass balance equations, i.e., first-order rate equations for both $c_1(t)$ and $c_2(t)$, with appropriate initial conditions.
(b) Convert the (a) results to a single ODE for $c_1(t)$ and then find solutions for both $c_1(t)$ and $c_2(t)$.
    Note: As a checkpoint,
    $$c_1(t) = c_0\left[\lambda_1 \exp(-\lambda_2 t) + (1-\lambda_1)\exp(-\lambda_3 t)\right]$$
    where λ_i's are constants.
(c) Plot $c_1(t)$ for $0 \le t \le 100$ h, $k_e = 3.8 \times 10^{-6}$ s^{-1}, and constants $\lambda_1 = 0.4$, $\lambda_2 = 5 \times 10^{-5}$ s^{-1}, and $\lambda_3 = 10^{-6}$ s^{-1}.

(d) Determine the half-times of the drug $t_{1/2}$ which is defined as the time when $c = c_0/2$, representing the distribution phase (see λ_2) and the elimination phase (see λ_3).

(e) Calculate $\int_{t_0}^{t_e} c_1(t)\,dt$ and interpret the result.

Sketch:	Assumptions:	Concepts:
Compartment I with volume ($\forall_1$) Compartment II with volume ($\forall_2$) Drug injection $\xrightarrow{}$ c_1 $\xrightarrow{k_1}$ c_2 $\xleftarrow{k_2}$ k_e Clearance (by inactivation and/or excretion)	• No drug in compartment II at $t=0$ • Chemical reactions are neglected in both compartments	• Two-component model • Mass balance equation • Modify Eq. (2.81)

Solution:

(a) According to this problem, the rate of mass transfer from compartment I to compartment II is equal to $k_1 c_1 V_1$, and the rate of mass transfer in the opposite direction is $k_2 c_2 V_2$, the rate of clearance is $k_e c_1 V_1$; therefore, the mass balance equations based on Eq. (2.81) are:

$$\frac{dc_1}{dt} = -k_1 c_1 + \chi_{21} k_2 c_2 - k_e c_1 \qquad\qquad \text{(E2.27-1)}$$

and

$$\chi_{21} \frac{dc_2}{dt} = k_1 c_1 - \chi_{21} k_2 c_2 \qquad\qquad \text{(E2.27-2)}$$

where $\chi_{21} = \forall_2/\forall_1$.

The initial conditions are:

$$c_1 = c_0 = \frac{M}{\forall_1} \quad \text{at } t = 0 \qquad\qquad \text{(E2.27-3)}$$

and

$$c_2 = 0 \qquad\qquad \text{at } t = 0 \qquad\qquad \text{(E2.27-4)}$$

(b) Rearranging Eqs. (E2.27-1) and (E2.27-2) to eliminate c_2, the governing equation for c_1 can be expressed as:

$$\frac{d^2c_1}{dt^2}+\left(k_1+k_2+k_e\right)\frac{dc_1}{dt}+k_2k_ec_1=0 \qquad \text{(E2.27-5)}$$

The initial conditions for Eq. (E2.27-5) are:

$$c_1=c_0=\frac{M}{\forall_1} \qquad \text{at } t=0 \qquad \text{(E2.27-6)}$$

and

$$\frac{dc_1}{dt}=-\left(k_1+k_e\right)c_0 \quad \text{at } t=0 \qquad \text{(E2.27-7)}$$

Solving Eq. (E2.27-5) gives:

$$c_1(t)=c_0\left[\lambda_1\exp\left(-\lambda_2t\right)+\left(1-\lambda_1\right)\exp\left(-\lambda_3t\right)\right] \qquad \text{(E2.27-8)}$$

where

$$\lambda_2=\frac{\left(k_1+k_2+k_e\right)+\sqrt{\left(k_1+k_2+k_e\right)^2-4k_2k_e}}{2} \qquad \text{(E2.27-9)}$$

$$\lambda_3=\frac{\left(k_1+k_2+k_e\right)-\sqrt{\left(k_1+k_2+k_e\right)^2-4k_2k_e}}{2} \qquad \text{(E2.27-10)}$$

and

$$\lambda_1=\frac{-\lambda_3+k_1+k_e}{\lambda_2-\lambda_3} \qquad \text{(E2.27-11)}$$

Substituting Eq.(E2.27-8) into Eq. (E2.27-1) yields:

$$c_2=c_0\frac{\lambda_1\left(1-\lambda_1\right)\left(\lambda_3-\lambda_2\right)}{\chi_{21}\left[\left(1-\lambda_1\right)\lambda_2+\lambda_1\lambda_3\right]}\left[\exp\left(-\lambda_2t\right)-\exp\left(\lambda_3t\right)\right] \qquad \text{(E2.27-12)}$$

(c) *Graph:*

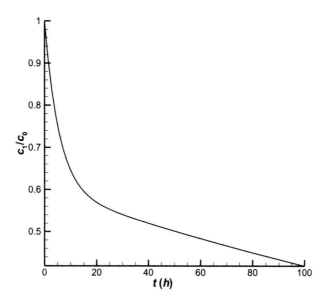

(d) The half-time of the drug, $t_{1/2}$, is defined as the time when $c = c_0/2$. When substituting the definition into Eq. (E2.27-8):

$$t_{1/2}\big|_{\lambda_1} = \frac{0.693}{\lambda_1}$$

and

$$t_{1/2}\big|_{\lambda_2} = \frac{0.693}{\lambda_2}$$

(e) $\displaystyle \int_{t_0}^{t_e} c_1(t)\,dt = c_0\left[\frac{\lambda_1}{\lambda_2} + \frac{1-\lambda_1}{\lambda_3}\right] = \frac{c_0}{k_e} = \frac{M}{k_e \forall_1}$

Example 2.28: Transient Species Concentrations in Open System with Linear Chemical Reaction

Consider a sudden pollutant input into a body of water with inflow and outflow streams, or a continuous-flow chemical reactor with species conversion. Set up the compartmental model and solve the system for two species input modes: impulse and step functions at time $t = 0$.

Sketch:	Assumptions:	Concept:
IN ↓ Q_{in}, c_{in} $\forall = ¢$, $k=¢$ $c=c(t)$ $Q_{out}=Q_{in}$ ↓OUT $c_{out}=c(t)$	• Well-mixed compartment with $\forall$ = constant • Reaction rate is constant • Constant flow rate	• Species mass: $m = c\forall$; $\forall$ = constant • Rate equation (2.81) with $n = 1$

Solution:

Equation (2.81) can be rewritten in terms of species mass, $m = c\forall$, as:

$$\frac{dm}{dt} = Q_{in}(t)C_{in}(t) - Q_{out}(t)C_{out}(t) \pm k(t)C(t)\forall(t) \qquad \text{(E2.28-1)}$$

In light of the assumptions (see the sketch), Eq. (E2.26a-1) reduces to:

$$\frac{dc}{dt} = \frac{Q}{\forall}C_{in}(t) - \frac{Q}{\forall}C(t) - kC(t) \qquad \text{(E2.28-2)}$$

where $\forall/Q \equiv \tau$ is the "residence time," i.e., the theoretical duration that a fluid element (or particle) resides in the compartment.

Species Input ① (Impulse): At $t = 0$ a finite amount of, say, a pollutant is dumped into a lake (compartment), i.e., the lake water starts with an initial contaminant concentration $c_0 > 0$. No species enters the lake thereafter, i.e., $c_{in}(t) = c_{in} = c_0$ because $\Delta t_{input} \approx 0$.

Species Input ② (Step Function): At $t = 0$ suddenly a fixed amount of a pollutant is discharged into the lake and that species input, $c_0 > 0$ and $c_{in}(t) = c_{in} = c_0$, stays constant.

The solutions to Eq. (E2.28-2) for the two different input conditions are of the form of Eq. (2.79). Specifically, for Case ① we have with $c_0 > 0$ and $c_{in}(t) = 0$, while the chemical reaction can be neglected, i.e., $k = 0$:

$$c(t) = c_0 e^{-t/\tau} \qquad \text{(E2.28-3)}$$

And for Case 2, with $c_{in}(t) = c_0 = $ constant (see Polyanin and Zaitsev, 1995):

$$c(t) = \frac{c_0}{1+k\tau}\left[1 - \exp\left(-t\left(k + \tau^{-1}\right)\right)\right] \qquad \text{(E2.28-4)}$$

Graph:

(a) Case 1 (impulse)

(b) Case 2 (step function)

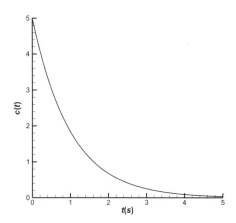

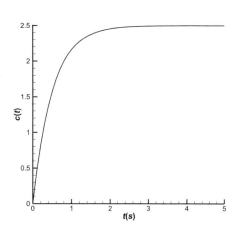

Comments:

Clearly, with an initial amount of pollutant residing in the lake, i.e., $c(t=0) = c_0 = 5\left[\text{M}/\text{L}^3\right]$ in Case 1, and no additional input, the contaminant is (exponentially) washed out via Q_{out}. Note the pollutant residence time $\tau \sim Q_{out}^{-1}$.

In Case 2, due to the constant species input (i.e., $c_{in}(t) = c_0$), an equilibrium concentration is reached due to the interplay of influx, efflux, and uptake.

Additional case studies using the compartmental transport approach may be found in undergraduate heat/mass transfer textbooks as well as in Middleman (1972), Cooney (1976), Godfrey (1983), Bird et al. (2002), Truskey et al. (2004), Kleinstreuer (2006), or Chandran et al. (2007), among other texts.

2.6 Homework Assignments

This section provides three types of homework assignments, i.e., text-illuminating questions (Category I), basic momentum/heat/mass transfer problems (Category II), and course-assigned homework sets (Category III).

2.6.1 Definitions, Concepts, and Physical Insight

2.1 Looking at Table 2.1: (a) list examples of transient flow in nature and applications in industry; (b) how can a realistic 3-D internal flow be transformed into an approximate 1-D flow; (c) depict the velocity vector components in spherical coordinates and cite a couple of flow applications; (d) for axisymmetric flow, why is the midplane symmetry condition better than the midplane u_{max}-condition; (e) why is entrance flow not unidirectional?

2.2 Equation (2.2c) describes a family of steady, laminar, unidirectional flows. Develop a table (or flowchart) listing all suitable solutions subject to basic BCs, i.e., velocity profiles, pressure drops, flow rates, and wall shear stresses, with sketches and parameter definitions.

2.3 Compare the math and physics of Poiseuille-type flows between parallel plates, on inclines, and in tubes.

2.4 Concerning Example 2.4, Cases A and B: (a) provide a physical explanation why the Poiseuille flow solution cannot be recovered from Eq. (E2.4-3); (b) derive an expression for $r = b$, the v_{max} location, and for v_{max}; (c) solve Case B for ω_0 clock-wise and ω_1 counter-clock wise; (d) analyze Case B for $aR \to 0$.

2.5 Assuming that the radial velocity v_r in Example 2.6 cannot be more than a small percentage ε of v_z, recast the (E2.6-4) solution, find $v_r(r, z)$, plot the profiles for $0 < \varepsilon < 10\%$, and comment.

2.6 Considering the graph in Example 2.8, provide physical insight to the axial flow development over time.

2.7 Derive Eqs. (E2.9-2) and (E2.9-3) in Example 2.9 and confirm the Poiseuille number Po $= f \times$ Re $= 57$, as given in the table. Why is Po$_{tube}$ > Po$_{duct}$?

2.8 Considering a saturated porous medium (Sect. 2.2), derive Eq. (2.9a) in 3-D. State correlations between κ (permeability), ε (porosity), K (hydraulic conductivity), and/or k (the Darcy coefficient), and provide physical explanations with sketches.

2.9 What are the basic equations for *unsaturated* porous media?

2.10 Concerning Sect. 2.3, discuss examples of natural and man-made 2-phase and 3-phase flow systems.

2.11 In Eqs. (2.16 and 2.17), why is the volume limit nonzero?

2.12 Analyze Eq. (2.21b) and provide the limiting conditions for $\kappa = c$ being correct.

2.13 Contrast Eqs. (2.28) and (2.29) and perform a literature search on v_{drift} correlations.

2.14 Extend Table 2.2 by including temperature-dependent correlations for the viscosities; starting with, say, Sutherland's law and Andrade's equation.

2.15 Concerning Example 2.14, find a mixture viscosity correlation valid for larger particle concentrations and plot Eq. (E2.14-2) plus WSS as a function of α.

2.16 Assuming homogeneous flow (Sect. 2.3.3), derive Eqs. (2.36) and (2.37). Discuss each term, and provide application examples.

2.17 Correlate pressure drop, WSS, and frictional pipe loss, and derive Eq. (2.41a).

2.18 What is the difference between (dilute or dense) homogeneous mixture flows and non-Newtonian fluid flows (see Sect. 2.3.4)? Provide contrasting examples, if any.

2.19 Replot the four types of fluids in Figure 2.6 in terms of "apparent viscosity over shear rate."

2.20 In Example 2.16, Eq. (E2.16-8) implies a small radial velocity. Determine v_r for a suitable $R(z)$-function, plot radial velocity profiles for different n-values, $0 < n < 1$, and comment.

2.21 Compare the graphs in Examples 2.14 and 2.16 and provide physical insight.

2.22 Concerning Example 2.17, given film thickness h, graph $Q(n)$ for $0 < n < 1$. How is m related to n for practical power law fluids?

2.23 In Sect. 2.3.5, derive Eqs. (2.51a-c) and discuss the meanings.

2.24 Determine and plot the particle acceleration in Example 2.19.

2.25 Considering nanoparticle transport, verify that the assumption of 6% particle volume fraction assures dilute suspension flow.

2.26 One of the computationally most efficient modeling approaches when dealing with nonspherical particles is the sphere-equivalent hypothesis. Review expressions for sphere-equivalent diameters and provide a couple of applications with suitable C_D correlations (for starters, see Hoelzer & Sommerfeld, 2008, *Powder Technology*, Vol. 184, pp. 361–365).

2.27 Concerning "momentum coupling" in fluid-particle dynamics, derive the $K_M = \kappa / (1 + \mathrm{St})$ correlation and discuss a couple of applications for one-way versus two-way coupled flows.

2.28 Concerning Eq. (2.60) in Sect. 2.4, why is there a minus sign and how would Fourier's law look for the conductivity varying in all directions?

2.29 In light of Figure 2.10, give a rationale for Eq. (2.66a,b).

2.30 Starting with the fact of similar, dimensionless velocity and temperature profiles in boundary-layer flow, derive the R-C analogy Eq. (2.70).

2.31 Employing an averaged form of Eq. (2.70), obtain the result commented on in Example 2.22.

2.32 Looking at Graph (a) in Example 2.24, why are the velocity profiles almost linear? What geometric/kinematic changes are required to generate more hyperbolic distributions (see Example 2.4B).

2.33 What are the math and physics reasons that the species convection-diffusion equation (2.72) looks very much like the heat transfer equation (2.61)?

2.34 Derive Eq. (2.76) from Eq. (2.72).

2.35 Plot Eq. (E2.26a-3) and discuss both the $k(\kappa)$ graph of Example 2.26 and the c/c_{sat} plot. Set up a problem solution for $c = c(t$ and $k)$!

2.36 Concerning Sect. 2.5.2, when encountering compartments in parallel, will Eq. (2.77) hold as well, or should the transient mass balance be reformulated? Provide a system sketch plus a sample solution.

2.6.2 Text Problems

2.37 For viscous fluid flow down a slope $S_0 = \tan \theta$ to be laminar and film thickness $h = $ constant, the Reynolds number has to be $\mathrm{Re}_h = v_{mean} hv < 500$. Find the average velocity $v_m = gh^2 S_0 / 3v$ and check the Re_h-value for crude oil ($v = 9.3 \times 10^{-5}$ m^2/s) on an incline of $S_0 = 0.02$ and with $h = 6$ mm. What is the ratio v_{max} / v_{mean}?

2.38 Comparing the maximum Poiseuille flow velocity in a tube and between parallel plates, one is $u_{avg.} = 0.5 u_{max}$ and the other $u_{avg.} = 0.66\overline{6} u_{max}$. Analyze and explain the difference.

2.39 Show that for Poiseuille flow the volumetric flow rate in an inclined tube of radius r_0 can be expressed as:

$$Q = \frac{\pi r_0^4}{8\mu}\left[-\frac{d}{dx}(p+\gamma z)\right]; \ \gamma = \rho g$$

Considering the flow in terms of average velocity v to be a streamline, as in a microtube of $D = 2r_0$, show that the extended Bernoulli equation

$$\frac{p_1}{\gamma} + z_1 = \frac{p_2}{\gamma} + z_2 + h_f$$

can be derived where $h_f \equiv (L/D)32\mu v/\gamma D$ encapsulates the frictional pipe resistance, or head loss.

2.40 Show that for Poiseuille flow between horizontal parallel plates, $0 \le y \le h$,

$$\frac{\partial p}{\partial x} = \frac{d\tau_{yx}}{dy} = \text{constant}$$

Use this equation as a starting point to find:

(a) $u(y) = \dfrac{h^2}{2\mu}\left(\dfrac{\partial p}{\partial x}\right)\left[\left(\dfrac{y}{h}\right)^2 - \left(\dfrac{y}{h}\right)\right]$

(b) An expression for τ_{yx}

(c) An expression for $\partial p/\partial x = -\Delta p/L$ as a function of

(d) An expression for v_{max}/v_{mean}

(e) An expression for the shear force exerted on either plate

2.41 A cylindrical piston ($D \times L = 25 \times 15$ mm) sits in a housing with $s = 5$ μm clearance, filled with oil ($\mu = 0.018$ kg/m·s). It is exposed to a pressure differential of 19 MPa. Find the leakage rate in [mm³/s] and check the Reynolds number $\mathrm{Re}_s = v_{avg}\, s/v$ when specific gravity $= 0.92$ for that oil.

2.42 Torque and power of an idealized journal bearing with shaft chamber D, angular velocity ω, and clearance s are occasionally analyzed in terms of simple Couette flow with plate spacing s and upper plate speed $U = D\omega$. For what s/D-ratio is this simple approach acceptable when the computed torque (T) error is not allowed to exceed 1%? Plot the exact and approximate $T(\omega, s/D, \mu)$ function using $n = 3000$ rpm, $D = 6$ cm, $s = 100$ μm, $L = 100$ cm, and $\mu = 0.02$ kg/m·s as a base case. Note that laminar flow prevails as long as:

$$\frac{\omega D^2}{4v} < \frac{40}{\kappa(1-\kappa)^{3/2}}$$

$$\text{where } \kappa = \frac{D_{shaft}}{D_{housing}} = \left(1 + 2\frac{s}{D}\right)^{-1} < 1$$

Note: Additional problems can be found as Homework Sets IIa and IIb, listed in Category III, as well as undergraduate fluid mechanics texts by Çengel & Cimbala (2006), Fox et al. (2008), Potter & Wiggert (2002), and White (2011).

2.6.3 Homework Sets

2.6.3.1 Homework Set Ia

<u>Notes:</u>

- For Part A, use additional sources plus images/sketches (e.g., internet) and physics/math descriptions.
- For Part B, follow the Format Sketch, Assumptions, and Concepts/ Approach.
- Reference sources (other than those found in this textbook).
- Record actual time spent (Part A versus Part B).

Part A: Insight:

1. (5pts) Discuss briefly the hypothesis that "PHYSICS is the Mother of All Science/Engineering Branches." Take not only chemistry and biology as disputable examples, but also neurology, etc.

2. (5pts) Plot the temperature-dependent viscosities of both liquids and gases and explain their (opposing) behavior.

3. (5pts) Certain bugs can walk on water. Explain this phenomenon and estimate the maximum weight of a typical bug crossing a quiescent water body at 20°C.

4. (5pts) Discuss internal energy versus enthalpy on both the macroscale and the nanoscale.

5. (5pts) Consider macro-, micro-, and nanoscale fluid flow applications in both nature and industry, i.e., provide three sample images (via book image scanning, website, etc.) each with brief descriptions.

Part B: Problems:

6. (10 pts) Consider a quiescent protein solution ($\mathcal{D} = 10^{-10}$ m^2/s) in a 100-μm- versus a 10-μm channel. How long does it take for the protein to diffuse across these channels? What is the time to cross a macrochannel of 1 m width?

7. (15 pts) Consider the velocity vector $\vec{v} = (0.5 + 0.8x)\hat{i} + (1.5 - 0.8y)\hat{j}$ in [m/s].

    (a) Classify this fluid flow field.

    (b) Determine the total (or material) acceleration vector field and find $\vec{a} = a_x\hat{i} + a_y\hat{j}$ at $x = 2$ m and $y = 3$ m.

    (c) Plot selected acceleration vectors ($\Delta x = \Delta y = 1$ m) plus streamlines and discuss the graphical results.

8. (10 pts) Recall the RTT (Reynolds Transport Theorem):

    (a) Discuss the RTT as a link of principles between solid mechanics and fluid mechanics.

    (b) Determine concisely the conservation laws in differential form from the general RTT.

9. (20 pts) For Poiseuille flow in straight channels of length L with different cross-sectional shapes, the hydraulic resistance is in general

$$R_{\text{hydr}} = \frac{\Delta p}{Q} \approx 2\mu L \frac{P^2}{A^3} \sim l^{-3} \qquad \text{(SP1)}$$

where P is the duct perimeter, A is the cross-sectional area, and l is the system's representative length scale, such as channel height or width, hydraulic diameter, etc. Prove Eq. (SP1) and discuss the implications for micro/nanosystems.

10. (25 pts) A long rod of radius R_1 moves horizontally with velocity U in a cylinder (inner radius R_2) which translates in the opposite direction with velocity V, separated by an oil film (ρ, μ).

   (a) Determine the velocity ratio V/U at which the wall shear stress is the same at both surfaces.

   (b) If the rod is heated (q_{wall} = const.), recast the governing equations in light of convection heat transfer with $\mu(T)$.

   (c) Define three dimensionless groups on which the Nu number would depend.

   (d) List an industrial application for both cases, i.e., with and without heat transfer.

2.6.3.2 Homework Set Ib

Notes:

- Follow the Format Graphics, Math, & Physics for Part A and Sketch, Assumptions, Concepts/Approach for Part B.
- Reference sources (other than those for Chapters 1 and 2).
- Record actual time spent (Part A versus Part B).

Part A: Insight:

1. (10 pts) *What is the difference between:*

   (a) Units and dimensions?

   (b) Flow system dimensionality and directionality?

   (c) Fluid kinematics and dynamics?

   (d) Newtonian and non-Newtonian fluids?

   (e) Entrance flow and fully developed flow?

   (f) Laminar and turbulent flows?

   (g) Transient and transitional flows?

   (h) Ideal and real fluid flows?

   (i) Conduction and convection heat transfer?

   (j) Integral and differential solution approaches?

2. (5 pts) Steady laminar incompressible flow in curved conduits (e.g., pipes, ducts, channels, etc.) is known as Dean's flow. Draw several axial velocity profiles in the straight/curved section for viscous fluid flow. Provide a real-world application/ occurrence!

3. (5 pts) Consider planar Couette flow where the upper (parallel) plate moves with a constant velocity u_0 *without* a pulling force. Draw the unique velocity profile and explain the underlying physics.

4. (5 pts) The long (blond) hair of a girl riding in a convertible is being pushed into her face. Explain!

5. (5 pts) Consider "carving out" 1-nm cubes from a 1-m cube without any material losses. What is the total surface area of all the resulting nanocubes approximately equal to in terms of a public space? Compare the total surface-area-to-volume ratios of the 1-m cube and all the nanocubes. What are the implications?

Part B: Problems:

6. (20 pts) A Venturi meter to measure flow rates (ρ, Q) consists of a smooth horizontal pipe constriction from D_1 via D_2 back to D_1 where a manometer (i.e., U-tube with fluid ρ_M) connects pipe D_1 with pipe throat D_2, exhibiting a fluid height difference h. Obtain an expression for $Q = Q(D_1, D_2, A_2; h, \rho, \rho_M)$. Plot Q versus D_2 / D_1 and comment!

7. (20 pts) Consider a horizontal solid cylinder of radius a moving axially with velocity U inside a housing of inner radius b which moves in the opposite direction at velocity V. The gap b-a is filled with oil (ρ, μ). Assuming constant pressure and neglecting end effects, find $v_z(r)$ and the ratio V/U at which the wall shear stress will be the same at both cylinder surfaces. Graph $v_z(r)$ for a couple of a/b ratios and comment!

8. (30 pts) Consider a (thin metal) tube of diameter $D = 60$ mm subject to a constant $q_s = 2000$ W/m^2. Pressurized water enters the tube at $\dot{m} = 0.01$ kg/s with $T_{mean, in} = 20°C$.

    (a) Estimate the thermal entrance length $x_{fd}/D \approx 0.05\,\mathrm{Re}_D\,\mathrm{Pr}$ to assure that fully developed conditions can be assumed.

    (b) What tube length L is required to obtain an exit temperature of 80°C?

    (c) Evaluate the tube surface temperature at the outlet, i.e., $T_s(x = L)$, and comment.

    (d) Calculate the total entropy generated when $T_{ambient} = 300$ K.

    (e) Provide an industrial application example for this model.

2.6.3.3 Homework Set IIa

Preparation:

Study the material in Chapters 1 and 2 and solve the given examples *independently.*

Part A: Insight

1. The long hair of a girl driving a convertible is being pushed into her face rather than swept back. Why? Provide quick answers (sketches, streamlines, math, physics).

2. Provide a physical interpretation of

$$\int_{C.V.} \rho \vec{f}_B \, d\forall + \int_{C.S.} \vec{T} \cdot d\vec{A} = \frac{\partial}{\partial t} \int_{C.V.} \rho \vec{v} \, d\forall + \int_{C.S.} \vec{v} \rho \vec{v} \cdot d\vec{A}$$

and sketch an example with all four terms appearing.

3. Employing the REV (representative elemental volume) approach, derive the transient 3-D continuity equation in cylindrical coordinates. Note, for incompressible fluids $\nabla \cdot \vec{v} = 0$; does that also hold for transient flows?

4. Discuss with examples boundary work vs. flow work, and explain why on a differential basis we start with δW rather than dW.

5. A velocity field is given/measured as:

$$\vec{v} = (ax + b)\hat{i} + (cx^2 - ay)\hat{j}$$

where a, b, and c are constants: (a) Classify this flow field; (b) check if the associated pressure field, $p = p(x, y)$, is a smooth function.

Hint: Cross-differentiation of the pressure gradients should generate the same results for "function smoothness," i.e.,

$$\frac{\partial}{\partial x}\left(\frac{\partial p}{\partial y}\right) = \frac{\partial}{\partial y}\left(\frac{\partial p}{\partial x}\right)$$

6. Explain the differences between thermodynamic (or static) pressure, dynamic pressure, and total (or stagnation) pressure. Draw a pressure probe which simultaneously can measure stagnation and static pressures.

7. Draw carefully velocity profiles in a pipe's entrance and fully developed regions for: (a) $\mathrm{Re}_D \approx 1800$ and (b) $\mathrm{Re}_D \approx 18{,}000$. Comment!

8. Develop a criterion which sets the limit for the "nearly parallel-flow" assumption.

9. How do heat conduction and convection heat transfer differ? How do $T_{surface}$, T_{wall}, T_{mean}, T_{bulk}, and T_∞ differ?

10. Derive $\frac{1}{2}C_f(x) = \mathrm{St}_x \mathrm{Pr}^{2/3}$ for $0.6 < \mathrm{Pr} < 60$ and comment on the advantages of heat momentum transfer and heat mass transfer analogies.

Part B: Problems

Solve Problems 11 to 15, following the Format.

11. Consider a jet plane with the engine mounted at the rear end discharging $\dot{m}_{gas} = 18$ kg/s of exhaust gases at $v = 250$ m/s relative to the plane. To shorten landing, a deflector vane (called a "thrust reverser") is lowered into the path of the exhaust stream, which deflects the gases and hence aids in braking. Analyze the effect of the vane angle, θ, on the braking force, i.e., a horizontal $\theta = 0°$ (no effect) to a $180°$ (full effect). Discuss the $F_{brake}(\theta)$ plot!

12. Consider a vertical concentric shaft rotating and translating (d, ω_0, u_0) inside a stationary cylinder (i.e., housing of inner diameter D) with a lubricant (or slurry) of properties ρ and μ. (a) What industrial operation can be modeled with this shaft/pipe annulus? (b) Set up the reduced N-S equations plus BCs based on suitable assumptions and postulates. (c) Can you solve for the velocity field and draw a representative velocity profile? (d) Estimate the amount of lubricant which has to be constantly supplied.

13. Consider steady fully developed air flow in a smooth tube where a Pitot static pressure arrangement measures p_{static} and $p_{stagnation}$ as shown in Figure 1.13. Estimate: (a) the centerline velocity; (b) the volumetric flow rate; and (c) the wall shear stress.

    Properties: $\rho_{air} = 1.2$ kg/m^3, $\mu_a = 1.8 \times 10^{-5}$ kg/(ms), $\rho_{water} = 998$ kg/m^3, $\mu_w = 0.001$ kg/(m·s).

14. A (wide, vertical) belt moving at velocity v_0 drags a viscous fluid layer of thickness h upwards. Develop expressions for the film's velocity profile, the average fluid velocity, the shear stress distribution, and the volumetric flow rate per unit width. What is the condition for the *minimum* belt speed in order to achieve net upward flow?

15. Consider a well-insulated counterflow heat exchanger (i.e., a thin-walled, double-pipe system) which connects a water heater to a shower. Specifically, cold water ($c_p = 4.18$ kJ/kg·°C) enters at 15°C at a rate of 0.25 kg/s and is heated to 45°C by hot water ($c_p = 4.19$ kJ/kg·°C) that enters at 100°C with 3 kg/s. Determine: (a) the rate of heat transfer and (b) the rate of entropy generation in the heat exchanger.

2.6.3.4 Homework Set IIb

Preparation:

Study the material in Chapters 1 and 2 and solve the given examples *independently*.

Part A: Insight (20 points for Tasks 1–7)

1. Considering p or $\tau_{normal} = F_{normal}/A_{surface}$, how does τ_{normal} differ physically from p; give an illustrative example.

2. Describe the math conditions for: (a) the continuum hypothesis and (b) thermodynamic equilibrium.

3. Explain the rationale for the "rate-of-deformation" equation $(\partial v_i/\partial x_j = \varepsilon_{ij} + \zeta_{ij})$, where mathematics is connected to physics: (a) prove that $2\vec{\omega} = \nabla \times \vec{v} \equiv \vec{\zeta}$ and (b) compare two circular flows, i.e., $v_\theta = \omega r$ and $v_\theta = C/r$ $(r \neq 0)$. Compute the vorticity fields and sketch them.

4. Provide basic answers (sketches, streamlines, math, physics) to the following counterintuitive fluid flow scenarios:

(a) When bringing a spoon near a jet, e.g., faucet stream, it gets sucked into the stream. Try it out and explain!

(b) The long hair of a girl driving a convertible is being pushed into her face rather than swept back. Why?

5. Although "inviscid flow" doesn't exist, why is the Bernoulli equation still quite popular and when is its application most suitable?

6. Explain the differences between thermodynamic (or static) pressure, dynamic pressure, and total (or stagnation) pressure. Draw a pressure probe which simultaneously can measure stagnation and static pressures.

7. Consider the following del operator expressions (solve in 2-D):

(a) Compare ∇s versus ∇v and provide an example of each.

(b) Expand $(v \cdot \nabla)v$ and $(v \cdot \nabla)s$ and provide an example of each.

Assume s is a scalar and v is a vector.

Part B: Problems (20 points each for Problems 8–11, following Format)

8. A skydiver ($m_{total} = m_{person} + m_{gear}$) jumps out of a plane and at terminal velocity v_T he/she opens at $t = 0$ a parachute ($F_{drag} = kv^2$) and lands with a final velocity v_F. Show that $k = mg/v_F^2$ and derive an equation for $v(t)$. Note that $v = v(t, v_T, v_F, g)$, but not m_{total}; why? Plot $v(t)$ for typical parameter values and comment.

9. Oil at $60°C$ ($\rho = 864$ kg/m³; $\mu = 72.5 \times 10^{-3}$ kg/m·s) is pushed ($\Delta p_{gage} = 1$ atm) between two horizontal parallel plates ($L = 1.5$ m, $W = 0.75$ m) a small distance h apart. Plot the Reynolds number, $Re_h = \rho u_{mean} h/\mu$, as a function of spacing h for $10 \ \mu m \le h \le 1$ mm and comment.

10. A (wide, vertical) belt moving at velocity v_0 drags a viscous fluid layer of thickness h upwards. Develop expressions for the film's velocity profile, the average fluid velocity, the shear stress distribution, and the volumetric flow rate per unit width. What is the condition for the minimum belt speed in order to achieve net upward flow?

11. Air flows through a tube of radius R with a measured steady fully developed velocity distribution of $u(r) = A[(R-r)/R]^{1/7}$ in m/s and a temperature profile $T(r) = B - C[(R-r)/R]^{1/7}$ in °C, where $A = 20$ m/s, $B = 70°C$, and $C = 40°C$.

(a) Plot both profiles and comment.

(b) Find the mean (or mixing-cup) temperature in the flow and comment.

References (Part A)

Batchelor, G.K., 1967, *An Introduction to Fluid Dynamics*, Cambridge University Press.

Batchelor, G.K., 1977, Journal of Fluid Mechanics, Vol. 128, pp. 240.

Beattie, D.R.H., Whalley, P.B., 1982, International Journal of Multiphase Flow, Vol. 8 (1), pp. 83-87 (Eng).

Bejan, A., 1996, *Entropy Generation Minimization, the Method of Thermodynamic Optimization of Finite-Size System and Finite-Time Processes*, CRC Press, Boca Raton, FL.

Bejan, A., 2002, International Journal of Energy Research, Vol. 26, pp. 545-565.

Beavers, G.S. Joseph, D.S., 1967, Journal of Fluid Mechanics, Vol. 30 (1), pp. 197-207.

Bird R.B., Armstrong, R.C., Hassager, O., 1987, *Dynamics of Polymeric Liquids*, 2nd ed., Wiley, New York.

Bird, R.B., Stewart, W.E., Lightfoot, E.N., 2002, *Transport Phenomena*, 2nd ed., Wiley, New York.

Brinkman, H.C., 1947, Physics, Vol.13 (8), pp. 447-448.

Brinkman, H.C., 1952, Journal of Chemistry Physics, Vol. 20, pp. 571-581.

Bruus, H., 2007, *Theoretical Microfluiodics,* Oxford University Press, Oxford, UK.

Buchanan, J.R., Kleinstreuer, C., Comer, J.K., 2000, Computers and Fluids, Vol. 29 (6), pp. 695–724.

Çengel, Y.A., Cimbala, J.M., 2006, *Fluid Mechanics: Fundamentals and Applications*, McGraw-Hill, Boston, MA.

Chandran, K.B., Rittgers, S.E., Yoganathan, A.P., 2007, *Biofluid Mechanics: The Human Circulation*, CRC Press, Boca Raton, FL.

Clift, R., Grace, J.R., Weber, M.E., 1978, *Bubbles, Drops and Particles*, Academic Press, New York.

Colin, S., 2010, *Microfluidics*, Wiley, Hoboken, NJ.

Cooney, D.O., 1976, *Biomedical Engineering Principles*, Marcel Dekker, New York.

Crowe, C.T. (editor), 2006, *Multiphase Flow Handbook*, CRC Press, Boca Raton, FL.

Crowe, C., Sommerfeld, M., Tsuji, Y., 1998, *Multiphase Flows with Droplets and Particles*, CRC Press, Boca Raton, FL.

Cussler, E.L., 2009, *Diffusion: Mass Transfer in Fluid Systems*, 3rd ed., Cambridge University Press, Cambridge.

Das, S.K., Choi, S.U.S., Yu, W., Pradeep, T., 2007, *Conduction Heat Transfer in Nanofluids: Science and Technology*, Wiley, Hoboken, NJ.

Einstein, A., 1906, Annual Physics, Vol. 19, pp. 289-306.

Ergun, S., 1952, Analytical Chemistry, Vol. 24 (2), pp. 388-393.

Faghri, A., Zhang, Y., 2006, *Transport Phenomena in Multiphase Systems with Phase Change*, Elsevier, Burlington, MA.

Forchheimer, P. 1901, Zeitschrift Vereine Deutscher Ingenieure, Vol. 45, pp. 1782–1788.

Fox, R.W., McDonald, A.T., Pritchard, P.J., 2008, *Introduction to Fluid Mechanics*, Wiley, Hoboken, NJ.

Godfrey, K., 1983, *Compartmental Models and Their Application*, Academic Press.

Graham, A.L., 1981, Applied Science Research, Vol. 37, pp. 275.

Greenberg, M.D., 1998, *Advanced Engineering Mathematics*, Pearson Education.

Hoffman, J.D., (2001), *Numerical Methods for Engineers and Scientists,* 2nd ed., Marcel Dekker, New York.

Hornyak, G.L., Moore, J.J., Tibbals, H.F., Dutta, J., 2008, *Fundamentals of Nanotechnology*, CRC Press, Boca Raton, FL.

Kleinstreuer, C., 1997, *Engineering Fluid Dynamics*, Cambridge University Press, New York.

Kleinstreuer, C., 2003, *Two-Phase Flow: Theory and Applications*, Taylor and Francis, New York.

Kleinstreuer, C., 2006, *Biofluid Dynamics: Principles and Selected Applications*, CRC Press, Boca Raton; part of the Taylor & Francis Group, London, New York.

Kleinstreuer, C., Zhang, Z., Kim, C.S., 2007, Journal of Aerosol Science, Vol. 38 (10), pp. 1047-1061.

Kirby, B.J., 2010, *Micro- and Nano-scale Fluid Mechanics: Transport in Microfluidic Devices*, Cambridge University Press, New York.

Macosko, C.W., 1994, *Rheology: Principles Measurements and Applications*, VHC Publishers, New York.

Michaelides, E.E., 1997, Journal of Fluids Engineering, Vol. 119 (2), pp. 233-247.

Middleman, S., 1972, *Transport Phenomena in the Cardiovascular System*, Wiley Interscience, New York.

Middleman, S., 1998, *An Introduction to Fluid Dynamics: Principles of Analysis and Design*, Wiley, New York.

Mitra, S.K., Chakraborty, S., 2011, *Microfluidics and Nanofluidics Handbook: Fabrication, Implementation, and Applications*, CRC Press, Boca Raton, FL.

Naterer, G.F., 2002, *Heat Transfer in Single and Multiphase Systems*, CRC Press, Boca Raton, FL.

Nguyen, N.T., Wereley, S.T., 2006, *Fundamentals and Applications of Microfluidics*, 2nd ed., Artech House, Boston.

Nichols, W.W., O'Rourke, M.F., et al., 1998, *McDonald's Blood Flow in Arteries: Theoretical, Experimental, and Clinical Principles,* Oxford University Press.

Ochoa-Tapia, J.A., Whitaker, S., 1995, International Journal of Heat and Mass Transfer, Vol. 34(14), pp. 2635–2646.

Panton, R.L., 2005, *Incompressible Flow*, 3rd ed., Wiley, Hoboken, NJ.

Papanastasiou, T.C., 1994, *Applied Fluid Mechanics*, Prentice-Hall, Englewood Cliffs, NJ.

Polyanin, A.D., Zaitsev, V.F. 1995, *Handbook of Exact Solutions for Ordinary Differential Equations*, CRC Press, Boca Raton, FL.

Potter, M.C., Wiggert, D.C., 2002, *Mechanics of Fluids*, Brooks Cole/Thompson Learning, Pacific Grove, CA.

Probstein, R.F., 1994, *Physicochemical Hydrodynamics: An Introduction*, Wiley, Hoboken, NJ.

Rogers, B., Adams, J., Pennathur, S., 2008, *Nanotechnology: Understanding Small Systems*, CRC Press, Boca Raton, FL.

Schlichting, H., Gersten, K., 2000, *Boundary-Layer Theory*, Springer, New York.

Shenoy A.R., Kleinstreuer C., 2010, Journal of Fluid Mechanics*,* Vol. 653, 463-487

Soo, S.L., 1990, *Multiphase Fluid Dynamics*, Science Press/Gower Technical.

Tabeling, P., 2005, *Introduction to Microfluidics*, Oxford University Press, New York.

Tanner, R.I., Walters, K., 1998, *Rheology: an Historical Perspective*, Elsevier, Amsterdam.

Truskey, G.A., Yuan, F., Katz, D.F., 2004, *Transport Phenomena in Biological Systems*, Pearson/Prentice Hall, Upper Saddle River, NJ.

White, F.M., 2011, *Fluid Mechanics*, 7th Edition, McGraw-Hill, Boston, MA.

Wilkes, J.O., 2006, *Fluid Mechanics for Chemical Engineers*, 2nd ed., with Microfluidics and CFD, Pearson Education.

Womersley, J.R., 1955, Journal of Physiology, Vol. 127, pp. 553-563.

Zapryanov, Z., Tabakova, S., 1999, *Dynamics of Bubbles, Drops and Rigid Particles*, Springer.

Zhang, Z.M., 2007, *Nano/Microscale Heat Transfer*, McGraw-Hill, New York.

Part B: MICROFLUIDICS

*Microfluidics deals with transport phenomena of low volumes of fluids in microsystems. A major application is the laboratory-on-a-chip (LoC) for the analysis of biological, chemical, and biomedical substances (e.g., DNA sequencing, drug/toxin assays, etc.). LoCs are part of mechanical or biomedical microelectromechanical systems (MEMSs or bio-MEMSs). As implied, the advantages of microsystems include the use of tiny material volumes, low production cost, compactness, low operating cost, and Controlled multifunctionality. The fluid transport occurs as either a <u>continuous</u>-flow operation or as a finite-volume, i.e., droplet-based, operation which is known as <u>digital microfluidics.</u> Fluid motion is achieved via displacement pumps and mechanical pumps as well as more esoteric driving forces based on electrostatic, magnetohydrodynamic, or surface tension effects. Another successful application of microfluidics is the inkjet print-head (also a MEMS example). An emerging field using microfluidics devices is "point-of-care" (PoC) diagnostics of human diseases as well as air/water pollution analysis. Thus, local real-time testing of tissue or fluid samples can detect quickly and cost-effectively major problems. In any case, main components of microsystems are, of course, the microchannel and micropumps as well as associated heaters, sensors, and actuators. **Microfluidics has a lot in common with macrofluidics (see Part A) as long as the continuum mechanics assumption holds.** However, as the characteristic length scale (e.g., the microchannel's hydraulic diameter or height/width) is reduced to the micrometer (10^{-6} m), or even the nanometer (10^{-9} m) range, quite significant changes in fluid properties and fluid flow pattern may occur. For example, considering the limit of a system's surface-area-to-volume ratio, it is evident that surface forces, fluxes, and processes become dominant. Furthermore, on the micro/nanoscale gases and liquids may have to be treated separately because of their very different molecular number density and hence their flow behavior. As a result, traditional conservation laws and associated boundary conditions have to be modified, or new modeling and simulation approaches are needed. Most working fluids are not pure. For example, the simulation of "mixture flows," such as nanofluids, non-Newtonian fluids, and living cell or drug suspensions, requires suitable apparent viscosity expressions or two-phase modeling approaches. Concerning "pumps," they may be mechanical or nonmechanical; examples of the latter include fluid displacement action due to capillary, electrostatic, or magnetohydrodynamic effects.*

CHAPTER 3

Microchannel Flow Theory

Microfluidics is the science and technology of systems that process/manipulate small amounts of fluids, say, 10^{-12} to 10^{-9} liters, and sort particles in microchannels with characteristic "length" scale L of, say, 10 to 500 μm, where L represents a hydraulic diameter or channel height or width. Hence, of special interest are gas and liquid flows as well as the fluid-particle dynamics in microchannels plus the associated drivers (i.e., "pumps"), mixers, and heaters. Other microsystem components, such as fluid-particle reservoirs, actuators, valves, sensors, and controllers, as well as microchannel materials and fabrication techniques are discussed elsewhere (see Lee & Sundararajan, 2010; Meng, 2011; among others). Clearly, the very high surface-to-volume ratio of microsystems forms the foundation of the many advantages over macroscale devices and processes. However, miniaturization can also be the cause of major problems; for example, those related to transport phenomena in microchannels as implied in Table 1.1 and discussed in Section 3.1.3 and in Chapter 4 (see Sect. 4.3).

3.1 Introduction

3.1.1 Microfluidic System Components

As mentioned, central to any microfluidics system is the microchannel with necessary accessories such as a pump, valve, mixer, and/or heater (see Figure 3.1).

A real-world example of a microfluidic system, known as a lab-on-a-chip (LoC), is depicted in Figure 3.2.

Microfluidic devices are manufactured using planar substrates such as silicon wafers, glass slides, ceramic blocks, stainless steel, polymer, or even paper chips.

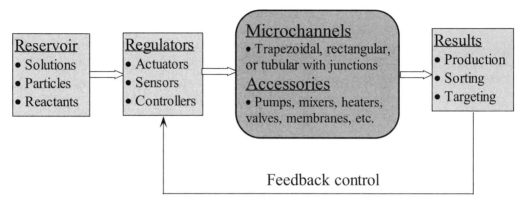

Figure 3.1   Key components and interactions of microfluidics systems

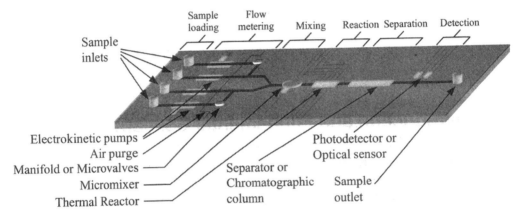

Figure 3.2   Microfluidic system example (after Chang & Yeo, 2010)

The type of material selected affects not only device architecture, cost, and fabrication method (e.g., micromachining or soft lithography) but also microscale phenomena (e.g., surface roughness or heat transfer effects). Polydimethylsiloxane (PDMS) is the most frequently used material to fabricate microsystems. When uncured, PDMS flows freely at room temperature, and once cured it produces a mechanically strong and optically transparent substrate.

3.1.2 Microfluidic System Integration

System integration implies the combination of independent components and their associated functions, mainly actuation, sensing, and analysis, to produce a desired outcome with, say, an LoC, a diagnostics system, a drug delivery device, or special MEMS. Figure 3.3 summarizes microsystem building, operation, microfluidic functions, and a few applications with end results. The underlying component functionalities may be based on mechanical, electrical, thermal, magnetic, optical, chemical, and/or biological transport phenomena, clearly, all obeying the three conservation laws. For example, electromechanical, say, piezoelectric, actuators are used for microvalve or micropump displacements, while electric motors or syringe pumps provide angular or linear displacements. In contrast, not having to deal with moving parts, electroosmosis allows for fluid pumping while electrophoresis allows for particle manipulation (see Sect. 3.2.4.3). Thermal energy transfer via electric heaters is used for temperature-driven gas expansion, finite liquid volume displacement, and/or control of chemical reactions (Zhang, 2007).

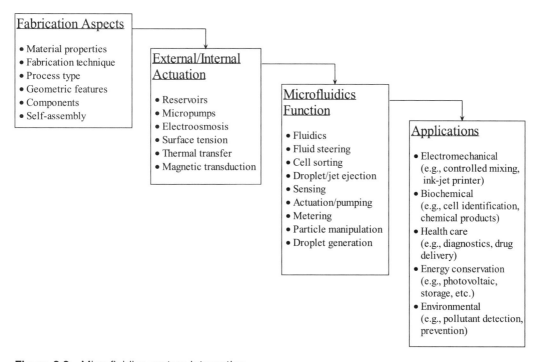

Figure 3.3 Microfluidics system integration

In summary, applications of microfluidic technologies can be most frequently found in chemistry, biology, and medicine as well as in chemical, biomedical, and mechanical engineering. As mentioned, advantages of microfluidic devices over conventional size equipment are largely based on the very high surface-to-volume ratio. In addition to the implied savings in space, material, energy, and hence cost, miniaturized process devices feature:

- higher heat and mass transfer rates;
- safer handling of hazardous/toxic chemicals;
- integrated analytic platforms (e.g., chips), such as LoC and micro-total analysis systems (μ-TAS) for biochemical and biomedical processing;
- simplified flow and screening process control;
- *in vivo* robotic diagnostics and treatment, for example, as in "image-guided surgery or drug delivery."

3.1.3 Microfluidics System Challenges

While microfluidics and microsystem applications can be very advantageous, they may also face a number of problems, as already alluded to with some Table 1.1 entries. Specifically, critical considerations associated with microfluidics systems include:

- Possible violation of the continuum mechanics hypothesis
- Slip flow for liquids on hydrophobic rough surfaces or more generally for rarefied gases
- Need for wall temperature jump condition in case of rarefied gases
- Impact of microchannel entrance flow in small devices
- Microchannel wall effects, such as roughness, electric charge, etc.
- Air bubbles and/or impurities in microchannel flow
- Particle/cell aggregation and biofouling leading to clogging
- Temperature rise due to Joule heating (see electroosmosis discussed in Sect. 3.4.2.1) or viscous dissipation
- Evaporation effects impairing especially digital microfluidics operation
- Macroscale-to-microscale flow connection problems
- Chaotic flow or turbulence effects
- Compressibility of rarefied gases
- Malfunctioning of accessories, e.g., tiny pumps, valves, heaters, actuators, sensors, controllers, etc.

As indicated, the last problem area includes pumps which are vital for any microfluidic device or system. While mechanical micropumps are needed for relatively high volumetric flow rates or any throughput in devices with strong hydraulic resistances (see Eq. (3.15)), numerous microsystems rely on nonmechanical driving forces generated by capillary or electrokinetic effects. After a brief summary of basic concepts and limitations (Sect. 3.2), Sections 3.3 and 3.4 deal specifically with liquid flows driven by surface tension and electroosmosis as well as charged particle transport via electrophoresis.

3.2 Basic Concepts and Limitations

3.2.1 Scaling Laws

Theory.   Fundamental to system, device, or process miniaturization is the scaling laws, starting with geometric ratio:

$$\kappa \equiv \frac{\text{surface area}}{\text{system volume}} \sim \frac{l^2}{l^3} = l^{-1} \tag{3.1}$$

Clearly, in the limit as $l \to 0$, $\kappa \to \infty$, *implying that surface-related forces, fluxes, and process characteristics become dominant in microfluidics and nanofluidics.* Table 3.1 lists a few fluidics-related scaling results. It should be recalled that material properties and most coefficients do not feature any length scales and that time t scales as $l^0 (=1)$.

Table 3.1   Scaling Laws

Quantity and Correlation	Result
System length: $L \sim l$	l
Conduit diameter, height or width: $L \sim l \left(\text{e.g., } L \equiv D_h = 4A/P \right)$	l
Surface area: S or $A \sim l^2$	l^2
Volume of body, system, or device: $\forall \sim l^3$	l^3
Heat flux: $\vec{q}\big|_{\text{macro}} = -k\nabla T \sim 1/l\,(k = \cancel{c})$	l^{-1}
Heat flux: $\vec{q}\big|_{\text{micro}} = -k\nabla T \sim l \cdot 1/l = l^0$	1
$F_{\text{gravity}} = \rho \forall g \sim l^3\,(g = \cancel{c})$	l^3
$F_{\text{dyn}} = ma \sim \forall \cdot a \sim l^3 \cdot l \to l^4\,(a \neq \text{const.})$	l^4
Pressure: $p = F/A \sim l^1 \left(F \sim l^3 \right)$	l
Shear stress: $\tau = F/A \sim l^1\ \left(F \sim l^3 \right)$	l
$F_{\text{viscous}} = \tau \cdot A \sim l \cdot l^2$	l^3
$F_{\text{spring}} = -k \cdot \Delta x \sim l^1\ (k = \cancel{c})$	l
$F_{\text{Coulomb}} = -\rho_{\text{electr.}}\,\nabla \phi \forall \sim l^3/l$	l^2
$F_{\text{capillary}} = \gamma \cdot L \sim l^1\ \left(\gamma = \cancel{c} \right)$	l
$F_{\text{electromagnetic}} \sim I^2_{\text{electr.}} A_{\text{cross-sect.}} \sim l^2 \cdot l^2$	l^4

Table 3.1 *Continued*

Quantity and Correlation	Result
Reynolds number: $\mathrm{Re} = vL/v \sim l \cdot l$	l^2
Molecular diffusion time: $\tau = <x^2>/2\mathcal{D}_{AB} \sim l^2$	l^2

When postulating scaling laws, the system's conditions may play an important role, e.g., see $F_{\text{gravity}} \sim l^3$ vs. $F_{\text{dyn.}} \sim l^4$. Another example on a macroscopic level is solid-body friction:

$$F_{\text{friction}} = \mu F_{\text{gravity}} \sim l^3 \text{ if } \mu = \text{constant} \tag{3.2}$$

On a microscopic level, adhesion may play a major role leading to stop-and-go combination known as striction; thus, the resulting force is dependent on the surface area alone, i.e.,

$$F_{\text{striction}} \sim l^2 \tag{3.3}$$

Scale variations can also be observed when considering the kinetic energy:

$$E_{\text{kin}} = \frac{m}{2}v^2 \sim l^3 \text{ if } v = \text{constant} \tag{3.4}$$

However, with variable velocity $v \sim l$,

$$E_{\text{kin}} \sim l^5 \tag{3.5}$$

It should be noted that in some cases a parameter's length-scale correlation can be directly used, e.g., $\tau_{\text{diffusion}} \sim l^2$ (see application of Eq. (3.12c)). However, when using equations it is their *relative importance* which counts, as demonstrated in Example 3.2 (see Eq. (E3.2-2)).

Example 3.1: Square Microchannel

Consider a square microchannel of width and height a with length $l = \mathcal{O}(a)$. From $\kappa = A/\forall$ and plot $\kappa(a)$ for $0 < a \leq 4$ mm:

Sketch:	Assumptions:	Concepts:
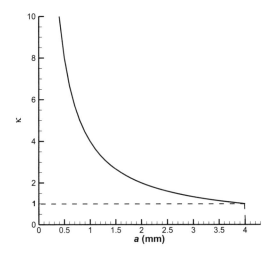	• Microchannel length $l = 6a$	• $\kappa = A / \forall$

Solution:

$$\kappa = \frac{A}{\forall} = \frac{24a^2}{6a^3} = 4a^{-1}$$

Graph:

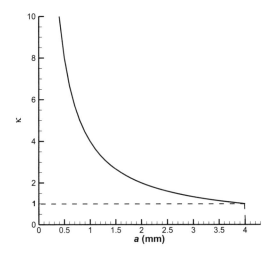

Comment:

As we move from the millimeter channel to the micrometer channel, the ratio κ shoots up and hence surface-related phenomena become dominant in fluid-particle flow.

Dimensionless Groups.   While all dimensionless groups are ratios of forces or fluxes or characteristic times, their length-scale dependence may be significant. For example, the Weber number scales as (see Table 3.1):

$$We = \frac{F_{inertia}}{F_{surface\ tension}} = \frac{\rho \forall (\vec{v} \cdot \nabla) \vec{v}}{\gamma \cdot L} = \frac{\rho v^2 L}{\gamma} \sim l^3 \qquad (3.6a\text{-}c)$$

for variable v and L but constant κ. In contrast, consider the capillary rise of height in a liquid column of diameter D:

$$\frac{F_{surface\ tension}}{F_{weight}} = \frac{(4D\pi)\gamma}{D^2 \pi h \rho g} = \frac{1}{hD} \frac{\gamma}{\rho g} \sim l^{-2} \qquad (3.7a\text{-}c)$$

The Stokes number, a ratio of particle and fluid residence times, can be scaled as:

$$St = \frac{\tau_{particle}}{\tau_{fluid}} = \frac{v \rho_p d_p^2}{18 \mu D} \sim \frac{l \cdot l^2}{l} = l^2 \qquad (3.8)$$

The Peclet number Pe = convection/diffusion scales as:

$$Pe = \frac{v D_h}{\mathcal{D}} \ or \sim \frac{v D_h}{v} \sim l^2 \qquad (3.9)$$

for a given diffusivity $\mathcal{D}$ or kinematic viscosity v. The Reynolds number $Re = F_{inertia} / F_{viscous}$ scales also as l^2, i.e.,

$$Re = \frac{v D_h}{v} \sim l^2 \qquad (3.10)$$

When capillary effects are dominant, the capillary number scales for constant viscosity and surface tension as:

$$Ca = \frac{viscous\ force}{surface\ tension} = \frac{\tau A}{\gamma L} \sim \frac{l^3}{l} = l^2 \qquad (3.11)$$

Clearly, the length-scale dependence of forces, fluxes, and dimensionless groups varies with the given system dynamics, e.g., which parameter can be kept constant or not. Furthermore, in micro/nanochannels as $l \to 0$, surface tension is more important than viscous forces or gravity (see Eqs. (3.6), (3.7) and (3.11)), and mass or momentum diffusion trumps convection or inertia (see Eqs. (3.9) and (3.10)).

Example 3.2: Evaluate the dominant forces in the Navier-Stokes equation for microchannel flow as $L_{\text{system}} \to 0$.

Sketch:	*Assumptions:*	*Concepts:*
Fluid Element $\vec{F}_{\text{pressure}}$ $\vec{F}_{\text{drag}}$ $\vec{F}_{\text{body}}$	• Constant fluid properties • Additive *net* surface and body forces	• Extended N-S equation • Characteristic length scaling

Solution:

Newton's second law of motion for fluid flow

$$m\frac{D\vec{v}}{Dt} = \vec{F}_{\text{pressure}} + \vec{F}_{\text{viscous}} + \vec{F}_{\text{gravity}} + \vec{F}_{\text{surface tension}} + \vec{F}_{\text{electro-static}} + ... \qquad (E3.2\text{-}1)$$

can be rewritten with relative length scales associated with each term as (see Table 3.1):

$$m\frac{\partial\vec{v}}{\partial t} + \underbrace{m(\vec{v}\cdot\nabla)\vec{v}}_{} = -\nabla p\forall + \nabla\vec{\vec{\tau}}\forall + m\vec{g} + \vec{\gamma}l + \underbrace{\rho_{el}\vec{E}\forall}_{} \; \Big|*L^{-4} \qquad (E3.2\text{-}2)$$

$$\begin{array}{ccccccc} L^4 & L^4 & L^3 & L^3 & L^4 & L^2 & L^2 \\ 1 & 1 & L^{-1} & L^{-1} & 1 & L^{-2} & L^{-2} \end{array}$$

Based on the relative L-scale of each term, for $L \to 0$ the implications are that four terms remain most important. Specifically,

- Fluid acceleration is negligible, i.e., $D\vec{v}/Dt \ll 1$.
- Gravity, being a body-force term, has a minimal effect, if any.
- Next to surface tension and electrostatic (or Coulomb) force, surface forces such as pressure and drag are most important.

Other illustrative examples demonstrating the potentially huge ratio of surface area to body volume are as follows.

➢ Volume-to-area changes: (i) A 1-m cube cut into N nanocubes and then spread out would cover the U.S. State of Delaware and (ii) the human lungs of $\forall \approx 6$ liters, spread out, would cover a tennis court.

➢ Creature heat loss: $Q = q_{wall} \cdot A \sim l^2$, while energy intake (eating) $E_{in} \sim \forall \sim l^3$. Thus, as $A/\forall = l^{-1} \to \infty$ and $E/Q = l$, we realize that creature-size reduction is limited because of $E \sim l \times Q_{loss}$. Hence, the smallest warm-blooded animal is the humming bird, while insects (1–2 mm) are all cold-blooded to keep food/energy intake at a minimum level.

➢ Speed gain vs. vehicle size: $E_{resist} \sim v^2 A_{body}$, where $E_{prop.} \sim \forall_{propulsion} \sim l^3$, and $A_{body} \sim l^2$. Thus, at dynamic equilibrium $E_{prop.} = E_{res.}$, or $l^3 \sim v^2 l^2 \succ v \sim \sqrt{l}$. Clearly, as the body/vehicle/propulsion system increases, the maximum speed gain is $v \sim l^{1/2}$.

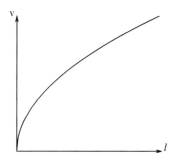

$$\text{Applications}: \begin{cases} \text{fruit fly} & v_{max} \approx 3\text{mph} \\ \text{bumble bee} & v_{max} \approx 12\text{mph} \\ \text{Boeing 747} & v_{max} \approx 360\text{mph} \end{cases}$$

Note : $\dfrac{v_{Boeing}}{v_{B\text{-}Bee}} = 30$ but $\dfrac{\forall_{Boeing}}{\forall_{B\text{-}Bee}} >>> 1$

➢ Diffusion scaling: In the nanometer range diffusional effects, e.g., molecular or nanoparticle transport, become important. The Stokes-Einstein equation reads:

$$\mathcal{D} = \frac{k_B T}{3\pi\mu d} \sim l^{-1} \tag{3.12a}$$

where $k_B = 1.38 \times 10^{-23}$ J/K, T in $[K]$, and μ in $[kg/m \cdot s]$, and particle diameter d in $[m]$.

Based on Einstein's random-walk analysis, the diffusion length of a molecule in solution is:

$$x = \sqrt{2\mathcal{D}\tau} \tag{3.12b}$$

from which the diffusion time can be estimated:

$$\tau = \frac{x^2}{2\mathcal{D}} \sim l^2 \left(\text{assuming } \mathcal{D} = \text{const.}\right) \tag{3.12c}$$

For a typical binary diffusivity of $\mathcal{D} = 10^{-9} \, \mathrm{m^2/s}$, the nanoparticle diffusion times decrease rapidly as, say, the channel width or height x decreases. Specifically, when:

- $x = 1\mathrm{mm} \Rightarrow \tau = 500s$
- $x = 10\mathrm{\mu m} \Rightarrow \tau = 0.05s$
- $x = 1\mathrm{\mu m} \Rightarrow \tau = 0.5ms$
- $x = 100\mathrm{nm} \Rightarrow \tau = 0.05ms$

Time scales.   Flow fields are transient due to (sudden) operational system changes. Examples include start-up, on-and-off switching, and periodic boundary conditions, such as pulsatile (blood) flow, inlet pressure waveforms, vibrating walls, etc. In general, all three transport fields, i.e., momentum, energy, and species mass, have to adjust to those time-dependent changes. The relevant characteristic time scales can be derived by setting the order of magnitudes of the time derivative terms of the conservation laws equal to one. As will be shown in Example 3.3, for the velocity field to adjust the characteristic time is:

$$\tau_v = \frac{\rho L^2}{\mu} \tag{3.13a}$$

For the temperature and species fields:

$$\tau_T = \frac{\rho c_p L^2}{k} \text{ and } \tau_c = \frac{L^2}{\mathcal{D}} \tag{3.13b,c}$$

where L is the system's characteristic length, k is the fluid conductivity, and $\mathcal{D}$ is the binary diffusion coefficient. For example, using water flow at $T = 20°C$ in a microchannel of $h \equiv L = 100$ μm, the adjustment times to reach a new equilibrium (i.e., steady state), after a sudden change in flow conditions, are:

$$\tau_v = 10\mathrm{ms}, \ \tau_T = 100\mathrm{ms}, \text{ and } \tau_c = 10\mathrm{s}$$

Clearly, the diffusion time is rate limiting. For periodically driven transport fields, the magnitude of the Strouhal number has to be considered. Specifically,

$$\mathrm{Str} = \frac{fL}{v} \tag{3.14}$$

where f is the frequency of the forcing function with period τ_p, L is a characteristic length, and v is a characteristic velocity. The time period of the forcing function, say, inlet pressure $p(t)$, wall heat flux $q_w(t)$, or input concentration $c_{in}(t)$, has to be compared with the τ-values of the system. For example, the fields are quasi-steady when the forcing period τ_p is much longer than the other timescales. In contrast, for periods much less than the timescales, the transport fields hardly respond. In-between, i.e., $\tau_{min} \leq \tau_p \leq \tau_{max}$, the transport fields are transient.

Before launching into illustrative examples, Tables 3.2 and 3.3 summarize aspects of flow system minimization as well as microscale liquid and gas flow problem solutions, respectively.

Table 3.2   Summary of Macro-to-Nano Flow Scales and Applications

A) Miniaturization of Liquid-Phase Devices:

- When L_{system} reduces (say, $L_s < 100\,\mu m$), the importance of:
  (i) Reynolds number and pressure drop go down; but,
  (ii) entrance length, surface roughness, and interfacial phenomena may go up

- Applications in:
  $\left\{\begin{array}{l} \text{Chem Eng (e.g.,lab-on-a-chip)} \\ \text{BME (bio-MEMS for assaying)} \\ \text{ME-EE (e.g.,MEMS and NEMS)} \end{array}\right\}$
  Actuators, pumps
  Microchannels, valves
  Heaters, sensors

- Knowledge/skills needed in:......$\left\{\begin{array}{l} \text{Fluid flow(mostly internal)} \\ \text{Fluid-particle dynamics} \\ \text{Heat transfer} \\ \text{Mass transfer} \end{array}\right.$

Note: For (rarefied) gases, consider also....$\left\{\begin{array}{l} \text{Velocity slip at walls} \\ \text{Wall-temperature jump} \\ \text{Compressibility effects} \end{array}\right.$

B) Length-Scale Examples:

Nature
- Human hair: $\approx 80\,\mu m$
- Air particle: $\approx 6\,\mu m$
- Bacteria: $100-200\,nm$
- Protein: $3\,nm$

Man made
- Nanomaterial $(1-10^3\,nm)$
- MEMS $(0.5-10^3\,\mu m)$
- NEMS $(10-500\,nm)$
- Pollutants $(1nm-mm)$

Table 3.3   Flow System Reduction and Simulation Approaches

Macroscale Flows ($l \geq 1\,\mathrm{mm}$)
- Classical Fluid Mechanics Where the Continuum Hypothesis Is Valid
- Conservation Laws (system of Navier -Stokes equations)

Microscale Flows ($10\,\mu\mathrm{m} \leq l_{\mathrm{channel}} \leq 1000\,\mu\mathrm{m}$)
(i) **Gases** (mean free path , e.g., $\lambda_{\mathrm{air}} \approx 65\,\mathrm{nm}$, $\delta_{\mathrm{intermolecular}} = 3\,\mathrm{nm}$)
- Continuum Mechanics Approach When $\mathrm{Kn} = \lambda/l \leq 0.01$ or $D_h \geq 100$ μm (i.e., employ N-S equations with no slip BC, as in macroscale flow)
- Slip Flow for $0.01 < \mathrm{Kn} \leq 0.1$ (i.e., N-S Eqs. with slip B.C., i.e., $v_{\mathrm{wall}} = v_{\mathrm{slip}} \neq 0$)
- Transitional Flow for $0.1 < \mathrm{Kn} < 10$ (i.e., use LBM or DSMC)
- Free Molecular Flow (e.g., rarefied gases with $\lambda > 10\,\mathrm{nm}$) for $\mathrm{Kn} \geq 10$ (Use molecular dynamics (MD) simulation)

(ii) **Liquids** (intermolecular distance, e.g., $\delta_{\mathrm{water}} \approx 0.3\,\mathrm{nm}$)
- Continuum Mechanics Approach When $\delta/l \leq 3 \times 10^{-5}$ or $D_h \geq 10$ μm (i.e., N-S equations with no-slip BC)
- Transitional Range $3 \times 10^{-5} < \delta/l \leq 10^{-3}$ (employ, e.g., DSMC simulation or LBM)
- Atomistic Flow Simulation When $\delta/l > 10^{-3}$ (employ MD simulation)

Nanoscale Flows ($1\,\mathrm{nm} \leq l_{\mathrm{channel}} \leq 10^2\,\mathrm{nm}$)
Gases/Liquids:
Atomistic/molecular flow simulations in nanochannels: the Lennard-Jones (L-J) potential (where its derivative is the particle interaction force) models the interaction of a collection of atoms/molecules, while the "particle" dynamics is described by Newton's second law of motion. Similarly, fluid and wall atom interactions are modeled by L-J as well with different length scales (i.e., atom diameter) and interaction energy strengths between the atoms $\Rightarrow$ MD simulation.

Example 3.3: Characteristic Time Scale for Momentum Transfer

Given transient liquid flow (ρ , μ) in a microchannel $(h,\ U)$. Derive the momentum time scale τ_v and calculate its value for $\rho = 10^3\,\mathrm{kg/m^3}$, m=$10^{-3}\,\mathrm{Ns/m^2}$, $h = 10$ μm, and $u_{\mathrm{avg.}} = U = 0.1\,\mathrm{m/s}$.

Sketch:	Assumptions:	Concepts:
	• Transient laminar 2-D flow with constant properties	• Reduced Navier-Stokes equation • Non-dimensionalization

Solution:

The N-S equation in vector form states:

$$\rho\left[\frac{\partial \vec{v}}{\partial t}+(\vec{v}\cdot\nabla)\vec{v}\right]=-\nabla p+\mu\nabla^{2}\vec{v} \qquad \text{(E3.3-1)}$$

With $\tilde{v}=\vec{v}/U$; $\tilde{p}=p/p_{0}$; $\tilde{t}=t/\tau$ and $\tilde{\nabla}=\nabla h$, we rewrite the momentum equation as:

$$\rho\frac{U}{\tau}\frac{\partial\tilde{v}}{\partial\tilde{t}}+\frac{\rho U^{2}}{h}\left(\tilde{v}\cdot\tilde{\nabla}\right)\tilde{v}=-\frac{p_{0}}{h}\tilde{\nabla}\tilde{p}+\frac{\mu}{h^{2}}U\tilde{\nabla}^{2}\tilde{v} \qquad \text{(E3.3-2a)}$$

or

$$\frac{\rho h^{2}}{\mu\tau}\frac{\partial\tilde{v}}{\partial\tilde{t}}+\frac{\rho Uh}{\mu}\left(\tilde{v}\cdot\tilde{\nabla}\right)\tilde{v}=-\frac{p_{0}h}{\mu U}\tilde{\nabla}\tilde{p}+\tilde{\nabla}^{2}\tilde{v} \qquad \text{(E3.3-2b)}$$

Setting the coefficient of the transient term equal to one (unity), we obtain

$$\tau=\tau_{v}=\frac{\rho h^{2}}{\mu} \qquad \text{(E3.3-3)}$$

while the coefficient of the inertial term is, as expected,

$$\text{Re}_{h}=\frac{Uh}{\nu} \qquad \text{(E3.3-4)}$$

With the given system values,

$$\tau_{v}=100\mu s \text{ and } \text{Re}_{h}=1.0\left(\text{Stokes flow}\right)$$

Comments:

A frequent application of transient microchannel flow is periodic flow due to changes in inlet pressure (e.g., in hemodynamics) or in electric potential (e.g., electrokinetic flow driven by AC fields). Then $\tau_v = f^{-1}$ and the dimensionless group is the Strouhal number, $\text{Str} = fh/U$, where f is the frequency (see Eq. (3.14)).

Flow Resistance   A very practical parameter to evaluate microchannel flow performance is the hydraulic resistance (Hagen-Poiseuille law) and its length-scale dependence:

$$R_{\text{hydr.}} \equiv \frac{\text{driving force}}{\text{flow rate}} := \frac{\Delta p}{Q} \tag{3.15}$$

Table 3.4 summarizes for *steady laminar fully developed flow* the velocity profiles, volumetric flow rates, and hydraulic resistances, considering common microconduits. For these Poiseuille-type flow conditions with $p(x)$ being linear, $F \sim l^2$ so that $p \sim l^0$ and hence $R_{\text{hydr}} \sim 1/l^3 = l^{-3}$

Table 3.4   Velocity Profiles, Flow Rates, and Hydraulic Resistances for Common Microchannels

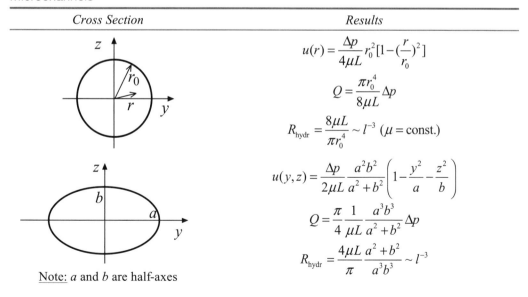

Cross Section	Results
	$u(r) = \dfrac{\Delta p}{4\mu L} r_0^2 [1 - (\dfrac{r}{r_0})^2]$
	$Q = \dfrac{\pi r_0^4}{8\mu L} \Delta p$
	$R_{\text{hydr}} = \dfrac{8\mu L}{\pi r_0^4} \sim l^{-3} \quad (\mu = \text{const.})$
	$u(y,z) = \dfrac{\Delta p}{2\mu L} \dfrac{a^2 b^2}{a^2 + b^2} \left(1 - \dfrac{y^2}{a} - \dfrac{z^2}{b}\right)$
	$Q = \dfrac{\pi}{4} \dfrac{1}{\mu L} \dfrac{a^3 b^3}{a^2 + b^2} \Delta p$
	$R_{\text{hydr}} = \dfrac{4\mu L}{\pi} \dfrac{a^2 + b^2}{a^3 b^3} \sim l^{-3}$

Note: a and b are half-axes

Table 3.4   *Continued*

Cross Section	Results
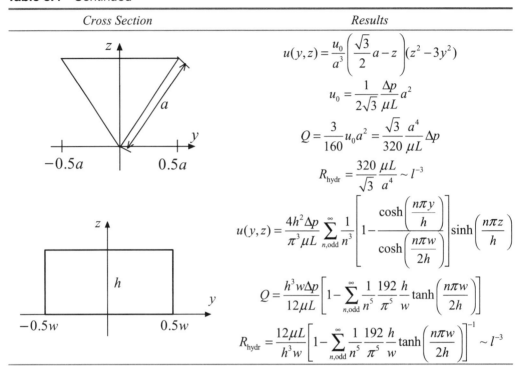	$$u(y,z)=\frac{u_0}{a^3}\left(\frac{\sqrt{3}}{2}a-z\right)(z^2-3y^2)$$ $$u_0=\frac{1}{2\sqrt{3}}\frac{\Delta p}{\mu L}a^2$$ $$Q=\frac{3}{160}u_0a^2=\frac{\sqrt{3}}{320}\frac{a^4}{\mu L}\Delta p$$ $$R_{\text{hydr}}=\frac{320}{\sqrt{3}}\frac{\mu L}{a^4}\sim l^{-3}$$
	$$u(y,z)=\frac{4h^2\Delta p}{\pi^3\mu L}\sum_{n,\text{odd}}^{\infty}\frac{1}{n^3}\left[1-\frac{\cosh\left(\frac{n\pi y}{h}\right)}{\cosh\left(\frac{n\pi w}{2h}\right)}\right]\sinh\left(\frac{n\pi z}{h}\right)$$ $$Q=\frac{h^3w\Delta p}{12\mu L}\left[1-\sum_{n,\text{odd}}^{\infty}\frac{1}{n^5}\frac{192}{\pi^5}\frac{h}{w}\tanh\left(\frac{n\pi w}{2h}\right)\right]$$ $$R_{\text{hydr}}=\frac{12\mu L}{h^3w}\left[1-\sum_{n,\text{odd}}^{\infty}\frac{1}{n^5}\frac{192}{\pi^5}\frac{h}{w}\tanh\left(\frac{n\pi w}{2h}\right)\right]^{-1}\sim l^{-3}$$

Example 3.4: Parallel-Plate Hydraulic Resistance

Consider Poiseuille flow between two horizontal parallel plates (spacing h, length L, and width w) forming a microchannel, where $\Delta p=p(0)-p(L)$.

Sketch:	*Assumptions:*	*Concepts/Postulates:*
	• Poiseuille flow • Constant properties	• Reduced N-S equation • $-\nabla p=\Delta p/L=\cancel{c}$ • $\vec{v}=[u(y);0;0]$

Solution:

- Based on the postulates, the x-momentum equation reduces to:

$$0 = \frac{\Delta p}{L} + \mu \frac{d^2 u}{dy^2} \qquad \text{(E3.4-1)}$$

  while continuity is satisfied. The BCs are:

$$u(0) = 0 \text{ and } u(h) = 0 \qquad \text{(E3.4-2a,b)}$$

  so that

$$u(y) = \frac{\Delta p}{2\mu L}(h - y)y = \frac{\Delta p}{2\mu L}h^2\left[\frac{y}{h} - \left(\frac{y}{h}\right)^2\right] \qquad \text{(E3.4-3a,b)}$$

- Flow rate $Q = \int_A u(y)\,dA$, $dA = w\,dy$; hence,

$$Q = \frac{h^3 w}{12\mu}\left(\frac{\Delta p}{L}\right) \qquad \text{(E3.4-4)}$$

- Hydraulic resistance $R_{\text{hydr.}} = \Delta p / Q$; thus,

$$R_{\text{hydr.}} = \frac{12\mu L}{wh^3} \sim l^{-3} \qquad \text{(E3.4-5)}$$

Comments:

- Clearly, for microchannel flow, let alone flow in nanochannels, the hydraulic resistance can be large, which may present a challenge to micropumping (see Sect. 3.3)
- It should be noted that in complex conduits, e.g., suddenly expanding/ contracting channels, the Hagen-Poiseuille law does not hold. Specifically, while for, say, three different flow regions in series $Q = Q_{\text{I}} = Q_{\text{II}} = Q_{\text{III}}$, $\Delta p \neq (R_{\text{I}} + R_{\text{II}} + R_{\text{III}})Q$.

Common Micro- and Nanoscales.   It now has been established that the surface-to-volume ratio $\kappa \equiv S/\forall = l^{-1}$ may imply a significant shift in the importance of certain forces, fluxes, and processes as $l \to 0$ (see Table 3.1). Thus, it is of interest to look at typical length scales in micro/nanofluidics and to consider natural and man-made applications. Table 3.5 lists some values of interest, where the human hair diameter of 80 μm and DNA being 2.5 nm can be considered as two yardsticks in microfluidics and nanofluidics, respectively. Conditions when the continuum assumption holds are given for liquid and gas flows as well.

Table 3.5   Reference Length Scales

In Nature:
- Human hair: 80 μm; Pollen: 20 μm; Red Blood Cell: 7 μm; Bacteria: ≥2 μm
- Rhinovirus: 20 nm; DNA: 2.5 nm; Silicon Atom: 0.25 nm

Man Made:
- MEMS and NEMS devices: mm to nm range
- LoC: mm to cm range, but with microchannels, etc.
- CNT: 1–50 nm in diameter and 1–4000 μm long
- Pin head: 1–2 mm

Fluidics:
- ➤ Fluid *continuum mechanics assumption* is applicable when:
  - $D_{\text{hydr.}}$ or $L_{\text{system}} \geq 10$ for liquid flow $\left(D_{\text{hydr.}} = 4A/P \right)$

  and/or

  - $\delta_{\text{IM}}/L_{\text{system}} \leq 3.0 \times 10^{-5}$ for liquid flow $\left(\delta_{\text{IM}} \triangleq \text{intermolecular spacing} \right)$

  -

  - $D_{\text{hydr.}}$ or $L_{\text{system}} \geq 100\,\mu\text{m}$ for gas flow

  and/or

  - Knudsen number $\text{Kn} = \lambda_{\text{mfp}}/L_{\text{system}} \leq 0.01$ for gases $\left(\lambda_{\text{mfp}} \triangleq \text{mean-free path} \right)$

  Note: $0.01 \leq \text{Kn} \leq 0.1$ is the "slip flow" regime where the N-S equations with velocity slip BCs hold.

  -

- ➤ Air and water molecular length scales at STP conditions
  - Air: Mean spacing $\delta_{\text{IM}} \approx 3.3\,\text{nm}$

    Mean free path $\lambda_{\text{mfp}} \approx 65\,\text{nm}$

    Typical diameter $d \approx 0.375\,\text{nm}\left(N_2 \right)$

    Number density $n \approx 2.69 \times 10^{19}\,\text{cm}^{-3}\left(\text{air} \right)$

Table 3.5 *Continued*

- Water: Intermolecular distance $\delta_{IM} \approx 0.3$–0.4nm

    Molecular diameter $d \approx 0.3$nm

    Number density $n \approx 3.34 \times 10^{19}$ mm^{-3}

Gas Flow. In *liquids*, groups of molecules move about each other, where the internal energy is a function of microscopic activities, including sensible and bonding energies. In *gases*, however, molecules exhibit random, highly temperature-dependent motion. Specifically, for gases the sensible (and hence a large part of the internal) energy is due to molecular translation, rotation, and vibration. The mean free path $\lambda_{mfp} = \lambda$ is the distance of gas molecules traveled before collision. Based on kinetic theory, assuming ideal sphere collision (Probstein, 1994) the mean-free path is:

$$\lambda_{mfp} = \frac{\mu}{\rho}\sqrt{\frac{\pi RT}{2}} = \frac{1}{\sqrt{2}\pi n_i d_c^2} = \frac{m_g}{\sqrt{2}\pi\rho\, d_c^2} \tag{3.16a-c}$$

as a statistical average. Here n_i is the number of molecules per unit volume, d_c is the molecular collision diameter, and m_g is the mass of a gas molecule. Associated ideal-gas expressions include:

$$p = \rho RT = n k_B T \text{ and } \delta_{IM} \approx n^{-1/3} \tag{3.17a-c}$$

where R is the specific gas constant for ideal gases, n is the number of molecules per cubic meter, d is the molecular diameter, δ_{IM} is the mean intermolecular distance, and κ_B is the Boltzmann constant $(1.38 \times 10^{-23}$ J/K$)$.

As indicated, the *Knudsen number value* for gas flow in micro/nanoconduits determines the "modeling regime." Specifically,

$$
\begin{array}{c}
\text{Fluid continuum} \\
\text{(Cauchy or N-S equations)}
\end{array}
\leftarrow
\left\{
\begin{array}{c}
\underbrace{0.01 \leq Kn = \frac{\lambda}{L} < 0.1}_{\substack{\text{Slip flow regime} \\ \text{(N-S equations with slip BCs)}}}
\end{array}
\middle\|
\begin{array}{c}
\underbrace{0.1 < Kn < 10}_{\substack{\text{Transitional flow regime} \\ \text{(lattice-Boltzmann or direct} \\ \text{simulation Monte Carlo methods)}}}
\end{array}
\right\}
\rightarrow
\begin{array}{c}
\text{Rarefied gases of free} \\
\text{molecular flow} \\
\text{(molecular dynamics} \\
\text{simulation)}
\end{array}
\tag{3.18}
$$

It should be noted that $Kn = \lambda/L$ depends on the definition of L, e.g., $L \triangleq D_{hydr.}$, $h_{channel}$, $\delta_{spacing}$, etc., and may vary with conduit length because λ varies with axial temperature changes.

Example 3.5: Gas Flow Characteristics

Consider nitrogen (N_2) gas flow at $v = 100$ m/s, p $= 200$ kPa, and $T = 350$ K in a microconduit of $D_h = 10\,\mu m$.

Given: $M_{N_2} = 28\,\text{kg}/\text{kmole}$, $d_{N_2} = 3.75 \times 10^{-10}\,\text{m}$, $R_u = 8.3145\,\text{kJ}/(\text{kmole} \cdot \text{K})$, $\kappa = c_p/c_v = 1.4$, $v = 2 \times 10^{-4}\,m^2/s$, and $k_B = 1.38 \times 10^{-23}$ J/K.

Find:
- δ/d to determine if the gas is rarefied, where $\delta/d \gg 1$ (say, at least $\delta/d > 7$) implies a "dilute" gas; $\delta \equiv \delta_{IM}$ is the intermolecular distance and d is the molecular diameter
- $Kn = \lambda/L$ where $\lambda = \lambda_{mfp}$ and $L \hat{=} D_h$ to determine the modeling regime (see Eq. (3.18))
- $Re = vD_h/v$ to determine the flow regime, recalling that $Re_{D_h} \leq 1200$ to assure laminar flow
- $Ma = v/c_{sound}$; $c = \sqrt{\kappa RT}$ to check if the gas is incompressible, i.e., $Ma < 0.3$

Sketch:	Assumptions:	Concept:
	• Ideal gas • Averaged, constant properties	• Correlations based on kinetics theory

Solution:

Gas constant $R_{N_2} = R_u/M_{N_2} := 0.2968\,\text{kJ}/(\text{kg} \cdot \text{K})$, where R_u is the universal gas constant.

Number of molecules $n = p/k_B T := 4.14 \times 10^{25}/m^3$ so that the intermolecular distance is:

$\delta \approx n^{-1/3} := 2.89 \times 10^{-9}\,\text{m}$

Molecular mean-free path $\lambda = \left(\sqrt{2}\pi n d^2\right)^{-1} := 3.9 \times 10^{-8}\,\text{m}$

Speed of sound in N_2 gas: $c_s = \sqrt{\kappa RT} := 381\,\text{m/s}$

Hence,
- $\delta/d := 7.7$, i.e., here N_2 is a rarefied gas because $\delta/d > 7$
- $Kn = \lambda/D_h := 0.004$, i.e., the N-S equations with no-slip wall condition are applicable, mainly because N_2 is a borderline "rarefied gas"
- $Re = vD_h/v := 92$, i.e., laminar flow
- $Ma = v/c_s := 0.26$, i.e., incompressible gas because $Ma < 0.3$

3.2.2 Fluid Properties and Surface Tension Effects

In thermal microfluidics the most important fluid properties are the viscosity, surface tension, thermal conductivity, and density. As indicated in Sect. 2.3.2 and further discussed in Sects. 3.4 and 6.3, the viscosity is measurably affected by the temperature and fluid microstructure (e.g., polymers versus water or gases) as well as additives forming mixtures (e.g., nanofluids). Surface tension of liquids in contact with solids (e.g., microtubes or capillaries) and gases (e.g., interfaces with air) may result in a net force which drives a finite liquid volume in microchannels. Thermal conductivities are typically regarded as a *bulk* transport property defined by Fourier's law $k = \vec{q}/\nabla T$. However, in micro/nanofluidics k may depend not only on the type of material and temperature level but also on the material's microstructure and the length scale of a device/system. Density best characterizes the difference between liquid $\left(\rho_{\text{water}} \approx 10^3\,\text{kg/m}^3\right)$ and gas $\left(\rho_{\text{air}} \approx 1\,\text{kg/m}^3\right)$. The "large" intermolecular distance of gases, $\delta_{\text{IM}} \approx 3\,\text{nm}$, leads to compressibility if $\text{Ma} \geq 0.3$ or $p \geq p_{\text{critical}}$. In contrast, $\delta_{\text{IM}} \approx 0.3\,\text{nm}$ for liquids and hence water, oil, etc., are basically incompressible. While fluid density and viscosity have been fairly well discussed in prerequisite courses, a brief review of thermal conductivity and surface tension and the associated capillary effects in microfluidics should be in order.

Thermal Conductivity.   Property values obtained from macrofluidics graphs and tables are applicable when the physical dimensions of the solid or fluid material/system are relatively large. However, on the micro- or nanoscale the material structure and boundaries have to be considered when evaluating, say, energy flow or heat transfer to determine thermal conductivity values. For example, in a *very thin* material film the "trajectories" of energy carriers, i.e., electrons or phonons (elastic lattice vibration/ waves), are greatly influenced by the film's physical boundaries. Specifically, energy carriers in the film have a smaller path length and encounter more frequently the interfaces, causing scattering which redirects their propagation. This in turn creates a directional dependence of the thermal conductivity k. For example, in a material layer of extent $0 \leq y \leq h$, where $h = \mathcal{O}(1\mu\text{m})$, $k_y \leq k_x \leq k_{\text{bulk}}$. Expressions for k_{micro} have been correlated with λ_{mfp} for gases and δ_{IM} for liquids and solids. While derivations of k_{micro}-values from k_{bulk}-values are about 5%, substantial increases in k-values have been reported for nanofluids, i.e., dilute suspensions of (metallic) nanoparticles in liquids (Kleinstreuer et al., 2012). Such elevated $k_{\text{nanofluid}}$-values may be due to Brownian-motion-induced micromixing and enhanced pathways for heat transfer (see Sect. 5.3).

Capillary Effects.   It is well known that water rises in a vertical microtube (capillary diameter d) against gravity, where the capillary height h due to surface tension $\gamma\left[\text{N/m}\right]$ has the dependence:

$$h \sim \frac{\gamma}{d} \qquad (3.19)$$

At the water-air interface the surface water molecules experience an inward pull because of the much stronger liquid-phase attraction forces than their interaction with the air molecules. In general, all material interfaces, e.g., liquid-air, solid-liquid, and solid-air, experience (surface) tension due to the presence of density discontinuities and resulting different cohesive forces between the molecules of the three substances involved. Figures 3.4a,b depict drops on hydrophilic and hydrophobic surfaces, respectively. For a *contact angle* of $\theta < 90°$ the liquid is wetting and nonwetting when $\theta > 90°$. Hydrophobic and rough surfaces may cause measurable velocity slip at solid walls.

Assuming immiscible phases, Young's equation relies on static equilibrium at the triple point (Figure 3.4a):

$$\gamma_{sl} + \gamma_{lg} \cos\theta - \gamma_{sg} = 0 \text{ or } \cos\theta = \frac{\gamma_{sg} - \gamma_{sl}}{\gamma_{lg}} \qquad (3.20a,b)$$

The Young-Laplace equation states that any curved surface at equilibrium separating phase 1 from phase 2 maintains a pressure drop across the surface (interfacial stress condition):

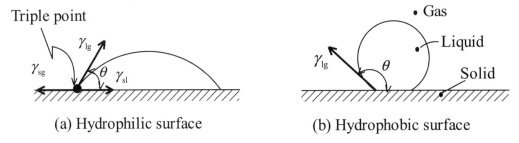

(a) Hydrophilic surface   (b) Hydrophobic surface

Figure 3.4   Contact angles of droplets on different surfaces

$$p_{interrior} - p_{exterrior} = p_1 - p_2 = \Delta p = \gamma_{12}\left(\frac{1}{R_1} + \frac{1}{R_2}\right) \qquad (3.21)$$

where R_1 and R_2 are the two principal radii of surface curvature. For an air bubble in water $R_1 = R_2 = R$ and hence:

$$\Delta p = 2\gamma_{water-air} / R \qquad (3.22)$$

Another example is the capillary rise already mentioned. Figure 3.5 shows wetting and nonwetting liquids, e.g., water and mercury, respectively.

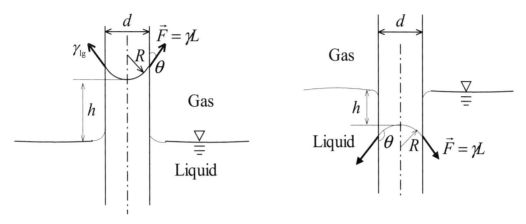

Figure 3.5   Capillary rise for $\theta < 90°$ (wetting) and $\theta > 90°$ (nonwetting)

Assuming a spherical interface for which $R = d/(2\cos\theta)$, the Young-Laplace equation (3.21) now reads:

$$\Delta p_{lg} = \frac{2\gamma_{lg}}{R} = \frac{4\gamma}{d}\cos\theta \tag{3.23}$$

With Eq. (3.20b) and the fluid static result:

$$\Delta p_{lg} = \rho gh \tag{3.24}$$

we obtain:

$$h = \frac{4}{\rho gd}\left(\gamma_{sg} - \gamma_{sl}\right) = \frac{4\gamma_{lg}}{\rho gd}\cos\theta \tag{3.25}$$

Clearly, in microconduits $h \sim d^{-1}$ can be substantial, a fact used for moving fluid volumes, known as "*capillary pumping.*" Applications of the Marangoni effect and electrowetting are two examples of "moving" finite fluid volumes, i.e., gas bubbles or liquid droplets. Such manipulation of small fluid volumes is termed *digital microfluidics*, as typically carried out with lab-on-a-chip systems (see Sect. 4.7).

Marangoni Effect.   So far, it was assumed that the surface tension γ is constant; however, spatial changes in surfactant concentrations and/or temperature cause gradients in the surface tension as well. Specifically, while $\gamma = F/L$, $\nabla\gamma$ implies a force per unit area, called the Marangoni or thermocapillary effect, which can move gas bubbles or liquid droplets in partially heated microchannels (see Figure 3.6a). The one-sided heat flux generates a temperature gradient across the gas bubble, where the local surface tension is lower at the higher temperature. As a result the Marangoni force reads:

$$f_M \equiv F_M / A = \nabla\gamma \tag{3.26}$$

which pushes the bubble towards the heat source. The inverse happens for the liquid plug (see Figure 3.6b). For example, γ_{water} is 72.75×10^{-3} N/m at 20°C but 62.6×10^{-3} N/m at 80°C.

A simple demonstration of the Marangoni effect can be accomplished when placing a drop of alcohol in the middle of a thin film of water on a smooth horizontal surface. Because $\gamma_{\text{alcohol}} < \gamma_{\text{water}}$ and hence a local surface tension gradient is created, the water will rush away from the region where the alcohol drop landed.

(a) Gas bubble moves towards heat source

(b) Liquid plug moves away from heat source

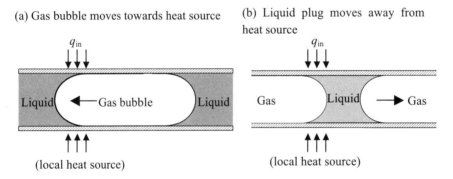

Figure 3.6 Thermocapillary surface tension effect

Electrowetting.   The phenomenon that the surface tension (and hence the contact angle) of a liquid droplet changes when applying an electrical potential across the liquid-solid interface is called electrowetting. Figure 3.7 depicts the impact of an electric field on droplet spreading. Specifically, without any voltage applied, the conductive liquid droplet rests on an electrically insulating (i.e., dielectric) layer with dielectric constant (or permittivity) ε, covering an electrode. While the droplet size is in the millimeter range, the dielectric coating is only 0.1 to 5 μm. The combination of an electrolyte droplet (say, salt water) on a dielectric surface (e.g., fluoropolymers) results in a high contact angle (see dashed droplet in Figure 3.7).

Applying a voltage up to 300 V reduces the contact angle significantly, causing surface wetting. Combined with well-placed electric currents, electrowetting can be employed to move (electrolyte) droplets on dielectric material (insulators) with hydrophobic surface coating (see Figure 3.8).

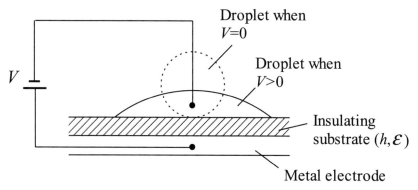

Figure 3.7 Generic electrowetting

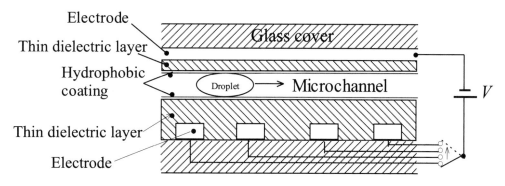

Figure 3.8 Droplet transport in microchannel due to electrowetting effect

Specifically, the applied electric field changes the droplets interfacial surface tension as described by the Lippman equation:

$$\gamma_{sl} = \gamma_{sl,o} - \frac{1}{2}CV^2 \tag{3.27}$$

where $\gamma_{sl,o}$ is the surface tension before applying the electric field, C is the interfacial capacitance per unit area, and V is the applied voltage across the interface. Dividing Eq. (3.27) by γ_{lg} and neglecting γ_{sg} we can express with Young's equation the associated contact angles as:

$$\cos\theta = \cos\theta_0 + \frac{1}{2\gamma_{\mathrm{lg}}}CV^2 \tag{3.28}$$

Now, this is a static result of contact angle change, assuming a uniform charge on the dielectric thin film wetted by the droplet. In order to move the droplet in the microchannel, an array of electrodes are activated to create a nonuniform charge distribution on the droplet surface, causing changes in contact angle and hence lateral droplet transport (see Figure 3.8). Clearly, by applying a sequence of electric pulses to different electrodes, different droplets can be made to disperse from reservoirs, move in microchannels, merge (and react) at junctions and form new products, test results, etc.

3.2.3 Wall Slip Velocity and Temperature Jump

Background Information.   The no-slip velocity and impermeable wall condition reads:

$$\vec{v}_{\mathrm{fluid}} = \vec{v}_{\mathrm{wall}} \tag{3.29}$$

It is correct for macroscale flows and is applicable for most microscale and even some nanoscale flows. However, in many natural and industrial microfluidics cases effective or apparent slip velocities can be observed. Examples include:

- Rarefied gases on actual solid surfaces
- Liquids on rough *and* hydrophobic surfaces, labelled "superhydrophobic"
- Liquid-molecule depletion-layer flow
- Electroosmotic flow of electrolytes with thin electric double layers (EDLs)
- Porous wall-layer flow

The underling physics for velocity slip differs greatly between rarefied gas molecules and liquid molecules. Due to the very low number density of *rarefied gas* molecules and the associated large mean free path, the near-wall molecules hardly collide with the solid surface which, statistically averaged, results in an effective slip velocity. The mean-free path, i.e., the distance travelled by gas molecules before collision, $\lambda_{\mathrm{mfp.}} = \mathrm{Kn}\, D_h$, is often taken as the key model parameter when gas flow is in the "slip range," i.e., $0.01 \leq \mathrm{Kn} \leq 0.1$ (see Eq. (3.18) and Figure 3.10a).

For *liquid flow* in micro- or nanochannels the average intermolecular distance δ_{IM} is employed to assure fluid continuum, i.e., $\delta_{\mathrm{IM}}/D_h > 3 \times 10^{-5}$. Nevertheless, the slip length, L_s, can be $\mathcal{O}(10\,\mathrm{nm})$ to $\mathcal{O}(100\,\mu\mathrm{m})$. Relative high apparent L_s-values may be encountered when rough hydrophobic surfaces with air/vapor pockets allow liquids to "glide" over the wall with minimum resistance (see Figures 3.9 and 3.10b). Alternatively, a depletion of liquid molecules, or particles in case of suspensions, causes much lower viscosities in a thin wall layer. That may lead to apparent velocity slip as has been demonstrated for liquid flow in carbon nanotubes. For non-Newtonian fluids, in addition to physical and chemical characteristics of the surface, the type of fluid or suspension and hence its properties, such as density, viscosity, yield stress, elastic modulus, particle concentration, and electric charge, determine the degree of true or apparent velocity slip.

In electroosmosis (see Figure 3.9b and Sect. 3.5) a near-wall layer of mostly positive (or negative) ions of an electrolyte is set into motion due to an applied electric field $E_x = V/L$. In 1-D microchannel flow, the electrostatic force $f_{coulomb}$ is balanced by f_{drag} which moves the fluid bulk at uniform velocity. In cases for which the charged wall layer (i.e., EDL) may be only a few nanometers thin, the electroosmotic velocity $u_{EO} = u_{bulk} \approx u_{slip}$.

(a) Effective u_{slip} due to low flow resistance

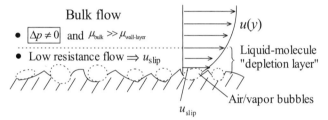

(b) Effective u_{slip} due to moving ion layer, driven by external electric field $E \sim V_{net}$

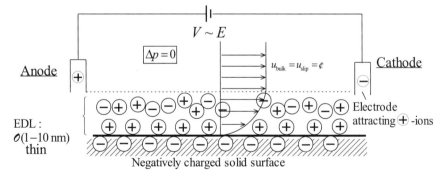

Figure 3.9   Examples of effective velocity slip of liquids at/near a solid surface: (a) rough hydrophobic wall and (b) electroosmotic flow inside EDL

Velocity Slip Models.   For rarefied *gas flow* Maxwell suggested a wall velocity slip dependent on the mean-free path and the wall velocity gradient. In contrast, for *liquid*

flow with wall slip, Navier postulated the "slip-length" L_s, a hypothetical linear extension of the velocity profile into the wall (Figure 3.10).

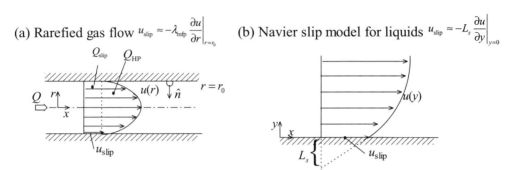

(a) Rarefied gas flow $u_{slip} \approx -\lambda_{mfp} \dfrac{\partial u}{\partial r}\Big|_{r=r_0}$ (b) Navier slip model for liquids $u_{slip} \approx -L_s \dfrac{\partial u}{\partial y}\Big|_{y=0}$

Figure 3.10: Slip flow schematics for gas flow (Maxwell) versus liquid flow (Navier)

Specifically, as a first-order approximation, Maxwell suggested for *rarefied gas flow* in the Knudsen number range of $0.01 \le \mathrm{Kn} = \lambda_{mfp}/D_h \le 0.1$:

$$u_{slip} = \lambda_{mfp} \frac{\partial u}{\partial y}\Big|_{\hat{n}=0} \tag{3.30}$$

where λ_{mfp} is the mean-free path, $D_h = 4A/P$ is the hydraulic diameter, and $\hat{n}$ is the wall normal vector. For *liquid flow* the Navier model envisions a linear extension of the velocity profile into the wall, generating a slip length L_s. Again,

$$u_{slip} = L_s \frac{\partial u}{\partial y}\Big|_{\hat{n}=0} \tag{3.31}$$

Clearly, such "slip flows" generate higher volumetric flow rates than no-slip flows. For example, in reference to basic Hagen-Poiseuille flow (see Sect. 2.1), i.e.,

$$Q_{HP} = \frac{\pi r_0^4}{8\mu}\left(\frac{\Delta p}{L}\right) \tag{3.32}$$

the flow rate enhancement ratio κ is greater than one (see Figure 3.10a):

$$\kappa = \frac{Q}{Q_{HP}}; \text{ where } Q = Q_{HP} + u_{slip}\left(\pi r_0^2\right) = Q_{HP} + Q_{slip} \tag{3.33a-c}$$

In fact, for Poiseuille-type flow in CNTs $\kappa \leq 8$ has been measured (see Sect. 5.4).

Temperature-Jump Model.   For cooled or heated rarefied gases, deviations from the *wall temperature* have been observed due to thermal nonequilibrium conditions in the wall region. Specifically, the relative few gas molecules near/at the wall cannot attain instantaneously the exact wall temperature. Hence, as a first-order approximation:

$$T_{\text{gas}}\left(x,\hat{n}=0\right)=T_{\text{wall}}+T_{\text{jump}}\approx T_{\text{wall}}+\frac{2\kappa\lambda_{\text{mfp}}}{\Pr(\kappa+1)}\frac{\partial u}{\partial y}\bigg|_{\hat{n}} \qquad (3.34\text{a,b})$$

where the heat capacity ratio $\kappa=c_p/c_v$ and the Prandtl number $\Pr=v/\alpha$.

Example 3.6: Nitrogen gas flow with viscous heating and velocity slip

Consider thermal micro-Couette gas flow with viscous heating in the slip regime. $0.01\leq \text{Kn}\leq 0.1$ (see Eq. (3.18)).
System Description:
- Nitrogen at $T_{\text{in}}=T_{\text{wall}}=300$ K and $\lambda_{\text{mfp}}\approx 64$ nm
- Channel parameters: $h=0.8\,\mu\text{m}$, $L=4\,\mu\text{m}$, and $U=100\,\text{m/s}$

Sketch:	Assumptions:	Concepts:
	• Slip regime: $\text{Kn}\equiv\lambda/h=0.08$ • 1-D conduction with viscous heating	• Reduced N-S equations • Viscous heating • Slip flow *Postulates:* • $u=u(y)$ only • $\Delta p\equiv 0$ • $\mu\varphi\approx\left(du/dy\right)^2$

Solution:

(i) *Fluid Flow:* Reduced x-momentum equation

$$0=\mu\frac{d^2u}{dy^2};\text{ or integrated } u(y)=C_1 y+C_2 \qquad (\text{E3.6-1a,b})$$

subject to $u(y=0) = u_{slip} = \lambda_{mfp}\, du/dy\big|_{y=0} = C_2$ and at $y=h$ (see negative unit vector for the moving surface):

$$u(y=h) = U - u_{slip} = U - \lambda \frac{du}{dy}\bigg|_{y=0} \; ; \text{ where } \lambda = h\,\text{Kn} \qquad \text{(E3.6-2a,b)}$$

With $du/dy = C_1$ we have:

$$C_2 = h\,\text{Kn}\,C_1 \text{ and } U - h\,\text{Kn}\,C_1 = C_1 h + C_2 \text{ or } C_1 = \frac{U}{h(1+2\,\text{Kn})} \qquad \text{(E3.6-3a-c)}$$

Hence,

$$u(y) = \frac{U}{1+2\,\text{Kn}}\left[\left(\frac{y}{h}\right) + \text{Kn}\right] \qquad \text{(E3.6-4)}$$

For $\text{Kn} \to 0$, i.e., $L_{system} \equiv h >>> \lambda_{mfp}$, $u(y) = U\,y/h$ (i.e., simple shear flow).

(ii) *Heat Transfer:*
Here,

$$0 = k\frac{d^2T}{dy^2} + \mu\left(\frac{du}{dy}\right)^2 \text{ or } \frac{d^2T}{dy^2} + \underbrace{\frac{\mu}{k}\left[\frac{U}{h(1+2\,\text{Kn})}\right]^2}_{\equiv A} = 0 \qquad \text{(E3.6-5a,b)}$$

subject to (see Eq. (3.34)):

$$T(y=0) = T_w + T_{jump} = T_w + \frac{2\kappa}{\kappa+1}\frac{h\,\text{Kn}}{\text{Pr}}\frac{dT}{dy}\bigg|_{y=0} \; ; \; \kappa = \frac{c_p}{c_v}; \; \text{Pr} = \frac{\nu}{\alpha} \qquad \text{(E3.6-6a,b)}$$

Employing symmetry, i.e., $dT/dy\big|_{y=h/2} = 0$, we have after double integration:

$$\frac{dT}{dy} = -Ay + C_1 \succ C_1 = A\frac{h}{2} \qquad \text{(E3.6-7a,b)}$$

and

$$T(y) = -\frac{A}{2}y^2 + C_1 y + C_2 \qquad \text{(E3.6-8)}$$

Hence, at $y = 0$:

$$T_w + \frac{2\kappa}{\kappa+1} \frac{h\,\text{Kn}}{\text{Pr}} \overbrace{\left(\frac{h}{2}A\right)}^{\equiv C_1} = C_2 \qquad (E3.6\text{-}9)$$

and finally,

$$T(y) = T_w + \frac{A}{2}h^2 \left[-\left(\frac{y}{h}\right)^2 + \left(\frac{y}{h}\right) + \frac{2\kappa}{\kappa+1}\frac{\text{Kn}}{\text{Pr}} \right] \qquad (E3.6\text{-}10)$$

(iii) _Numbers:_ For nitrogen (N_2): $\kappa = 1.4$, Pr $= 0.72$, $\mu = 1.656 \times 10^{-5}\,\text{N} \times \text{s}/\text{m}^2$, and Kn $= 0.08$; $U = 100\,\text{m}/\text{s}$ and $h = 0.8\,\mu\text{m}$.

(iv) _Graph:_ The theoretical temperature jump $T_{\text{jump}} = T_{\text{gas}} - T_{\text{wall}} \approx 0.3\,\text{K}$ is minimal as confirmed by DSMC (direct simulation Monte Carlo) results of Fang & Liou (2002).

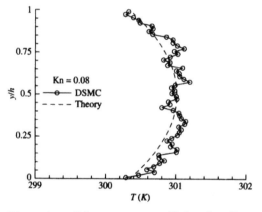

*Comparison of the temperature profile for micro-Couette flow (DSMC Source: Fang & Liou, 2002)

In addition to surface tension effects, other driving forces of special fluids (or particle suspensions) in microchannels are electrokinetic and magnetohydrodynamic phenomena. The interactions and resulting forces are summarized in Figure 3.11 and discussed in the next two sections.

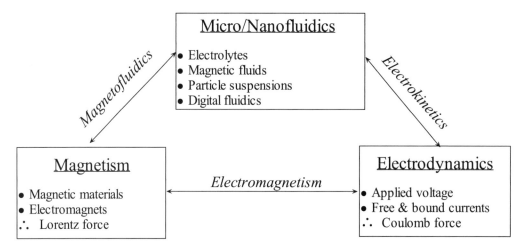

Figure 3.11   Interactive phenomena and resulting forces in micro/nanofluidics

3.2.4 Electrokinetic Phenomena

Overview.   As mentioned and further discussed in Chapters 4 and 6, surface-modulated driving forces such as surface tension (see Sect. 3.2.2) and especially electroosmosis are often employed for liquid flow in micro- and nanochannels in the Stokes flow regime ($\mathrm{Re} \leq 1.0$). The advantages of such "pumps" include minimal space requirements, no moving parts, and low costs. Movement of charged particles is accomplished via electrophoresis. In fact, there are four distinct electrokinetic phenomena as depicted in Figure 3.12.

3.2.4.1 Electroosmosis

When a solid surface, e.g., plastic or glass, contacts on electrolyte, such as salt water, the solid surface becomes naturally charged. This occurs due to the difference of electron (or ion) affinities between the solid surface and the solution. As a result, the surface charge, i.e., being either negative or positive depending on the type of solid-electrolyte pairing, causes a special layered ion structure at/near the interface. It is labeled the EDL which has a net mobile electric charge. Now, applying externally an electric field, $\vec{E}$, between channel inlet and outlet induces a Coulomb force which causes a major part of the EDL to move, e.g., the positive ions moving towards the cathode (see Figure 3.12(ii)). This moving fluid (or diffuse) layer of typically nanometer thickness λ_D (known as the *Debye length*) near the wall drags via viscous effects the adjacent layer, practically the entire bulk fluid, with it. This phenomenon is known as *electroosmotic flow* (see Figures 3.12(i) and 3.13). Thus, electroosmosis

(i) **Electroosmosis:** Bulk liquid flow over a stationary surface with EDL can be achieved via applied net voltage $\Delta V \sim E_x$, i.e., electroosmotic (EO) flow.

(ii) **Electrophoresis:** Particle with EDL migrates towards the cathode in quiescent liquid, known as electrophoresis.

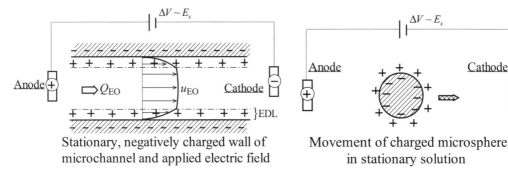

Stationary, negatively charged wall of microchannel and applied electric field

Movement of charged microsphere in stationary solution

Note: In contrast to electroosmosis or electrophoresis requiring an external electric field, when applying a net surface force (pressure drop) or body force (gravity) to get the bulk fluid or the microspheres in nano/microchannel moving, an electric field is generated due to electric charge advection. The "streaming potential" in pressure-driven microchannel flow and the "sedimentation potential" using charged particles subjected to gravity are the other two electrokinetic phenomena.

(iii) **Streaming Potential**

(iv) **Sedimentation Potential**

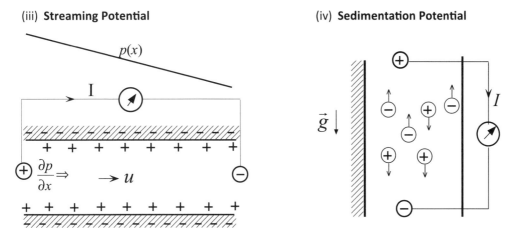

Figure 3.12   Electrokinetic phenomena

works like a (nonmechanical) *pump* (see Sect. 4.2.1). In devices where the length scales are micro to nanometers, it is most desirable for the flow rates to be very low ($\text{Re} \leq 1.0$), or the fluid volumes conveyed to be in the nano- and picoliter range as in digital microfluidics. Hence, the required pumping device is very small and pumping has to be reliable and cost-effective.

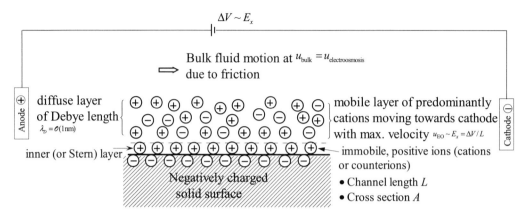

Figure 3.13   Moving diffuse layer generating microchannel flow rate $Q_{EO} = u_{EO} A$

Definitions.   Before discussing the underlying physics, i.e., electrostatics of electroosmosis, a few definitions are in order:

- Ion: An atom or molecule for which the total number of electrons (-) does not match the total number of protons (+) will result in a net positive charge (i.e., cations) or net negative charge (anions). The process of gaining or losing electrons from a neutral atom/molecule is called ionization. This electron transfer process is driven by the tendency to attain a stable electronic configuration, i.e., a "closed shell."

    For example, sodium is stable when losing an (extra) electron and hence becoming a cation:

$$\text{Na} \rightarrow \text{Na}^+ + e^- \nearrow \tag{3.35}$$

The opposite holds for the chlorine atom where gaining an electron, i.e., $\text{Cl} + e^- \rightarrow \text{Cl}^-$, yields an anion. As an aside, being oppositely charged, ionic bonds are formed:

$$\text{Na}^+ + \text{Cl}^- \rightarrow \text{NaCl (Salt)} \tag{3.36}$$

- Electrolyte: It is a substance containing free ions that make the medium electrically conductive. For example, salt in water forms an ionic solution, i.e., solid salt dissolves into component elements:

$$\text{NaCl}\big|_{\text{solid}} \rightarrow \text{Na}^+_{\text{aq.}} + \text{Cl}^-_{\text{aq.}} \tag{3.37}$$

Another example is CO_2 dissolving in water, producing H_3O^+, CO_3^{2-}, and HCO_3^- ions. As an aside, all higher life forms require a subtle electrolyte balance between intracellular and extracellular milieu for proper body hydration, blood pH, and nerve and muscle functions. Key electrolyte ions are sodium (Na^+), potassium (K^+), chlorine (Cl^-), calcium (Ca^{2+}), hydrogen phosphate (HPO_4^{2-}), and hydrogen carbonate (HCO_3^-). Imbalance of certain ions may cause dehydration (i.e., lack of Na^+ or K^+) or weak muscle function (i.e., lack of Ca^{2+}, Na^+ and K^+).

- Electrical Double Layer (EDL): When a naturally charged object, e.g., a solid particle or surface or porous plug, gas bubble, or liquid droplet, is placed into an electrolyte (e.g., salt water) a double layer of ions is formed around/on the object. For example, a negatively charged surface attracts and immobilizes cations (also called counterions), forming an inner sublayer, i.e., the Stern layer of net electric surface potential $\phi_{wall} = \phi_{stern} - \phi_{bulk}$. The adjacent second layer contains also mainly cations, known as the movable diffuse layer of thickness λ_D with typically an exponentially decaying electric potential (see Figure. 3.14). In 1-D:

$$\phi(y) \approx \phi_w e^{-x/\lambda_D} \tag{3.38}$$

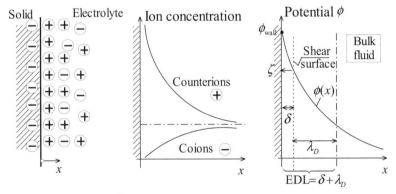

Notes:
- $0 \leq x \leq \delta$: Immobile Stern layer
- $\delta \leq x \leq \lambda_D$: Mobile diffuse layer
- $\phi_{wall} \approx \zeta$ (zeta potential) because $\delta <<< 1$

Figure 3.14   Ions, the resulting electric potential, and the electrical double layer (EDL) with zeta potential $\zeta \leq \phi_{wall}$

where λ_D is the Debye length in the range $1 \leq \lambda_D \leq 50$ nm, with typically $25 \leq \phi_w \leq 100$ mV. For example, ion concentrations of 0.04 to 4 mM may generate a λ_D-value of 10 to 1 nm. Clearly, there is a "slipping plane" (i.e., the shear surface) that separates potentially moving fluid, i.e., the electrolyte, from the ion (or Stern) layer attached to the (charged, typically solid) surface.

3.2.4.2 Electrostatics

Although electroosmosis deals with *moving* fluid layers driven by forces, the laws of *electrostatics* (i.e., Coulomb, Gauss, Poisson, etc.) are applicable. In other words, although resulting observations/laws hold strictly only for static charges (e.g., ions in an electrolyte), they actually describe slowly moving charges as well. Hence they are conducting liquids at Re ≤ 1.0. In general, electrostatics deals with phenomena involving time-independent distributions of electric point charges and resulting electric fields. Fundamental is Coulomb's law concerning the interactive force between charged bodies at rest.

<u>Electrostatic Force:</u> Two point charges q_i and q_j a distance r_{ij} apart cause a static interaction force $F_{ij} \sim r_{ij}^2$ and $\sim q_i q_j$ (see Figure 3.15). Specifically,

$$\vec{F}_{ji} = -\vec{F}_{ij} = \frac{1}{4\pi\varepsilon_0} \frac{q_i q_j}{r_{ij}^2} \hat{e}_{ij} \quad \text{(Coulomb)} \tag{3.39}$$

where $\vec{F}_{ji}$ is the force exerted on q_j by charge q_i, ε_0, is the permittivity in a vacuum, and $\hat{e}_{ij}$ is the unit vector. Note, the *relative permittivity*, or <u>dielectric constant</u>, is:

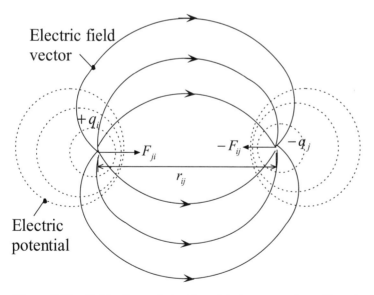

Figure 3.15   Distributions of electric point charges and resulting electric fields

$$\varepsilon_r = \frac{\varepsilon}{\varepsilon_0} = \begin{cases} 1 & \text{in a vacuum} \\ 1.005 & \text{in dry air} \\ 6 & \text{in glass} \\ 12 & \text{in silicon} \\ 80 & \text{in water} \end{cases}$$

where ε in $\left[C/(V \cdot m)\right]$ and $\varepsilon_0 = 8.85 \times 10^{-12}\,C/(V \cdot m)$. Clearly, 1 Coulomb $C = 1\,N \cdot m/V$.

The electric permittivity indicates how much electric field is caused by a source charge (see Gauss' law).

As Figure 3.15 indicates attractive forces, equally charged particles generate repulsive forces. For example, the repulsive electrostatic force increases as $+q_1$ moves towards $+q_2$. The associated electric fields are due to outward traveling electromagnetic waves at the speed of light.

Rewriting Coulomb's law as:

$$\left|\vec{F}\right| = F_{\text{Coulomb}} = k\frac{|q_1||q_2|}{r^2} \tag{3.40a}$$

where $k = 8.99 \times 10^9\,N \times m^2/C^2$ is the electrostatic constant, Eq. (3.40a) is similar to Newton's gravitational force between two objects of masses m_1 and m_2, i.e.,

$$F_{\text{Newton}} = G\frac{m_1 m_2}{r^2} \tag{3.40b}$$

<u>Electric Field and Potential:</u> The electric field vector $\vec{E}\left[N/C \hat{=} V/m\right]$ is defined as:

$$\vec{E} \equiv \frac{\vec{F}_{ji}}{q_j} = \frac{1}{4\pi\varepsilon_0}\frac{q_i}{r_{ij}^2}\hat{e}_{ij}; \text{ where unit vector } \hat{e}_{ij} \equiv \frac{\vec{r}_{ij}}{r_{ij}} \tag{3.41a-c}$$

Thus, for a continuous spatial distribution of charges:

$$\vec{E} = \frac{1}{4\pi\varepsilon_0}\int\frac{\vec{r}}{r^3}dq := \frac{1}{4\pi\varepsilon_0}\int_{\forall}\frac{\rho_{\text{el}}\vec{r}}{r^3}d\forall \tag{3.42a,b}$$

where ρ_{el} is the volume charge density, i.e.,

$$\rho_{el} \equiv \lim_{\Delta\forall\to 0} \frac{\Delta q}{\Delta\forall} \quad \left[\frac{C}{m^3}\right] \tag{3.43a}$$

For an ensemble of point charges q_i, as in an electrolyte, we have:

$$\sum q_i = \int_{\forall} \rho_{el} d\forall \tag{3.43b}$$

The electric field $\vec{E}$ being conservative (see Chang and Yeo, 2010) and irrotational (i.e., $\nabla \times \vec{E} = 0$), it can be derived from a scalar potential function ϕ [V]:

$$\vec{E} = -\nabla\phi \left[\frac{V}{m} \text{ or } \frac{N}{C}\right] \tag{3.44}$$

Here, ϕ is the electric potential (or net voltage) and typical values for the electric field $\vec{E}$ are:

- 3×10^{21} N/C at the surface of a uranium nucleus
- 5×10^{4} N/C within a hydrogen atom
- 10^{3} N/C   near a charged plastic rod
- 10^{-2} N/C   inside a copper wire

The flux of the electric field through a (spherical) surface A surrounding a point charge is (see Figure 3.16):

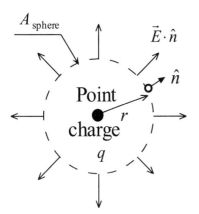

Figure 3.16   Gradient of electric point charge

$$\int_S \vec{E} \cdot d\vec{A} = \int_S \vec{E} \cdot \hat{n} \, dA = 4\pi r^2 \left(\frac{1}{4\pi\varepsilon_0} \frac{q}{r^2} \right) = \frac{q}{\varepsilon_0} \qquad (3.45\text{a-c})$$

For an ensemble of point charges in an electrolyte, i.e., $\sum q_i$, we have:

$$\int_S \vec{E} \cdot \hat{n} \, dA = \frac{1}{\varepsilon_0} \sum q_i = \frac{1}{\varepsilon_0} \int_\forall \rho_{el} \, d\forall \quad (\text{Gauss}) \qquad (3.46)$$

Applying the Divergence Theorem (Sect. A.3.1), $\int_S \vec{E} \cdot \hat{n} \, dA = \int_\forall \nabla \cdot \vec{E} d\forall$, we have:

$$\int_\forall \nabla \cdot \vec{E} d\forall = \int_\forall \frac{\rho_{el}}{\varepsilon_0} \, d\forall \text{ or } \int_\forall \left[-\nabla^2 \phi - \frac{\rho_{el}}{\varepsilon_0} \right] d\forall = 0 \qquad (3.47)$$

Hence,

$$\nabla \cdot \vec{E} = -\nabla^2 \phi = \frac{\rho_{el}}{\varepsilon_0} \quad (\text{Poisson}) \qquad (3.48)$$

or for a heterogeneous dielectric, i.e., insulating medium:

$$-\nabla \cdot (\varepsilon \nabla \phi) = \rho_{el} \xrightarrow[\text{uniform}]{\varepsilon = \ell} -\varepsilon \nabla^2 \phi = \rho_{el}; \ \varepsilon = \varepsilon_r \varepsilon_0 \qquad (3.49\text{a-c})$$

Note: Actual microfluidics materials are typically "weak electric conductors" and hence have finite permittivities ε ranging from low (e.g., insulators like glass or polymers or silicon) to high (e.g., electrodes), with aqueous solutions being in between.

Now, from the definition $\vec{E} = \vec{F}/q$ and with $\rho_{el} = q/\forall$ we have the Coulomb electric force per unit volume:

$$\vec{f}_C \equiv \vec{f}_{el} = \rho_{el} \vec{E} = -\left(-\varepsilon \nabla^2 \phi \right) \nabla \phi \qquad (3.50)$$

Thus, when an electric field $\left(\vec{E}_{ext.} \right)$ is externally applied to an electrolyte $\left(\rho_{el} \right)$ in, say, a microchannel, a new body force $\vec{f}_{el}$ has to be added to the N-S equation:

$$\rho \frac{\partial \vec{v}}{\partial t} + \rho (\vec{v} \cdot \nabla) \vec{v} = -\nabla p + \mu \nabla^2 \vec{v} + \rho \vec{g} + \rho_{el} \vec{E}_{ext.} \qquad (3.51)$$

Poisson-Boltzmann Equation: The expression for the new body force is then:

$$\vec{f}_{el} = \rho_{el}\vec{E} = -\varepsilon\nabla^2\phi\vec{E} \tag{3.52}$$

We recall:

- An applied voltage difference or electric potential, $\Delta V = \phi_{ext,}$ results in an external electric field $\vec{E}_{ext.} = -\nabla\phi_{ext.}\left[\text{V/m}\right]$
- For a continuous medium Maxwell showed that the charge density is:

$$\rho_{el} = -\nabla\cdot\left(\varepsilon\vec{E}\right)\xrightarrow{\varepsilon=\ell} = -\varepsilon\nabla^2\phi_{ext.}\left[\text{C/m}^3\right]$$

where again $\varepsilon = \varepsilon_r\varepsilon_0$ is the dielectric constant (or permittivity).

- The charge density ρ_{el} along the wall is confined to the EDL $\approx\lambda_D$ thick, i.e., $\rho_{el}\sim\phi_{ext.}$. Specifically, based on statistical thermodynamics analysis (Boltzmann) and following the Debye-Hückel linear approximation:

$$\rho_{el} = -\frac{\varepsilon}{\lambda_D^2}\phi_{ext.} \tag{3.53}$$

Hence,

$$-\frac{\varepsilon}{\lambda_D^2}\phi_{ext.} = -\varepsilon\nabla^2\phi_{ext.} \text{ or } \nabla^2\phi = \frac{\phi}{\lambda_D^2}\left(\text{Poisson-Boltzmann}\right) \tag{3.54a,b}$$

so that

$$\vec{f}_{el} = \rho_{el}\vec{E} = -\varepsilon\nabla^2\phi\vec{E} = -\frac{\varepsilon}{\lambda_D^2}\phi\vec{E} \tag{3.55a-c}$$

For axial (1-D) flow:

$$f_x^{el} = -\varepsilon\frac{d^2\phi}{dy^2}E_x \tag{3.55d}$$

1-D Electric Potential Solutions: For the 1-D case, the Poisson-Boltzmann equation reads:

$$\frac{d^2\phi}{dy^2} = \frac{\phi(y)}{\lambda_D^2} \tag{3.56}$$

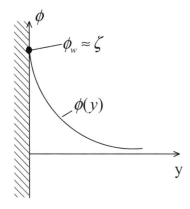

Case I: Equation (3.56) is subject to two BCs, i.e., $\phi(y=0)=\phi_{\text{wall}}\approx\zeta$ (zeta potential) and $\phi(y\rightarrow\infty)=0$ (zero bulk potential), resulting in:

$$\phi(y)=\phi_w\exp\left(-\frac{y}{\lambda_D}\right) \tag{3.57a}$$

Note: Equation (3.57a) is suitable for a single charged surface or very wide channels where the focus is on one charged wall.

Case II: For the sketched microchannel flow we impose $\phi(y=h)=\phi_{\text{wall}}\approx\zeta$ and $d\phi/dy\big|_{y=0}=0$ and the solution of Eq. (3.56) is:

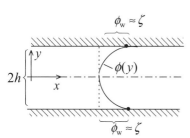

$$\phi(y)=\phi_w\frac{\cosh\left(\dfrac{y}{\lambda_D}\right)}{\cosh\left(\dfrac{h}{\lambda_D}\right)} \tag{3.57b}$$

Notes:

- These 1-D solutions hold for relatively low $\phi_{\text{wall}}\approx\zeta$ values, say, $\zeta\le 50$ mV.
- To avoid Joule heating, the applied electric current $I\sim E_{\text{ext.}}$ should be low.

Example 3.7: Electroosmotic Flow of an Electrolyte in a Microchannel with Charged Surfaces

Consider saltwater (μ) in a microchannel (height h, length L, and permittivity ε) with zeta potential $\zeta\approx\phi_{\text{wall}}$ and Debye length λ_D subject to an electric field $E_x=-\Delta\phi/\Delta x=\Delta V/L$.

Find the induced velocity $u(y)$ and plot several nondimensional velocity profiles. Discuss the interpretation of $u(y=\lambda_D)=u_{\text{EO}}\approx u_{\text{slip}}$ when $\lambda_D/h<<<1$.

Sketch:	*Assumptions/Postulates:*	*Concepts:*
	• Steady laminar fully developed flow • Constant properties/ parameters • $\phi_{\text{wall}} \approx \zeta$ and $\phi(y \geq \lambda_D) \approx 0$ • $\nabla p = 0$ and $\vec{v} \to u(y)$	• Reduced N-S equation • $\vec{f}_{\text{coulomb}} = \rho_{\text{el}}(\vec{E}_{\text{ext.}} = -\varepsilon)\nabla^2\phi\vec{E}_{\text{ext.}}$ $\therefore f_x = -\varepsilon \ d^2\phi/dy^2 \ E_x$ • $\phi(y) \approx \zeta \exp(-y/\lambda_D)$ • $f_{x,\text{electric}} = f_{x,\text{drag}}$

Solution:

Based on the assumptions and postulates, Eq. (3.51) in conjunction with Eqs. (3.52) and (3.55d) yields the *x*-momentum equation:

$$0 = \mu\frac{d^2u}{dy^2} - \varepsilon\frac{d^2\phi}{dy^2}E_x \quad \text{or} \quad \mu\frac{d^2u}{dy^2} = \varepsilon\frac{d^2\phi}{dy^2}E_x \qquad \text{(E3.7-1a,b)}$$

where $E_x = -d\phi/dy \approx \Delta V/L$.

Integrating twice each side, subject to $u(y=0)=0$ and $du/dy\big|_{y=h/2} = 0$ or $u(y=h)=0$, as well as $\phi(y=0)=\varphi_w \approx \zeta$ and $\phi(y\to\infty, \text{ or } \lambda_D)=0$ (see Case I, Eq. (3.57a)) which may be valid for $h \gg \lambda_D$, we obtain:

$$u(y) = \frac{\varepsilon E_x}{\mu}\phi(y) - \zeta = \frac{\varepsilon}{\mu}\zeta E_x\left(1 - e^{-y/\lambda_D}\right) \qquad \text{(E3.7-2a,b)}$$

Nondimensionalization yields:

$$\hat{u} \equiv \frac{u(y)}{\varepsilon\zeta E_x/\mu} = 1 - \exp[-2P\hat{y}] \ ; \ P \equiv \frac{h}{2\lambda_D} \text{ and } \hat{y} = \frac{y}{h} \qquad \text{(E3.7-3a-d)}$$

Comments (see graph):

- For very thin EDLs, i.e., $\lambda_D \ll 1$ and hence $P \gg 1$, the exponential term approaches zero within a thin wall layer. Hence $u(y=\lambda_D) \equiv u_{\text{EO}} \hat{=} u_{\text{slip}} = u_{\text{bulk}} \equiv \mu_{\text{EO}}E_x$, where $\mu_{\text{EO}} \equiv \varepsilon/\mu\phi_w \approx \varepsilon/\mu\zeta$ is the *electroosmotic mobility*.
- Clearly, measuring u_{EO} and knowing $E_x = \Delta V/L$ yield μ_{EO} and hence $\phi_{\text{wall}} \approx \zeta = \mu_{\text{EO}}\mu/\varepsilon$.
- For $P < 5.0$, the Case II boundary conditions, i.e., Eq. (3.57b), would be more appropriate.

Graph:

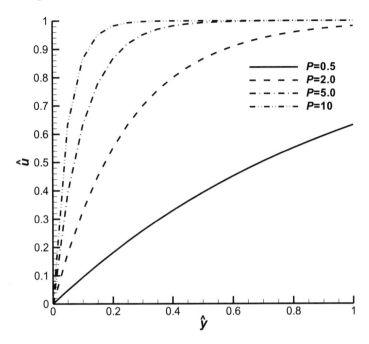

Summary of Thermal Electroosmotic Fluid Flow Equations. Assuming steady electrolyte flow with constant fluid properties and uniform zeta potential, we have shown:

- Continuity: $\nabla \cdot \vec{v} = 0$ (3.58)

- Momentum Transfer: $\rho(\vec{v} \cdot \nabla)\vec{v} + \nabla p - \mu \nabla^2 \vec{v} - \rho_{el} \vec{E} = 0$ (3.59)

- Energy Transfer: $\rho c_p (\vec{v} \cdot \nabla) T + \nabla \cdot (k \nabla T) - \mu \Phi - \sigma_{el}(\vec{E} \cdot \vec{E}) = 0$ (3.60)

where the term $\sigma_{el}(\vec{E} \cdot \vec{E})$ indicates Joule heating.
    The applied electric potential can be obtained from:

$$\nabla^2 \phi = -\frac{\rho_{el}}{\varepsilon}$$ (3.61)

For a microchannel of height $2h$, where the driving Coulomb force balances the viscous resistance force, we have for $\phi_{\text{wall}} \approx \zeta$:

$$u = u_{\text{EO}}\left(1 - \frac{\phi}{\zeta}\right) \tag{3.62}$$

and the electric potential $\phi(y)$ is (Dutta & Beskok, 2001):

$$\phi = \frac{4k_B T}{ez}\tanh^{-1}\left[\tanh\left(\frac{ez\zeta}{4k_B T}\right)\exp\left[\frac{|y| - h}{\lambda_D}\right]\right] \tag{3.63}$$

Horiuchi & Dutta (2004) showed that the ratio of Joule heating to viscous dissipation for 1-D microchannel flow is:

$$\kappa \approx \frac{\sigma_{\text{el}}\delta h}{\zeta^2}\frac{\mu}{\varepsilon^2} \tag{3.64}$$

where typical parameter values are $\sigma_{\text{el}} = 10^{-3}\,\text{S/m}$, $\delta \leq 5\lambda_D$ is the wall layer where $\phi(y) \neq 0\,(0 \leq y \leq \delta)$ $\lambda_D = -100\text{mV}$, $h \gg \delta$, $\mu = 10^{-3}\,\text{Ns/m}^2$, and permittivity $\varepsilon = \varepsilon_0\varepsilon_r$ (see Eq. (3.40)).

It should be noted that $\kappa = \kappa\left(h, \lambda_D, \zeta^{-2};\text{ medium parameter values}\right)$ but is not explicitly dependent on $E_x = \Delta V / L$.

3.2.4.3 Electrophoresis

Fluid flows and particles move because of external forces, typically grouped into surface and body forces of which pressure, viscous, and gravitational forces dominate macrofluidics. When dealing with particle transport in micro/nanoconduits, diffusion may become more important or even more desirable than convection. When particles are charged (e.g., ions, molecules, chemical compounds, nanoparticles, etc.) and subjected to an electric field (see Sect. 3.2.4.2), they migrate towards the opposite electrode. Such a resulting particle motion in a quiescent liquid is known as electrophoresis (see Figure 3.11(ii)). Depending on the size of the EDL surrounding the particle, we distinguish between electrophoresis of "pointparticles," such as ions where $\lambda_D \gg r_0$, and larger "charged particles" with very thin EDLs.

Point Particle Electrophoresis. As in electroosmosis (see Comment in Example 3.7), we introduce an *electrophoretic mobility* $\mu_{\text{EP}}\left[\text{m}^2/(\text{V}\cdot\text{s})\right]$ so that (see Figure 3.17) the particle velocity vector is:

$$\vec{v}_{EP} = \mu_{EP}\vec{E} \tag{3.65}$$

A charged ion (actually a point charge without an EDL of influence) subject to an electric field experiences Coulomb's force, i.e.,

$$\vec{F}_c \equiv q\vec{E} = ze\vec{E} \tag{3.66}$$

where q is the point charge, z is the ion's charge number or valence (e.g., $Na^+ = 1$, $Ma^{2+} = 2$, $OH^- = -1$, $SO_4^{2-} = -2$, etc.), and $e = 1.6 \times 10^{-19}$ C is the charge of an electron, i.e., the unit charge.

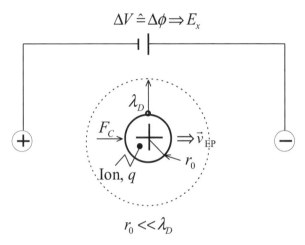

Figure 3.17   Positive ion in water

Example 3.8: Consider a sodium ion $\left(Na^+ \rightarrow z = 1\right)$ in a water reservoir at $t = 0$ and $u = 0$ subject to $E_{external} = 10^4$ V/m. The measured electrophoretic mobility is $\mu_{EP} = 5.2 \times 10^{-8}$ m^2/(V·s) and the sodium ion mass $m = 3.8 \times 10^{-26}$ kg.

•   The electric force exerted on the ion is:

$$F_{Coulomb} = zeE = 1 \times 1.6 \times 10^{-19} \times 10^4 \, C \cdot V/m \tag{E3.8-1}$$

$$\therefore \qquad\qquad F_c = 1.6 \times 10^{-15} \, N$$

where in 1-D $F_c = m\,du/dt = ma$.

- The terminal ion velocity is:

$$u_{\text{term}} = \mu_{\text{EP}} E = 5.2 \times 10^{-4} \, \text{m/s} = 520 \, \mu\text{m/s} \tag{E3.8-2}$$

- The constant ion acceleration is:

$$a = F/m = \frac{1.6}{3.8} 10^{-15} \times 10^{26} \, \frac{\text{N}}{\text{kg}} \tag{E3.8-3}$$

$$\therefore \qquad\qquad a = 4.2 \times 10^{10} \, \text{m/s}^2$$

- The time required to reach $u_{\text{term}} \hat{=} \Delta u$ assuming $a = \text{¢}$:

$$\Delta t = \frac{\Delta u}{a} = 1.3 \times 10^{-14} \, \text{s} \tag{E3.8-4}$$

Comments:

(i) Clearly, the Na ion due to its tiny mass reaches the terminal velocity instantaneously.

(ii) Under steady-state conditions, i.e., $ma = \sum F = 0$, the electric force is opposed by Stokes' drag:

$$F_C = F_d \tag{E3.8-5a}$$

or

$$zeE = 3\pi\mu u_{\text{particle}} d_p \tag{E3.8-5b}$$

Hence,

$$d_p = \frac{zeE}{3\pi\mu u_{\text{EP}}} = 0.33 \mu\text{m for an Na ion} \tag{E3.8-5c}$$

Particle Electrophoresis. Charged (larger) particles, in contrast to ions, are surrounded by a (thin) EDL. This is because they can attract/support enough counterions to create a continuous EDL. The underlying physics and math describing particle electrophoresis is the same as in electroosmosis when $r_0 \gg \lambda_D$, i.e., a very thin EDL exists (see Figure 3.18). Otherwise the effect of interface curvature, especially for smaller particles, cannot be ignored because $\vec{E}_{ext.}$ is nonuniform in any *thick* EDL. Also, the EDL exhibits two-way coupling between particle motion and the electrolyte velocity field when the surface potential is large.

In summary, the electrophoretic particle velocity is:

$$\vec{u}_{EP} = f \frac{\varepsilon \phi_w}{\mu} \vec{E}_{ext.} \tag{3.67}$$

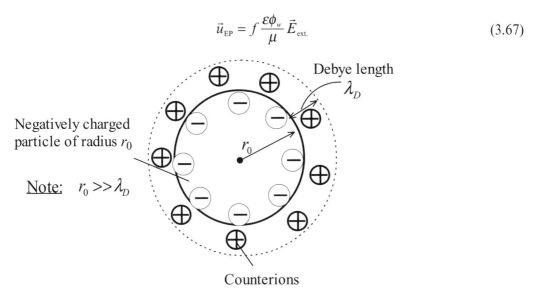

Figure 3.18   Charged particle and the corresponding EDL

where the correction factor $f = \mathcal{O}(1)$, i.e., $f = 1$ when $r_0 \gg \lambda_D$ <Smolochowski limit>.

Again, analogous to electroosmosis, we introduce the electrophoretic mobility $\mu_{EP} \left[m^2 / (V \cdot s) \right]$ so that

$$\vec{u}_{EP} \equiv \vec{u}_{particle} = \mu_{EP} \vec{E}_{ext.} \tag{3.68}$$

i.e.,

$$\mu_{EP} = \mu_{EO} = \frac{\varepsilon \phi_w}{\mu} \tag{3.69}$$

Dielectrophoresis.   In electrophoresis the moving particles are charged and $\vec{u}_{particle} = \vec{u}_{EP} = \varepsilon \zeta / \mu \cdot \vec{E}_{ext.}$, while in dielectrophoresis (DEP) polarizable particles (or cells) translate in an electrolyte due to a net electric field gradient, i.e., $\vec{u}_{DEP} \sim \nabla E_{net}$.

A major DEP advantage is that the particles can be nonconductors (i.e., dielectric), where generally the strength of the DEP force depends on the electrolyte, the particle's electric properties, its size and shape as well as the type of electric field (see Figure 3.19). One outcome is that if the permittivity of the medium is greater than that of particles, they move towards the region of the lesser DC or AC electric field. Also, DEP is most effective for particles in the diameter range of $1\,\mu m < d_p < 1\,mm$. With DEP migration or retention, such microspheres can be separated, guided, and manipulated.

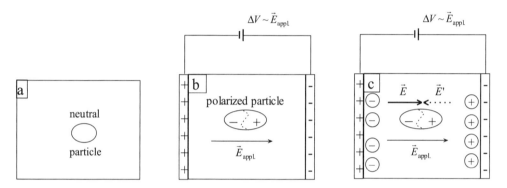

(a) A nonpolar dielectric particle in an electrolyte　(b) Particle polarization via an applied electric field　(c) The particle surface charges set up an electric field $\vec{E}'$

Figure 3.19　Polarization effect in DEP and net electric field $\vec{E} = \vec{E}_{applied} - \vec{E}'$

Particle magnetophoresis (see Sect. 3.2.5) is similar to DEP; however, the outcome is vastly different because of the large variations available in magnetic properties of particles and media as well as magnetic fields (see Sect. 3.2.5).

3.2.4.4 Nernst-Planck Equation

As discussed, chemical species and biological particles may move in a base fluid due to convection, diffusion, and/or electrophoresis. Thus, considering charged ion i the total velocity could be

$$\vec{v}_i = \vec{v}_{fluid} + \vec{v}_{diff} + \vec{v}_{EP} \tag{3.70}$$

where $\vec{v}_{fluid}$ is the (pressure-driven) fluid flow velocity (i.e., species convection), $\vec{v}_{diff}$ is the species transport due to Brownian motion and/or thermal diffusion, and $\vec{v}_{EP}$ is the electrophoretic velocity due to an applied electric field $\vec{E} = -\nabla\phi$. Furthermore, an electrolyte solution may contain n ionic species of concentrations $c_i (i = 1, 2, \ldots, n)$ which all influence the electrostatic force, i.e.,

$$\vec{f}_{\text{Coulomb}} = F\nabla\phi\sum_{i=1}^{n} z_i c_i \tag{3.71}$$

where F is the Faraday constant and z_i is the valence of the i_{th} ion species.

The fate of dissolved species i of concentration c_i in a fluid flow field, subject to diffusion and electrophoresis (no chemical reaction), is described by the Nernst-Planck equation (see Eq. (1.96) and Table 1.2 for comparison):

$$\underbrace{\frac{\partial c_i}{\partial t}}_{\substack{\text{Local time rate} \\ \text{of change in } c_i}} = -\nabla \cdot \left[\underbrace{-\mathcal{D}_i\nabla c_i}_{\text{Species diffusion}} + \underbrace{\left(\vec{u}_{\text{fluid}} + \vec{u}_{\text{EP}}\right)c_i}_{\substack{\text{Species convective} \\ \text{transport}}} \right] \tag{3.72}$$

where $\mathcal{D}_i$ is the species diffusivity coefficient, $\vec{u}_{\text{fluid}}$ is the solution to the Navier-Stokes equation, and $\vec{u}_{\text{EP}} = \mu_{\text{EP},i}\vec{E}$. Assuming $\mathcal{D}_i = \text{constant}$ and setting $\vec{u}_{\text{fluid}} + \vec{u}_{\text{EP}} = \vec{u}_i$, we obtain:

$$\frac{\partial c_i}{\partial t} = \mathcal{D}_i\nabla^2 c_i - \nabla \cdot \left(\vec{u}_i c_i\right) \tag{3.73a}$$

or

$$\frac{\partial c_i}{\partial t} + \vec{u}_i \cdot \nabla c_i + c_i\nabla \cdot \vec{u}_i = \mathcal{D}_i\nabla^2 c_i \tag{3.73b}$$

In more compact form:

$$\frac{\partial c_i}{\partial t} + c_i\nabla \cdot \vec{u}_i =_i \mathcal{D}\nabla^2 c_i \tag{3.74}$$

Clearly, Eq. (3.74) can be interpreted as the conventional "transient convection-diffusion" equation with an extra transport term, i.e.,

$$c_i\nabla \cdot \vec{u}_{\text{EP}} = -c_i\mu_{\text{EP}}\nabla^2\phi \tag{3.75}$$

3.2.5 Magnetohydrodynamics

Overview. When describing the flow of an electrically and/or magnetically conducting fluid of velocity $\vec{v}$, say, electrolytes, charged particle/magnetic suspensions, and/or liquid metals, the Navier-Stokes equation carries additional body force terms, summarized as the electromagnetic force $\vec{f}_{EM}$. Assuming that the externally imposed magnetic field $\vec{B}$ is unperturbed by the flow, i.e., $\text{Re}_{magnetic} \ll 1$ (see Eq. (3.91)), we have (Griffith, 1999):

$$\vec{f}_{EM} = \vec{f}_{electrostatic} + \vec{f}_{magnetic} + \vec{f}_{\underset{\text{-phoretic}}{\text{magneto}}} = \vec{f}_{Lorentz} + \vec{f}_{m-p} \qquad (3.76a,b)$$

By definition, the Lorentz force is the force on a point charge of a moving particle due to electromagnetic fields, i.e., caused by the induced current density $\vec{J}$ and external fields $\vec{B}$ and $\vec{E}$:

$$\vec{f}_L = \vec{J} \times \vec{B} \; ; \; \vec{J} = \sigma\left(-\nabla\phi + \vec{v} \times \vec{B}\right) \; ; \; -\nabla\phi = \vec{E}_{ext.} \qquad (3.77a\text{-}c)$$

In case the ions or fluid particles are magnetic, we have to consider:

$$\vec{f}_{m-p} = \frac{\Delta\chi}{\mu_0}\left(\vec{B} \cdot \nabla\right)\vec{B} \qquad (3.78)$$

where $\Delta\chi = \chi_{particle} - \chi_{fluid}$ is the difference in magnetic susceptibilities and permeability $\mu_0 = 4\pi \times 10^{-7}$ Tm/A.

Evaluation of the electrostatic (i.e., Coulomb) force for an electrolyte solution with n ionic species of concentrations $c_i (i = 1, \ldots, n)$ requires first the solution of the Nernst-Planck equation (see Eq. (3.72)) because

$$\vec{f}_{Coulomb} = F\nabla\phi \cdot \sum_{i=1}^{n} z_i c_i \qquad (3.79)$$

Here z_i is the valence of the i_{th} ion species, F is the Faraday constant, and ϕ is the applied electric potential. Clearly, the applied electric potential and magnetic field are uniform, $\vec{f}_{Coulomb}$ and $\vec{f}_{m-p}$ are zero, respectively.

In summary, the system of the extended Navier-Stokes equations read:

$$\nabla \cdot \vec{v} = 0 \qquad (3.80)$$

$$\frac{\partial \vec{v}}{\partial t} + \left(\vec{v} \cdot \nabla\right)\vec{v} = -\frac{1}{\rho}\nabla p + \nu\nabla^2\vec{v} + \vec{f}_{EM} \qquad (3.81)$$

$$\rho c_p \left[\frac{\partial T}{\partial t} + (\vec{v} \cdot \nabla) T \right] = k \nabla^2 T + \frac{1}{\sigma_{el}} \vec{J} \cdot \vec{J} + \mu \Phi \tag{3.82}$$

where σ_{el} is the electric conductivity of the solution, the term $\sigma_{el}^{-1} \vec{J} \cdot \vec{J}$ indicates possible Joule heating, and $\mu \Phi$ is the viscous dissipation function.

Charged Particle in a Magnetic Field. As a charged plastic rod produces an electric field $\vec{E}$ around the rod, permanent magnets (or electromagnets) generate a magnetic field $\vec{B}$ at all points in the space around them. One is familiar with a permanent magnet which holds up pictures on a refrigerator door and with an electromagnet, i.e., when an electric current is sent through a wire coil wound around an iron core, which can collect, sort, and transport scrap mental. Thus, magnetic fields come about in two ways: (i) elementary particles (e.g., electrons) have an intrinsic magnetic field around them adding up to a net magnetic field in certain materials such as permanent magnets and (ii) moving electrically charged particles (e.g., due to a current in a wire or ions in an electrolyte) create magnetic fields.

As shown in Sect. 3.2.4.1, a particle of charge q in an electric field $\vec{E}$ experiences an electrostatic (Coulomb) force $\vec{F}_C = q\vec{E}$. Somewhat similarly, when a charged particle, i.e., an ion of charge q, subject to a magnetic field $\vec{B}$ moves with velocity $\vec{v}$, a magnetic force $\vec{F}_B$ acts on that particle:

$$\vec{F}_B = q \left(\vec{v} \times \vec{B} \right) \tag{3.83}$$

Its magnitude is:

$$F_B = |q| vB \sin \phi \tag{3.84}$$

where ϕ is the angle between the magnetic field $\vec{B}$ and the direction of velocity vector $\vec{v}$. Clearly, for $\vec{B} \perp \vec{v}$ (i.e., $\phi = 90°$) F_B has its maximum value and it always acts perpendicular to $\vec{v}$ and $\vec{B}$. For example, a charged particle in an axial flow field subject to a magnetic field, acting in a horizontal plane normal to the velocity, may get deflected downwards (see Example 3.9).

Example 3.9: Force on and Acceleration of a Proton Subject to a Uniform Magnetic Field

Consider a uniform magnetic field $|\vec{B}| = 1.2 \, \text{mT} = 1.2 \times 10^{-3} \, \text{N}/(\text{A} \cdot \text{m})$ directed perpendicular to a channel. A proton $(q = 1.6 \times 10^{-19} \text{C} \, ; \, m = 1.67 \times 10^{-27} \, \text{kg})$ enters the horizontal channel with $E_{kin} = 8.48 \times 10^{-13}$ Nm. What is the magnetic force deflecting the particle and what is its acceleration? Indicate the particle trajectory.

Sketch:	Assumptions/Postulates:	Concepts:
N Magnet + $\vec{v}$ $\vec{B}$ m,q S Magnet	• Ignore viscous, gravity, and earth magnetic effects, i.e., $\vec{F}_B$ is the only force acting on the proton	• $\vec{F}_B$ is in the direction of the cross product $\vec{v} \times \vec{B}$ • $E_{kin} = m/2 \cdot v^2$

Solution:

$$\left|\vec{F}_B\right| = qvB \sin \phi; v = \left(2E_{kin}/m\right)^{1/2} := \sqrt{\frac{16.96 \times 10^{-13}}{1.67 \times 10^{-27}}} := 3.2 \times 10^7 \text{ m/s}$$

With $v = 3.2 \times 10^7$ m/s and $\phi = 90°$,

$$F_B = 1.6 \times 10^{-19} \times 3.2 \times 10^7 \times 1.2 \times 10^{-3} \cdot 1 = 6.1 \times 10^{-15} \text{ N}$$

which appears to be miniscule. However, with $m_{proton} = 1.67 \times 10^{-27}$ kg, the steady particle acceleration is:

$$a = \frac{F_B}{m} = 3.7 \times 10^{12} \text{ m/s}^2$$

Graph (side view):

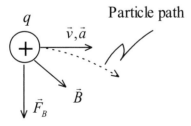

Electromagnetism Background.   As mentioned, the coupled interaction of a magnetic field (or flux density) $\vec{B}\left[\text{N}/(\text{A}\cdot\text{m})\right]$ and a conducting flow field of velocity $\vec{v}$, being subject to an electric current $\vec{I}\left[\text{A}\right]$ (or current flux $\vec{J}\left[\text{A}/\text{m}^2\right]$), gives rise to the Lorentz force moving the fluid. Thus, for example, a strong electrolyte can be pumped through a microchannel when equipped with magnets and electrodes. Such a basic system is a microfluidics example of *magnetohydrodynamics (MHD)*. As outlined in Davidson (2001), this fluid transport process consists of three steps:

(i)   The relative movement of a conducting fluid and a magnetic field causes an electromagnetic field (EMF $\sim\left|\vec{v}\times\vec{B}\right|$) due to induction (Faraday's law), which in turn is causing an electrical current flux, i.e., $\vec{J}=\vec{I}/A\sim\sigma_{el}\left(\vec{v}\times\vec{B}\right)$, where σ_{el} is the electrical conductivity.

(ii)   The induced current flux (or density) generates a second, induced magnetic field (Ampere's law).

(iii)   The two magnetic fields (i.e., applied and induced) combine and interact with the electrical current in the conducting fluid stream, resulting in the Lorentz force $\vec{F}_L/\forall=\vec{f}_L=\vec{J}\times\vec{B}$.

Clearly, the moving conductor (e.g., an electrolyte) can "drag" magnetic field lines (see Step (ii)), and the magnetic field can pull on (i.e., pump) the conducting liquid. If in addition an electric field $\vec{E}\left[\text{V}/\text{m}\right]$ is applied, the current flux (or density) reads (Ohm's law):

$$\vec{J}=\sigma_{el}\left(\vec{E}+\vec{v}\times\vec{B}\right);\ \vec{E}=-\nabla V \tag{3.85a,b}$$

where $\left[\vec{E}+\left(\vec{v}\times\vec{B}\right)\right]$ is the total electromagnetic force per unit charge q. Figure 3.20 depicts schematically a moving conductor, i.e., a collection of charged fluid elements/particles/ions, subject to both electric and magnetic fields.

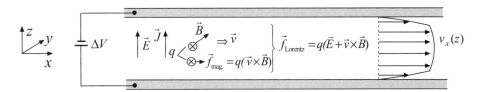

Notes:
- carrier of charge q subject to an electric field $\vec{E}$ and a magnetic field $\vec{B}$ causing Lorentz force $\vec{f}_L=\vec{F}_L/\forall$
- $\vec{f}_{mag.}=q(\vec{v}\times\vec{B})$ is the magnetic force on a single particle with charge q
- permanent magnets at $y=-w/2$ and $y=+w/2$, where w is the channel width

Figure 3.20   Schematic of MHD pumping process

As shown in Sect. 3.2.4, a conducting liquid (say, salt water) subject to an (external) electric field generates a body force (named after Coulomb) which has to be included in the momentum equation. Similarly, a magnetohydrodynamic force (named after Lorentz) is imposed when an electrolytic fluid of charge q flows with velocity $\vec{v}$ in an electric field $\vec{E}$ and a magnetic field $\vec{B}$. As outlined, the Lorentz force per unit volume is:

$$\vec{f}_{\text{Lorentz}} = q\left(\vec{E} + \vec{v} \times \vec{B}\right) \tag{3.86a}$$

In case $\vec{E} = 0$ and using now the current flux (or density) $\vec{J} = \sum q\vec{v}$ in $\left[\text{A}/\text{m}^2\right]$, which of course differs from Eq. (3.77b) or Eq. (3.85a), we have the magnetic body force acting in the flow field as:

$$\vec{f}_{\text{mag.}} = \vec{J} \times \vec{B} \left[\frac{\text{N}}{\text{m}^3}\right] \tag{3.86b}$$

Maxwell Equations.   It should be recalled that a magnetic field is induced by the free electric current $\vec{J}_f$ causing the local magnetization $\vec{M}$ and the bound electric current $\vec{J}_b$, which is related to the magnetic field strength $\vec{H}$. The Maxwell equations describe the magnetic field as (Davidson, 2001):

$$\nabla \times \vec{H} = \vec{J}_f \text{ and } \nabla \times \vec{M} = \vec{J}_b \tag{3.87a,b}$$

$$\nabla \times \vec{B} = \mu_0\left(\vec{J}_f + \vec{J}_b\right) = \mu_0\vec{J} \text{ and } \nabla \cdot \vec{B} = 0 \tag{3.88a-c}$$

Combining Eqs. (3.87a,b), the magnetic field can be expressed as:

$$\vec{B} = \mu_0\left(\vec{H} + \vec{M}\right) = \mu_0(1-\kappa)\vec{H} \tag{3.89a,b}$$

The flux density $\vec{B}$ is measured in Tesla $\left[\text{T} = \text{N}/(\text{m} \cdot \text{C}/\text{s}) \equiv \text{N}/(\text{m} \cdot \text{A})\right]$, the magnetic field strength $\vec{H}$ in $\left[\text{A}/\text{m}\right]$, the local magnetization $\vec{M}$ in $\left[\text{A}/\text{m}\right]$, and the (vacuum) permeability μ_0 in $\left[\text{N}/\text{A}^2\right]$. The material parameter κ (or χ) is the susceptibility, where for paramagnets $\kappa > 0$ and for ferromagnetic materials $\kappa \gg 1$. In case $\vec{E} = 0$ and $\vec{J}_b = 0$, but a magnetic field of magnetic energy density $e_m = -\vec{M} \cdot \vec{B}$ exists, the resulting magnetic force is:

$$\vec{f}_m = -\nabla \cdot e_m = -\nabla\left(\vec{M} \cdot \vec{B}\right) \tag{3.90a,b}$$

MHD Application.   Various dimensionless groups, e.g., the magnetic Reynolds number, Re_m, the Hartmann number, Ha, and the interaction parameter, IP, with their value ranges determine dominant forces in MHD. For example, in micromagnetofluidics:

$$\text{Re}_m = vl/\lambda = \mathcal{O}(10^{-8}) \tag{3.91}$$

where v is the average velocity, l $(\text{say } D_h)$ the characteristic length scale, and $\lambda \equiv (\varepsilon\sigma_{el})^{-1}$ is the magnetic field diffusivity with ε being the permittivity and σ_{el} the electric conductivity. Typical microchannel flow values are $v = 100\,\mu\text{m/s}$, $l = 100\,\mu\text{m}$, and $\lambda = 1\text{m}^2/\text{s}$, which is much higher than momentum $(10^{-6}\ \text{m}^2/\text{s})$ and species $(10^{-9}\ \text{m}^2/\text{s})$ diffusivities (see Nguyen, 2012; among others). An important dimensionless group in MHD is the Hartman number:

$$\text{Ha} = Bl\sqrt{\sigma_{el}/\mu} \tag{3.92}$$

where $(\text{Ha})^2$ is the Lorentz force over the viscous force and μ is the dynamic viscosity. Clearly, as a microchannel's $D_h \equiv\to 0$, $f_{\text{Lorentz}} \to 0$ and the liquid flow ceases.

A basic application of the extended momentum equation, assuming constant properties, is Poiseuille-type flow between parallel plates a distance $2d$ apart. In the absence of an electric field, the N-S equation reads:

$$\rho\frac{D\vec{v}}{Dt} = -\nabla p + \mu\nabla^2\vec{v} + \rho\vec{g} + \vec{f}_{\text{Lorentz}} \tag{3.93}$$

where here $\vec{f}_{\text{Lorentz}} = \vec{f}_{\text{magnetic}} = \vec{J}\times\vec{B}$. At low magnetic Reynolds numbers, i.e., $\text{Re}_{\text{mag.}} = \varepsilon\sigma_{el}vl \ll 1$, a steady magnetic field $\vec{B}_0$ can be assumed and hence

$$\vec{J} = \sigma_{el}\vec{v}\times\vec{B}_0 \tag{3.94}$$

so that in 1-D

$$f_L = \sigma_{el}uB^2 \tag{3.95}$$

where u is the axial velocity. Hence, the N-S equation reduces to (see Figure 3.21):

$$0 = \frac{\Delta p}{L} + \mu\frac{d^2u}{dy^2} - \sigma_{el}uB^2 \tag{3.96}$$

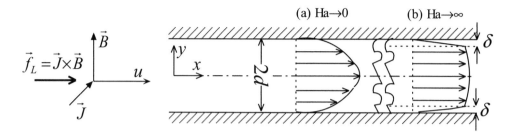

Figure 3.21   Magnetically induced current in the conducting liquid and resulting MHD pumping for low and high Hartmann numbers (Ha = δ/d)

subject to $\left(du/dy \right)\Big|_{y=0} = 0$ and $u(y = \pm d) = 0$. Before solving Eq. (3.96), it should be pointed out that the impact of $\vec{f}_L = \vec{J} \times \vec{B}$ on a conducting liquid causing, say, microchannel flow is actually a "wall shear-layer" effect, somewhat similar to a "slip velocity" as shown for electroosmotic flow (Sect. 3.2.4.1).

Considering *quasi-fully-developed flow of a conducting liquid with uniform approach velocity on a horizontal wall* (i.e., $u_\infty = \cancel{c}$, $\partial p/\partial x = 0$) and subject to a constant magnetic field $B(B \perp u)$, the reduced N-S equation reads:

$$0 = \mu \frac{d^2 u}{dy^2} - \sigma_{el} B^2 u \tag{3.97}$$

Introducing the Hartmann layer

$$\delta = \left[\mu/\sigma_{el} B^2 \right]^{1/2} = l/\text{Ha} \tag{3.98}$$

Equation (3.90) can be rewritten as:

$$\frac{d^2}{dy^2}\left(u - u_\infty \right) - \frac{u - u_\infty}{\delta^2} = 0 \tag{3.99}$$

Double integration subject to $u(y = 0) = 0$ and $u(y \to \infty) \to u_\infty$ yields:

$$u(y) = u_\infty \left[1 - e^{-y/\delta} \right] \tag{3.100}$$

Clearly, high velocity gradients exist near the wall, i.e., within the Hartmann layer $0 \le y \le \delta$ (see Figures 3.21 and 3.22).

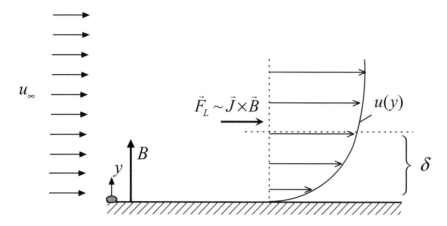

Figure 3.22   Constant shear-layer flow caused by a magnetic field and induced current

In contrast, for *fully developed flow in a duct*, Davidson (2001) showed that

$$u(y) = u_0 \left[1 - \frac{\cosh(y/\delta)}{\cosh(d/\delta)} \right]$$

(3.101)

where

$$u_0 = \frac{\Delta p}{L} \left(\sigma_{el} B^2 \right)^{-1}$$

(3.102)

Recasting the solution in terms of $\mathrm{Ha} = d/\delta = dB \left(\sigma_{el}/\mu \right)^{1/2}$ we have:

$$u(y) = u_0 \left[1 - \frac{\cosh\left[(\mathrm{Ha}) y/d \right]}{\cosh(\mathrm{Ha})} \right]$$

(3.103)

This allows the emergence of two limiting velocity profiles (see Figure 3.20):

$$u(y) = u_0 \left[1 - \left(\frac{y}{d} \right)^2 \right] \text{ for } \mathrm{Ha} \to 0 \quad \text{(Poiseuille flow)}$$

and

$$u(y) \approx u_0 \text{ for } \mathrm{Ha} \to \infty \qquad \text{(plug flow)}$$

When including a constant applied electric field E_0, the reduced N-S equation reads:

$$0 = -\frac{\partial p}{\partial x} + \mu \frac{d^2 u}{dy^2} - \sigma_{el} B^2 u - \sigma_{el} B E_0 \tag{3.104}$$

Equations (3.88) and (3.89) hold again when defining u_0 as:

$$u_0 = \frac{\Delta p}{L}(\sigma_{el} B)^{-1} - \frac{E_0}{B} \tag{3.105}$$

MHD Devices. Clearly, for devices based on the MHD phenomenon only large Ha number flows are of interest. Specifically, with the results:

$$u(y) \approx u_0; \ u_0 B = \frac{1}{\sigma_{el} B} \frac{\Delta p}{L} - E_0 \ \text{and} \ J = \sigma_{el}(E_0 + u_0 B)$$

we can freely set E_0-values and hence generate different effects/devices. Here are two scenarios:

(i) $E_0 = -u_0 B$, i.e., $J = 0$, $f_{\text{Lorentz}} = 0$, and hence no pressure gradient is created. Measuring E and knowing B we can determine u, i.e., we have a flow meter (see Figure 3.23).
   Note: When $E_0 = 0$, and hence $J = \sigma_{el} u_0 B$ and $|\Delta p / L| = u_0 \sigma_{el} B^2 \approx BJ$, a current is induced, but associated with a Lorentz force $(J \times B)$ opposite to the pressure gradient and the moving fluid.

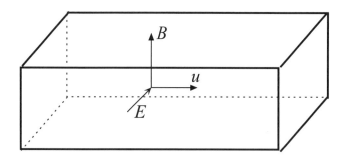

Figure 3.23 MHD flow meter

(ii) If the supplied $E^0 < 0$ and $|E_0| > u_0 B$, $|\vec{J}| < 0$ and hence $\vec{J} \times \vec{B}$ is being reversed, acting in the flow direction. The resulting $\Delta p / L = u_0 \sigma_{el} B^2 - |\sigma_{el} B E_0| = \sigma_{el} B[u_0 B - |E_0|]$ is negative and the liquid is being pumped (see Figure 3.24).

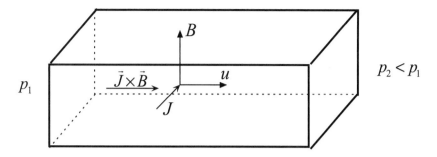

Figure 3.24   MHD pump

Magnetophoresis. In a magnetic field gradient, $\nabla \vec{B}$, a magnetic (spherical) particle, driven by the magnetic force $F_m \sim \Delta\chi (\vec{B} \cdot \nabla) \vec{B}$, moves along the gradient. Specifically,

$$\vec{F}_m = \frac{\forall_p (\chi_p - \chi_f)}{\mu_0} (\vec{B} \cdot \nabla) \vec{B} \tag{3.106a}$$

or

$$\vec{F}_m = \mu_0 \forall_p (\vec{M}_p \cdot \nabla) \vec{H}_a \tag{3.106b}$$

where $\forall_p$ is particle volume, $\chi_{p,f}$ are the particle and fluid susceptibilities, $\mu_0 = 4\pi \times 10^{-7} \text{ N/A}^2$ is the vacuum permeability, and $\vec{M}_p \stackrel{\frown}{=}$ particle magnetization and $\vec{H}_a \stackrel{\frown}{=}$ applied magnetic field strength, both in A/m.

Note: Neglecting magnetic (spherical) particle acceleration in Stokes flow, and considering only magnetic and drag forces in dynamic equilibrium, i.e., $F_{\text{drag}} = F_{\text{mag}}$, we have:

$$3\pi\mu d_p \vec{v}_p = \frac{\pi d_p^3}{6\mu_0} (\chi_p - \chi_f)(\vec{B} \cdot \nabla) \vec{B} \tag{3.107}$$

from which the particle velocity vector can be deduced:

$$\vec{v}_p = \frac{d\vec{x}_p}{dt} = \frac{\zeta}{\mu_0} (\vec{B} \cdot \nabla) \vec{B} \tag{3.108}$$

where $\zeta \equiv d_p^2 / 18\mu (\chi_p - \chi_f)$ is the *magnetophoretic mobility*.

Clearly, with $\zeta = \zeta\left(d_p, \Delta\chi\right)$ particles can be separated/sorted according to size and magnetic type, i.e., paramagnetic where $\chi_p > 0$ and diamagnetic where $\chi_p < 0$.

To solve for the particle velocity and then obtain its trajectory, a model for the magnetic response of the particle is needed (see Furlani, 2001). Recalling that

$$\vec{B} = \mu_0\left(\vec{H} + \vec{M}\right) = \mu_0\left(1 + \chi\right)\vec{H} \tag{3.109}$$

a linear magnetization model with saturation is assumed. As outlined by Furlani and Ng (2006), with $\mu_{\text{fluid}} \approx \mu_{\text{vacuum}}$ we have:

$$\vec{F}_m = \mu_0 \forall_p \frac{3\left(\chi_p - \chi_f\right)}{3 + \left(\chi_p - \chi_f\right)}\left(\vec{H}_a \cdot \nabla\right)\vec{H}_a \tag{3.110}$$

The necessary applied field distribution $\vec{H}_a(\vec{x})$ and its gradient field have to be set up on a system-specific basis (see Sect. 3.3 and Example 4.2).

3.3 Homework Assignments

In this section two types of suggested homework problems, i.e., text-illuminating questions (Sect. 3.3.1) and basic microscale engineering problems (Sect. 3.3.2), are given. The problems in Sect. 3.3.2 were taken in modified and extended form from the open literature as well as previous homework and test assignments of the author's fluid mechanics courses.

3.3.1 Physical Insight

3.1 Using Figure 3.1 as a guide, develop a flowchart for a microfluidic system, such as a lab-on-a-chip (see Figure 3.2), or a MEMS (e.g., ink-jet printer), or a PoC (e.g., blood analysis).

3.2 Select a specific microfluidics system and provide images and descriptions of major system aspects from fabrication to application, as given in Figure 3.3.

3.3 Discuss LoC versus μ-TAS functions and applications.

3.4 Concerning Sect. 3.1.3 and Table 1.1, select five bulleted problem areas and provide real-world microsystem examples.

3.5 Most dimensionless groups discussed in Sect. 3.2.1 scale with l^2, i.e., appearing to be "irrelevant as $l \to 0$." However, show that when focusing on individual forces, fluxes, time-scales or transport phenomena, Eqs (3.6) to (3.11) are most useful for micro/nanosystem analysis.

3.6 Redo Example 3.2 for: (a) momentum transfer Eq. (1.80); (b) heat transfer Eq. (1.85); (c) species mass transfer Eq. (1.96); and (d) particle dynamics Eq. (2.49).

3.7 Contrast the Strouhal number $\text{Str} = fL/v$ to the Womersley number $\text{Wo} = L\left(\omega\rho/\mu\right)^{1/2}$ and provide applications.

3.8 Considering Table 3.3, illustrate and discuss one flow device with important applications on all three geometric system scales, i.e., from macro to nano.

3.9 Show that for steady fully developed flow in an ellipsoidal tube (see $u(y,z)$ in Table 3.4) the hydraulic resistance is again $R_{\text{hydr.}} \sim l^{-3}$.

3.10 Equation (3.18) provides definite gas flow modeling regimes in terms of the Knudsen number $\text{Kn} = \lambda_{\text{mfp}}/l_{\text{system}}$. Why isn't there a sound theory for liquid flows in micro/nanoconduits? Discuss ideas for developing such liquid flow modeling regimes.

3.11 Obtain a specific data set for argon from the open literature and redo Example 3.5.

3.12 Review gas versus liquid properties (i.e., density, viscosity, and conductivity) contrasting macroscale with nanoscale phenomena and resulting property values.

3.13 Elaborate on the nanoscale physics of Eq. (3.34a).

3.14 Research the differences between electrophoresis and dielectrophoresis.

3.15 Research the impact of the diffuse layer on electroosmosis in microchannel flow.

3.16 Derive, illustrate, and explain Eqs. (3.41b), (3.42b), (3.45c), and (3.46).

3.17 Derive, illustrate, and explain Eqs. (3.50) and (3.52).

3.18 Redo Example 3.7 for $P<5$, i.e., using Eq. (3.57b).

3.19 Review the paper by Dutta and Beskok (2001), derive Eq. (3.63) and contrast it to Eq. (3.57a).

3.20 Review the paper by Horiuchi and Dutta (2004) and derive/discuss Eq. (3.64). Cite practical applications where $\kappa > 1$, i.e., Joule heating is important.

3.21 Research the Nernst-Planck Equation and provide an illustrative (i.e., tutorial) example (see Sect. 3.2.4).

3.22 Research magnetohydrodynamics and provide an illustrative (i.e., tutorial) example (see Sect. 3.2.5).

3.23 Provide from the open literature two tutorial examples contrasting the cases of $\mathrm{Ha} \to 0$ and $\mathrm{Ha} \to \infty$ (see Figure 3.21).

3.24 Provide from the open literature a tutorial example of an MHD pump (see Figure 3.24).

3.25 Research and derive Eqs.(3.107) to (3.110).

3.3.2 Text Problems

3.26 Consider a quasi-spherical drop (D_d, σ, θ) about to fall from the tip of a capillary (D_c). It is assumed that the contact angle $\theta = 0$ when the drop actually falls. Based on a 1-D force balance, i.e., $F_z \cos\theta = (\sigma \pi D_c)\cos\theta = m_d g$, find the falling drop volume $\forall_{max}$ and an expression for the diameter ratio D_d / D_c. Introducing the Bond number $\left(\mathrm{Bo} \equiv D_c^2 \rho g / \sigma\right)$, show that $D_d / D_c = 1.82\mathrm{Bo}^{-1/3}$.

3.27 As a correction to $\forall_{max}$ of problem 3.26, Harkins and Brown (1919) suggested $\forall_{drop} = \forall_{max} C$, where $C = 0.6 + 0.4\left(1 - 0.488\mathrm{Bo}^{1/3}\right)^{2.2}$. Estimate $\forall_{drop}$ for $D_c = 1\,\mathrm{mm}$, $\sigma = 0.06\,\mathrm{N/m}$, and $\rho = 1000\,\mathrm{kg/m}^3$.

3.28 Consider spherical microbubble growth (radius R) in a very viscous fluid.

   (a) Using the reduced continuity equation and the dynamic boundary condition $v_r \left[r = R(t)\right] = dR/dt \equiv \dot{R}$, show that $v_r(r) = \dot{R}R^2 / r^2$.

   (b) With only the pressure and viscous forces being dominant show that the internal bubble pressure is:

$$p(t) = p_\infty + 4\mu \dot{R}R$$

   (c) Find $R(t)/R_0$, where $R_0 = R(t=0), \Delta p = p - p_\infty$ and compute/graph $R(t)/R_0$ for $R_0 = 100\mu\mathrm{m}$, $\Delta p = 900\mathrm{kPa}$, $\rho = 1000\mathrm{kg/m}^3$, and $\mu = 100\mathrm{Pa}\cdot\mathrm{s}$

3.29 Consider a rod $(\kappa R, L)$ falling with velocity v_0 in a tall, liquid-filled cylinder (radius R) with a closed bottom. Assuming axisymmetric steady laminar fully developed flow between these cylinders, find the velocity distribution $v_z(r)$ in the microgap.

3.30 Consider radial flow between horizontal parallel microdisks of inner radius R_i, outer radius R_o, and spacing $2h$. Given inlet flow rate $Q = 4\pi R_i h u_{mean}$, show that:

(a) $v_r(r,z) = \left(\dfrac{3Q}{8\pi h} \right) \dfrac{1}{r} \left[1 - \left(\dfrac{z}{h} \right)^2 \right]$

(b) $\Delta p = -\dfrac{3\mu Q}{4\pi h^3} \ln\left(R_i / R_o \right)$

(c) Plot v_r and Δp using the base values $Q = 10^{-4}\,\mathrm{m}^3/\mathrm{s}$, $R_i = 5\mathrm{mm}$, $R_o = 50\mathrm{mm}$, $h = 200\mu\mathrm{m}$, $\rho = 1000\,\mathrm{kg/m}^3$, and $\mu = 10\mathrm{Pa}\cdot\mathrm{s}$. Check if $\mathrm{Re}_h < 1.0$!

3.31 Consider planar flow in a microchannel (h, L) with the bottom wall being porous where fluid is being withdrawn at velocity as $v_{wall}(x)$.

(a) Show that for $\varepsilon \equiv h/L \ll 1$ and $\mathrm{Re}_h = u_{in} h/\nu \leq 1.0$ (Stokes flow) the N-S equations reduced to:

$\dfrac{\partial u}{\partial x} + \dfrac{\partial v}{\partial y} = 0; \dfrac{\partial p}{\partial y} = 0;$ and $0 = -\dfrac{\partial p}{\partial x} + \nu \dfrac{\partial^2 u}{\partial y^2}$

(b) Using the dimensionless variables $\hat{x} = x/L, \hat{y} = y/h, \hat{u} = u/u_{in}, \hat{v} = v/(\varepsilon u_{in})$, and $\hat{p} = (p - p_{ref.})/[\mu u_{in}/(\varepsilon h)]$, show that $\hat{u}(\hat{x}, \hat{y}) = 6\hat{u}_{avg.}(\hat{x})[\hat{y} - \hat{y}^2]$, where $\hat{u}_{avg.}(0) = 1$.

(c) Employing the continuity equation, show that $\hat{v}(\hat{x}, \hat{y}) = v_w(\hat{x})[1 - 3\hat{y}^2 + 2\hat{y}^3]$.

(d) Plot the velocity profiles and comment.

3.32 Assume Poiseuille flow with velocity slip between parallel plates a distance h apart or in a tube of radius r_0.

(a) Show that $u(y) = 4u_c\left[(y/h) - (y/h)^2\right] + u_{slip}$ for microtube flow. Interpret u_c and v_c!

(b) Derive expressions for the mass flow rates $\dot{m}_{plate}$ and $\dot{m}_{tube}$, plot $\dot{m}_p(h)$ and $\dot{m}_t(r_0)$, and comment.

3.33 Consider thermal micro-Couette flow of a gas $(U, h; \Delta p = 0; T(y=0) = T_w = \cent$ and $q(y=h) = q_w = 0)$ in the slip flow regime $0.01 < \mathrm{Kn} \equiv \lambda/2h < 0.1$. Assume 1-D fluid flow and only thermal diffusion with viscous heating.

(a) Show that, indeed, $\mathrm{Kn} = \lambda/2h$.

(b) Derive the velocity profile and temperature distribution.

(c) Graph $T(y/h)$ for $U = 100\,\mathrm{m/s}$, $T_{wall} = 300\mathrm{K}$, $h = 500\mu\mathrm{m}$, $\kappa = 1.4$, and $\mathrm{Pr} = 0.7$ by selecting Kn=0.0, 0.01, 0.1

(d) Comment and explain!

3.34 Consider electroosmotic microchannel flow characterized by h, λ_D, μ, ε, ζ, and E_x. Derive and graph $\hat{u} = u(y)/u_{\text{bulk}}$ as a function of y/λ_D for two different sets of BCs with regard to the linearized $\phi(y)$-distribution. Comment!

3.35 Consider pressure-driven flow $(\Delta p/L = ¢)$ of a particle suspension, which can be described by a power law fluid (m, n) in a microchannel of height of $2h$.

(a) Obtain the velocity profile and check your answer for $n = 1$ and $m = \mu$, i.e.,

$$u(y) = \frac{h^2}{2\mu}\left(\frac{\Delta p}{L}\right)\left[1 - \left(\frac{y}{h}\right)^2\right]$$

(b) Graph $u(y)$ for $n = 0.2$, 1, where u_{max} equals 0.8 and 1 m/s, respectively.

3.36 Consider a tube of diameter $D = 5\mu\text{m}$ and length $L = 10$ cm, filled with an electrolyte $(\varepsilon = 7.08 \times 10^{-10}\,\text{C}/(\text{V}\cdot\text{m})$; $\mu = 10^{-3}\,\text{Pa}\cdot\text{s})$. The wall potential, i.e., approximately the zeta potential is $\zeta = -100\text{mV}$, based on a voltage of 1 kV which is applied across the tube length. Calculate the pressure drop required to achieve the same volumetric flow rate as Q_{EO}. Plot the velocity profiles and comment!

CHAPTER **4**

Applications in Microfluidics

4.1 Introduction

Fluid flow and particle transport in microchannels, being part of basic MEMS, lab-on-chip (LoC), and point-of-care (PoC) devices, generally deal with aqueous solutions and nano/microsphere suspensions which require micropumps. Thus, relying on Part A and Chapter 3 material, the applications of fluid-particle flow in microchannels and the workings of micropumps are illustrated.

4.2 Micropumps and Microchannel Flow

Continuous flow and discrete (i.e., slug or droplet) flow in microconduits, such as channels, tubes, and slits, as well as associated fluid-particle dynamics, convection-conduction heat transfer, and species convection-diffusion are essential transport phenomena in microfluidics devices. Two standard examples include high-heat-flux removal via an array of microchannels as part of high-power electronic equipment (e.g., computer chips or MEMS) and efficient chemical process and/or biological cell analysis with LoCs. All microfluidics devices require fluid driving forces as summarized in Figure 4.1 and discussed in Sect. 4.2.1.

Section 3.2 contains a few basic sample problem solutions providing background information for Sects. 4.2.2 and 4.2.3, which feature solved examples illustrating liquid and gas flows in microchannels.

4.2.1 Micropumps

Conventional pumps, e.g., centrifugal, rotary, reciprocal-sealed-piston, etc., are less useful in micro/nanofluidics applications. The main reasons include difficulties in microfabrication and unfavorable scaling efficiencies with decreasing Reynolds numbers. Thus, the present R&D focus is on electromechanical membrane displacement and nonmechanical micropumps. They may employ piezoelectric actuators as well as electrodynamic, magnetic, surface tension, or acoustic driving forces, suitable for very small fluid volume as well as continuous-fluid transport over very short distances. Such micropumps are attractive because of the ease of miniaturization, lack of moving parts, and relatively high reliability. Still, conventional (external) micropumps are needed for relatively large Reynolds number flows as well as for non-electrolytic and/or high-resistance working fluids. Prime examples include cooling of heat-generating microscale devices and high-production chemical microreactors.

There are various ways of grouping micropumps. Traditionally they are categorized as mechanical pumps and nonmechanical ones, i.e., those without moving parts. Micropumps utilizing unconventional forces, such as electrostatic, magnetic, or capillary, are of special interest. Nevertheless, linear or nonlinear piston movement in a syringe and vibration of a diaphragm (or flexible membrane), as part of a fluid chamber with synchronized valves, are basic examples of "displacement" pumps, set into motion by motors or piezoelectric actuators. Alternatively, fluid driving forces, and hence pressure differentials, can be generated via electrohydrodynamics (EHD), electroosmosis (EO), or magnetohydrodynamics (MHD), as summarized in Figure 4.1.

An alternative classification for pumping methods in microfluidics is *on-device pumps* and *external pumps*. The first category includes piezoelectric, electrostatic, electromagnetic, acoustic, and rotary micropumps which are fabricated with the device itself, i.e., built in. External pumping is typically accomplished with powered syringes.

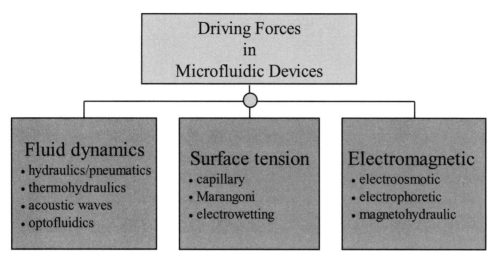

Figure 4.1   Typical driving forces in microfluidic devices

In any case, the overall flow resistance and hence the required pumping power for a given flow rate depend on the following microchannel parameters:

- Geometric features, such as circular, trapezoidal, or rectangular cross sections (captured by the hydraulic diameter $D_h = 4A/P$), channel length, possible entrance/exit effects as well as curvature, bifurcations, and changing cross-sectional areas, and surface roughness, the latter becoming less of a problem due to refined fabrication techniques.
- The flow regime, i.e., laminar vs. transitional vs. turbulent flow.
- The type of working fluid, e.g., dense particle suspensions, shear-rate-dependent fluids, or slippage over rough and hydrophobic solid surfaces.

Assuming that the continuum theory is valid for microscale flow (see Table 3.5), the pumping power can be expressed as:

$$P_{\text{pump}} \equiv \dot{W} = \rho g Q h_{\text{pump}} / \eta_{\text{pump}} \tag{4.1}$$

where Q is the volumetric flow rate, η_{pump} is the efficiency, and $h_{\text{pump}} \sim h_{loss}$ can be obtained from the extended Bernoulli equation. For example, for tubular flow (i.e., average velocity v, diameter D, length L) without elevation changes Δz:

$$h_{\text{loss}} = \frac{f \cdot L}{D} \cdot \frac{v^2}{2g} \tag{4.2a}$$

where the Darcy friction factor can be expressed as:

$$f = \frac{2D_h}{\rho v}\left(\frac{dp}{dx}\right)$$
(4.2b)

Here, the pressure gradient is a function of μ, v (or Q), and D_h. Very useful for laminar flow comparisons is the Poiseuille number $Po = f\,Re$ (see Tables 1.3 and 4.1). Interestingly enough, Hetsroni et al. (2011) compared experimental and theoretical Poiseuille flow results for both macroconduits and microconduits. They concluded that for both conduit size categories the Poiseuille number, axial velocity distribution, and critical Reynolds number are the same, i.e., independent of duct size. For rough pipes and channels, the Poiseuille number increases measurably for both macro- and microconduits. For fully developed *turbulent* flow in smooth conduits of cross-sectional area A and for a Reynolds number based on the length scale $\sqrt{A}$, Duan (2012) proposed:

$$f\,Re_{\sqrt{A}} = \left[3.6\log\left(\frac{6.115}{Re_{\sqrt{A}}}\right)\right]^{-2} Re_{\sqrt{A}}$$
(4.3)

   In summary, micropumps generate a pressure differential causing fluid flow or finite volume/boundary displacements. Examples include diaphragm displacement or vibration, peristaltic pumps as well as ferrofluid, phase-change, and gas boundary pumps.

Diaphragm Displacement Pump.   Quite attractive are diaphragm micropumps made out of elastic composite/polymer materials and characterized by low power requirements (see Figure 4.2). The actuator (or driving force) for diaphragm displacement can be piezoelectric, electrostatic, magnetic, thermal, or pneumatic.

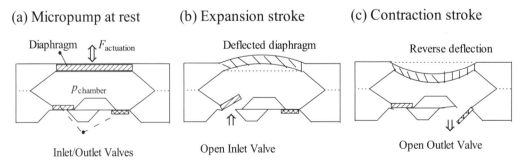

(a) Micropump at rest   (b) Expansion stroke   (c) Contraction stroke

Figure 4.2   Diaphragm displacement pump

When the fluid inlet pressure is higher than the chamber pressure, i.e., $p_{in} > p_{chamber}$, the inlet valve opens and the expanding chamber deflects the membrane (Figure 4.2b). Triggered by thermo/pneumatic or electro/magnetic actuation, the diaphragm (or membrane) contracts, pushing the liquid through the open outlet valve (Figure 4.2c). In order to obtain flow rectification from oscillatory membrane motion, actively controlled valves are incorporated into the micropump design and hence its operation.

Magnetohydrodynamic (MHD) Pumps.   MHD pumps rely on the Lorentz force that is induced when a conducting fluid, subject to an electric current I, is placed in a magnetic field of strength B (see Sect. 3.2.5). For example, a rectangular microchannel of width w and length l is sandwiched between two electrodes and two magnets. The resulting force moving a liquid with current-carrying ions is perpendicular to both the electric and magnetic fields (see Figure 4.3). Specifically, an external electric field, $\vec{E} = -\nabla V$, causes the current across the microchannel between the electrodes, while electromagnets generate the magnetic field, i.e., $\vec{B} \perp \vec{I}$.

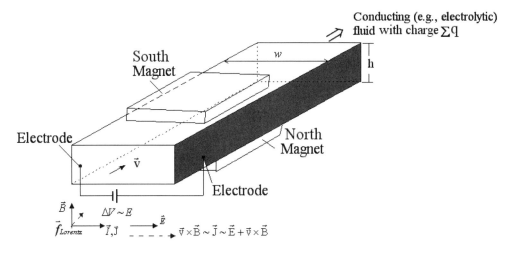

Figure 4.3   MHD micropump driving a conducting fluid in a microchannel of $\forall = w \times h \times l$

Based on vector analysis (see also Sect. 3.2.5),

$$\vec{F}_{mag.} = \vec{J} \times \vec{B}(w \cdot h \cdot l) \tag{4.4}$$

In terms of the current density $\vec{J} = \sum q\vec{v}$, where q is the charge of the fluid of velocity $\vec{v}$, the maximum pressure across microchannel length l, i.e., effective pressure drop, is $(\vec{E} = 0)$:

$$p_{max} = \left| \vec{J} \times \vec{B} l \right| \tag{4.5}$$

Thus, using the Hagen-Poiseuille solution of Chapter 2, the associated flow rate is:

$$Q = \left| \vec{J} \times \vec{B} \right| \frac{\pi D_h^4}{128 \mu} \tag{4.6}$$

where for a rectangular microchannel $\left| \vec{J} \times \vec{B} \right| = J_x B_y \triangleq \Delta p / L$ (see Sect. A.1.2 for expanding a vector cross product). Equation (4.6) should be compared to the Hagen-Poiseuille solution given in Chapters 1 and 2. As mentioned, the strength of the magnetic field, e.g., 1 T equals $1\ N/(m \cdot A)$, may cause Joule heating (see Eq. (3.75)). Also, the scaling with the conduit's hydraulic diameter, i.e., $B \sim D_h^4$, limits the applicability of MHD micropumps.

Magnetic particles in suspensions, e.g., ferromagnetic nano/microparticles in liquids, have numerous microfluidics applications in MEMS as well as bio-MEMS, as reviewed by Ganguly & Puri (2010), Gijs et al. (2010), and Nguyen (2012) (see Example 4.2).

Example 4.1: MHD Pumping Illustration

Consider a microchannel ($h = 400\ \mu m$, $w = 1$ mm, $L = 40$ mm) with a conducting liquid influenced by uniform magnetic and electric fields. Assuming a negligible magnetic Reynolds number (i.e., $Re_m = \sigma_{el} \mu_{mag} U L \approx 0$) and constant fluid properties, find the induced pressure drop as a function of applied current and magnetic flux density. A suitable liquid is seawater ($T = 20°C$, $p = 1$ atm), where the density $\rho = 1025\ kg/m^3$, viscosity $\mu = 1.09 \times 10^{-3}\ kg/(m \cdot s)$, electric conductivity $\sigma_{el} = 4\ A/(m \cdot V)$, relative permeability $\mu_{mag} = 1$, and the relative permittivity $\varepsilon = 72$ (see Jang & Lee, 2000).

Sketch:	*Assumptions/Postulates:*	*Concepts:*
	• Poiseuille-type flow • Constant magnetic and electric fields • Constant properties	• Reduced N-S equation • Ohm's law • $f_{Lorentz} \sim \nabla p \approx \dfrac{p_2 - p_1}{L_e}$ • L_e is the electrode length • $\vec{v} = [v(y); 0; 0]$

Note: The direction of $\vec{f}_L \sim \vec{J} \times \vec{B}$ can be reversed by switching the permanent magnets; hence $\vec{B} \rightarrow -\vec{B}$, so that the water is pumped from the inlet to the outlet tube.

Solution:

As shown in Sect. 3.2.5.1, the extended N-S equation (3.86) can be reduced to:

$$0 = \mu \nabla^2 \vec{v} + \vec{J} \times \vec{B} \qquad \text{(E4.1-1)}$$

which is the MHD micropump equation where the Lorentz force $\vec{f}_L \sim \vec{J} \times \vec{B}$ is equivalent to a pressure gradient $-\nabla p$.

For unidirectional flow ∇p is reduced to:

$$\nabla p = -\frac{\Delta p}{L} := -\frac{p_2 - p_1}{L_e} \qquad \text{(E4.1-2a,b)}$$

where L_e is the length of the top/bottom electrodes equal to the channel length, i.e., typically $L_e = L$.

In the present 1-D case, when solving the cross product

$$\vec{J} \times \vec{B} = \begin{vmatrix} \hat{i} & \hat{j} & \hat{k} \\ J_x & J_y & J_z \\ B_x & B_y & B_z \end{vmatrix}$$

only the product $-J_y B_x \hat{k}$ is nonzero. Thus, we can set:

$$\Delta p / L = -J_y B_x \qquad \text{(E4.1-3)}$$

Employing the hydraulic diameter ($D_h = 4A/P$) concept, Eq. (E4.1-1) reduces to:

$$0 = \frac{\mu}{r} \frac{d}{dr} \left(r \frac{dv_z}{dr} \right) - J_y B_x \text{ or } 0 = \frac{\Delta p}{L} + \frac{\mu}{r} \frac{d}{dr} \left(r \frac{dv_z}{dr} \right) \qquad \text{(E4.1-4a,b)}$$

Its solution is given with Example 2.1 where $r_0 \equiv D_h / 2$ and hence:

$$\bullet \ u(r) = u_{max} \left[1 - \left(\frac{r}{r_0} \right)^2 \right]; \ u_{max} = \frac{|J_y B_x|}{4\mu} r_0^2 \qquad \text{(E4.1-5a,b)}$$

$$\bullet \; Q = \frac{\pi r_0^4}{8\mu}\left|J_y B_x\right| \tag{E4.1-6}$$

and

$$\bullet \; \Delta p = L\frac{2\tau_{\text{wall}}}{r_0} \tag{E4.1-7}$$

Comments:

Assuming a constant magnetic flux of, say, $B_x = 0.44$ T and varying the applied current in the range of $0 \le I \le 40$ mA, a linear pressure drop $\Delta p(I)$ is expected. However, when external voltage exceeds 20 V, bubbles may occur which dramatically increase the flow resistance, i.e., Δp, as observed by Jang & Lee (2000), among others.

Example 4.2: Magnetic Particle Transport in Microchannel

Consider an iron oxide $\left(\text{Fe}_3\text{O}_4\right)$ particle $\left(R_p = 250 \text{ nm}, \; \rho_p = 5\times10^3 \text{ kg/m}^3, \; \mu = 1.2\times10^{-3} \text{ N}\cdot\text{s/m}^2\right)$ in water flowing fully developed in a microtube $\left(R = 75 \text{ μm}\right)$ adjacent to which is a stationary cylindrical (bias N-S) magnet of radius R_M located at a variable distance $d = 1+\kappa R_M$ with $0 \le \kappa \le 1$. Along the channel axis the normal and axial force components vary due to the induced magnetic field $\vec{B}$. Releasing nanospheres of different sizes at the microtube inlet, i.e., upstream of the magnet location, of interest are the particle trajectories and capture (i.e., deposition) criteria as a function of magnet distance d and particle radius R_p.

As indicated by Furlani & Ng (2006), among others, this analysis can be relevant to drug targeting of tumors located close to the body's surface.

Sketch:	*Assumptions/Postulates:*	*Concepts:*
	• Poiseuille flow without particle impact • Stokes drag, gravity, and simplified magnetic force components • Spherical particles	• Newton's second law $m_p \, d\vec{v}_p/dt = \vec{F}_D + \vec{F}_M$ • $\vec{F}_D = 6\pi\mu R_p\left(\vec{v} - \vec{v}_p\right)$ • $\vec{F}_M = \text{fct}(\vec{H})$, $\vec{H}$ being the magnetic field distribution

Solution:

- Fluid velocity (from Chapter 2):

$$v = 2v_{\text{avg.}}\left[1 - (r/r_0)^2\right]$$

(E4.2-1)

- Forces:

$$\vec{F}_g = (\rho - \rho_g)\forall_p \vec{g}$$

(E4.2-2)

$$\vec{F}_{\text{drag}} = 6\pi\mu R_p(\vec{v} - \vec{v}_p)$$

(E4.2-3)

and $\vec{F}_{\text{magn.}} \equiv \vec{F}_M$ is given as Eq. (3.110). In component form (see Furlani & Ng, 2006), with respect to the sketch, we have:

$$F_{M,x} = -\mu_0\forall_p \text{fct}(H_a)M_{sp}^2 R_M^4 \frac{(x+d)}{2\left[(x+d)^2 + z^2\right]^3}$$

(E4.2-4)

Equation (E4.2-4) is key to particle capture. In contrast,

$$F_{M,z} = -\mu_0\forall_p \text{fct}(H_a)M_{sp}^2 R_M^4 \frac{z}{2\left[(x+d)^2 + z^2\right]^3}$$

(E4.2-5)

while the drag force decelerates the particle in the axial direction. For magnetite nanoparticles (Fe_3O_4) $M_{sp} = 4.78 \times 10^5$ A/m and assuming $\chi_p \gg 1$,

$$\text{fct}(H_a) = \begin{cases} 3 & \text{for } H_a < M_{sp}/3 \\ M_{sp}/H_a & \text{for } H_a \geq M_{sp}/3 \end{cases}$$

(E4.2-6)

- Newton's second law with $\vec{v}_p = (v_x, v_y, v_z)$:

$$m_p\frac{dv_x}{dt} = -\mu_0\forall_p \text{fct}(H_a)M_{sp}^2 R_M^4 \frac{(x+d)}{2\left[(x+d)^2 + z^2\right]^3}$$
$$-6\pi\mu R_p v_x + (\rho_p - \rho)\forall_p g_x$$

(E4.2-7)

$$m_p\, dv_y/dt = -6\pi\mu R_p v_y := 0$$

(E4.2-8)

$$m_p \frac{dv_z}{dt} = -\mu_0 \forall_p \ \text{fct}\left(H_a\right) M_{sp}^2 R_M^4 \frac{z}{2\left[\left(x+d\right)^2 + z^2\right]^3}$$

$$-6\pi\mu R_p \left[v_z - 2v_{avg.}\left(1 - \frac{x^2 + y^2}{r_0^2}\right)\right]$$

(E4.2-9)

- Introduce the dimensionless parameter,

$$\overline{x} = \frac{x}{r_0}, \ \overline{z} = \frac{z}{\left(1+\kappa\right)R_M}, \ \overline{t} = \frac{2tv_{avg}}{\left(1+\kappa\right)R_M}, \ \overline{v}_x = v_x \frac{\left(1+\kappa\right)R_M}{2v_{avg}R_M}, \ \overline{v}_z = \frac{v_z}{2v_{avg}}$$

- Using these definitions, Eqs. (E4.2-7) and (E4.2-8) can be rewritten in nondimensional form:

$$\overline{m}_p \frac{d\overline{v}_x}{dt} = -C \frac{\beta\overline{x}+1}{\left[\left(\beta\overline{x}+1\right)^2 + \overline{z}^2\right]^3} - \overline{v}_x + \left(\overline{m}_p - \overline{m}\right)g_x$$

(E4.2-10)

and

$$\overline{m}_p \frac{d\overline{v}_z}{dt} = -C\beta \frac{\overline{z}}{\left[\left(\beta\overline{x}+1\right)^2 + \overline{z}^2\right]^3} - \left[\overline{v}_z - \left(1-\overline{x}^2\right)\right]$$

(E4.2-11)

where

$$C = \frac{\mu_0 R_p^2 \ \text{fct}\left(H_a\right) M_{sp}^2}{18\eta r_0 v_{avg} \left(1+\kappa\right)^4}$$

(E4.2-12)

$$\overline{m}_p = \frac{4v_{avg}\rho_p R_p^2}{9\eta\left(1+\kappa\right)R_M} = m_p \cdot \frac{v_{avg}}{3\pi\eta\left(1+\kappa\right)R_M R_p}$$

(E4.2-13a)

$$\overline{m} = \frac{4v_{avg}\rho R_p^2}{9\eta\left(1+\kappa\right)R_M} = m \cdot \frac{v_{avg}}{3\pi\eta\left(1+\kappa\right)R_M R_p}$$

(E4.2-13b)

and

$$\beta = r_0 \big/ \left(1+\kappa\right)R_M$$

(E4.2-14)

- For practical application, it is easy to verify that $\bar{m}_p \ll 1$, $\bar{m} \ll 1$, and $\beta \ll 1$. Thus, Eqs. (E4.2-10) and (E4.2-11) reduce to:

$$d\bar{x}/d\bar{t} = -C \cdot 1 \Big/ \left(1+\bar{z}^2\right)^3 \tag{E4.2-15}$$

and

$$d\bar{z}/d\bar{t} = \left(1-\bar{x}^2\right) \tag{E4.2-16}$$

- We divide Eq. (E4.2-15) by Eq. (E4.2-16) to obtain:

$$\frac{d\bar{x}}{d\bar{z}} = -C\frac{1}{\left(1+\bar{z}^2\right)^3 \left(1-\bar{x}^2\right)} \tag{E4.2-17}$$

- The analytical equation for the particle trajectory is:

$$\left(\bar{x} - \frac{\bar{x}^3}{3}\right) = \left(\bar{x}_0 - \frac{\bar{x}_0^3}{3}\right) - C\left[\frac{3}{8}\left[\tan^{-1}\left(\bar{z}\right) - \tan^{-1}\left(\bar{z}_0\right)\right]\right.$$
$$\left. + \frac{1}{8}\left(\frac{\bar{z}\left(3\bar{z}^2+5\right)}{\left(1+\bar{z}^2\right)^2} - \frac{\bar{z}_0\left(3\bar{z}_0^2+5\right)}{\left(1+\bar{z}_0^2\right)^2}\right)\right] \tag{E4.2-18}$$

where $\bar{x}_0$ and $\bar{z}_0$ are the initial normalized coordinates of the particle.

Graph of Eq. (E4.2-18) after Furlami & Ng (2006):

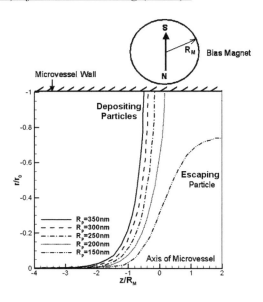

Comments:

- The particle trajectories are strongly dependent on particle size and the magnetic force.
- The critical radius of the barely captured particle is $R_p \approx 170$ nm.

Electrohydrodynamic (EHD) Pumps.   EHD pumps utilize electrostatic force fields acting on dielectric liquids, generating axial flow. There are several types of EHD pumps depending on the mechanism for space-charge generation, i.e., induction or injection, causing varying electric properties of the working fluid:

- *Induction EHD pumps* require either a gradient in the electrical conductivity or permittivity of the liquid. For example, induced transverse fluid temperature or property gradients can cause an electrical conductivity. A sinusoidal voltage applied to electrodes at the beginning and end of short microchannels creates a traveling wave through the fluid, inducing varying charges in the liquid. The fluid element charges are attracted (or repelled in reverse flow) by the spatially and temporally changing electric field, dragging with them the bulk liquid due to viscous effects. Typical mean velocities range from 10 to 100 μm/s, depending on the electrical properties of the working fluid, traveling-wave frequency, and induced temperature/property gradient.
- *Injection EHD pumps* rely on the injection of ions, say, via electrochemical reactions at the electrodes, into the dielectric liquid. The Coulomb force drives, say, cations to the anode and drags surrounding fluid elements along, which results in a net axial velocity field.

In addition to surface tension, electric fields, and magnetic forces, other microscale phenomena employed to generate fluid flow include acoustic wave and optoelectrostatic vortex pumps. It should be noted that all *nonmechanical micropumps* are only applicable in conjunction with specific fluids and/or devices.

4.2.2 Liquid Flow in Microchannels

Sections 1.3.3 and 3.2 both contain a few internal flow problem solutions directly applicable to microfluidic systems with straight long microchannels, i.e., when Poiseuille-type flow can be assumed. The examples in Sect. 3.2.4.1 can be reviewed as "*microchannel flow with built-in electroosmotic micropumps.*"

In this section more realistic microchannel flow problems are considered, featuring heat transfer, velocity slip, microchannel array, particle transport, and entrance effects.

Example 4.3: Laminar Fully Developed Thermal Flow in a Microtube

Consider steady laminar flow of water with velocity slip in an insulated microtube of diameter $D = 2R$. Assuming hydrodynamically and thermally fully developed flow, determine the friction factor from the temperature rise due to viscous dissipation heating. Discuss, in general, the relationship between pressure drop and temperature rise due to viscous dissipation heating in microtubes (see EI-Genk & Yang, 2008; among others).

Sketch:	*Assumptions/Postulates:*	*Concepts:*
	• Hydrodynamically and thermally fully developed laminar flow • Constant fluid properties • Adiabatic wall • $L \gg D$, i.e., no end effects	• Reduced Navier-Stokes equation • Simple energy balance • Navier's slip model

Solution:

Based on the postulates $\bar{v} = [0;0;v_z(r)]$ and $\nabla p \Rightarrow \Delta p / L$, the axial Navier-Stokes momentum equation for fully developed flow reads:

$$\frac{\mu}{r}\left[\frac{d}{dr}\left(r\frac{dv_z}{dr} \right) \right] = \frac{dp}{dz} = -\frac{\Delta p}{L} = \mathcal{C} \qquad \text{(E4.3-1)}$$

The following boundary conditions can be applied, i.e., the fluid velocity at the wall $(r = R)$ is proportional to the "shear stress" and the velocity profile is symmetric:

$$v_z(r = R) = -\beta \frac{dv_z}{dr}\bigg|_{r=R} \quad \text{and} \quad \frac{dv_z}{dr}\bigg|_{r=0} = 0 \qquad \text{(E4.3-2a,b)}$$

where typically $0 \le \beta \le 1\,\mu\text{m}$.

The Navier slip length β is the distance into the wall at which the liquid velocity extrapolates to zero. Solving Eq. (E4.3-1) subject to the BCs (see Eq. (E4.3-2a,b)) yields:

$$v_z(r) = \frac{D^2 \Delta p}{16\mu L}\left[1 - \left(\frac{2r}{D} \right)^2 + \left(\frac{4\beta}{D} \right) \right] \qquad \text{(E4.3-3)}$$

The average flow velocity in the microtube is then given as:

$$\overline{v} = \frac{D^2 \Delta p}{32 \mu L}\left[1 + 8\left(\frac{\beta}{D}\right)\right] \tag{E4.3-4}$$

The corresponding tube Reynolds number is:

$$\mathrm{Re} = \frac{\rho D \overline{v}}{\mu} = \frac{\rho D^3 \Delta p}{32 \mu^2 L}\left[1 + 8\left(\frac{\beta}{D}\right)\right] \tag{E4.3-5a,b}$$

Now, neglecting thermal diffusion, the energy balance for fluid flow with viscous dissipation in microtubes can be expressed as:

$$\rho c_p v_z(r)\left(\frac{\partial T}{\partial z}\right) = \mu\left(\frac{\partial v_z}{\partial r}\right)^2 \tag{E4.3-6}$$

The right-hand side of Eq. (E4.3-6) is evaluated using Eq. (E4.3-3). When integrating Eq. (E4.3-6) over the total volume of the liquid in the microtube, the rearranged results can be expressed in terms of the average flow velocity as:

$$\frac{8 \pi \mu L}{\left(1 + 8\dfrac{\beta}{D}\right)^2}\overline{v}^2 = \left(\frac{\pi}{4}D^2\right)\rho \overline{v} c_p \Delta T \tag{E4.3-7}$$

Inserting the average flow velocity from Eq. (E4.3-4) and rearranging the results provide the following expression for the rise in liquid temperature across the microtube due to viscous dissipation:

$$\Delta T = \frac{64}{\left(1 + 8\dfrac{\beta}{D}\right)^2}\left(\frac{\mu^2}{2\rho^2 L^2 c_p}\right)\left(\frac{L^3 \mathrm{Re}}{D^3}\right) \tag{E4.3-8}$$

Equation (E4.3-8) can be used to determine the friction number, c_f, in terms of a modified Reynolds number Re^* and dimensionless temperature rise, ΔT^*, in the microtube:

$$\Delta T^* = \frac{64}{\left(1 + 8\beta/D\right)^2}\mathrm{Re}^* = c_f \mathrm{Re}^* \tag{E4.3-9}$$

where $\mathrm{Re}^* \equiv \mathrm{Re}(L/D)^3$ and $\Delta T^* \equiv (\Delta T/\mu^2)(2\rho^2 L^2 c_p)$. The friction factor, c_f, can also be obtained from a similar equation written in terms of the dimensionless pressure drop and the modified average Reynolds number for laminar flow in a smooth microtube:

$$\Delta p^* = c_f \, \mathrm{Re}^* \qquad \text{(E4.3-10)}$$

where again $c_f = 64/[1+8\beta/D]^2$. A comparison between Eq. (E4.3-9) and Eq. (E4.3-10) gives the following relationship between the pressure drop and the temperature rise due to viscous dissipation in microtubes:

$$\Delta p^* = \Delta T^* \qquad \text{(E4.3-11)}$$

Graphs:

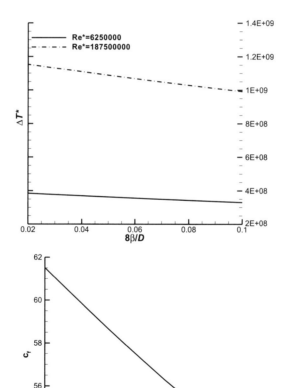

Comments:

With the no-slip condition for flow in a microtube (i.e., $\beta = 0$), the friction factor, c_f, from Eqs. (E4.3-9) and (E4.3-10) is the same as the classical Hagen-Poiseuille solution for smooth microtubes (i.e., $f = 64$). On the other hand, with velocity slip at the wall, i.e., $\beta > 0$, the friction factors for microtube flows are lower than that of the Hagen-Poiseuille case for smooth microtubes. They almost exponentially decrease as the diameter of the microtube decreases. It approaches that of the Hagen-Poiseuille solution when the microtube diameter is large enough, i.e., $8(\beta/D) \ll 1$.

While Example 4.3 reviewed a couple of interesting features of microtubular flow with velocity slip and heat source, Example 4.4 highlights a more practical application. Specifically, microelectronic devices, ranging from computers to play stations, may require 200 to 1000 W/cm^2 of heat dissipation. Effective device or processor cooling can be achieved via:

- large heat transfer surface areas (i.e., consider finned microchannel heat sinks);
- high fluid flow rates (i.e., turbulent flow);
- special coolants (e.g., nanofluids).

Additional cooling consideration includes the type of thermal wall condition, microchannel cross section, array arrangement, and wall roughness. Optimal Nusselt number $\left(Nu = hD_h/k\right)$ and Poiseuille number $\left(Po = f\,Re_{D_h}\right)$ values to be determined for system-specific cases indicate most efficient heat dissipation and minimal pumping power requirement, respectively.

Again, if fully developed flows can be assumed, Nu and Po numbers and/or correlations are readily available for macro-, mini-, and microchannels (see Kandlikar et al., 2006). Table 4.1 summarizes such values for steady laminar fully developed flow of Newtonian fluids.

Clearly, use of such average dimensionless group values lead to quick solutions but only for a host of restrictive assumptions (see Example 4.3). In order to assure fully developed flow, the condition $L_{entrance} \ll L_{channel}$ has to be fulfilled before using Table 4.1. Again, for laminar flow:

- Hydrodynamic entrance length $L_h = (0.05 \sim 0.09)\,Re_{D_h}\,D_h$
- Thermal entrance length $L_{th} = 0.1\,Re_{D_h}\,Pr\,D_h$

Table 4.1:   Summary of Poiseuille and Nusselt Numbers for Straight Conduits

Duct Shape		Aspect Ratio/Angle	Po=f·Re	Nu_H	Nu_T
	Circular		64.00	4.36	3.66
	Flat channel		96.00	8.24	7.54
a b	Rectangular channel aspect ratio b/a	1	56.92	3.61	2.98
		2	62.20	4.13	3.39
		3	68.36	4.79	3.96
		4	72.92	5.33	4.44
		6	78.80	6.05	5.14
		8	82.32	6.49	5.60
		∞	96.00	8.24	7.54
	Hexagon		60.20	4.00	3.34
θ	Isosceles triangle apex angle θ	10°	49.88	2.45	1.61
		30°	52.28	2.91	2.26
		60°	53.32	3.11	2.47
		90°	52.60	2.98	2.34
		120°	50.96	2.68	2.00
a b	Ellipse major/minor axis b/a	1	64.00	4.36	3.66
		2	67.28	4.56	3.74
		4	72.96	4.88	3.79
		8	76.60	5.09	3.72
		16	78.16	5.18	3.65
		Aspect Ratio		$Nu_{H,3}$	$Nu_{H,4}$
a b	Rectangular channel aspect ratio b/a	1		3.991	3.740
		2		4.505	4.111
		2.5		4.885	4.457
		3.33		5.393	4.969
		5		6.072	5.704
		10		6.939	6.700
		∞		8.235	8.235

Notes: Po $= f \, \mathrm{Re}_{D_h}$, where f is Darcy friction factor $f = 8\tau_w / \rho v^2$ and $\mathrm{Re} = \rho v D_h / \mu$; the Nusselt number Nu $= h D_h / k$, subscript H means heat flux boundary condition, T means temperature boundary condition, 3 or 4 means 3 or 4 sides were heated by constant heat flux.

While some semiempirical Nu and Po correlations exist for developing and turbulent flows, numerical simulations (using engineering software) of real-world systems are the preferred choice.

Example 4.4: Microchannel Array for Microprocessor Cooling with Water (Kandlikar et al. 2006)

Consider multiple fins forming (with an adiabatic cover plate) an array of microchannels with width $w = 50$ μm, height $h = 350$ μm, and a spacing $s = 40$ μm, which is also the fin thickness. This array has a 10 mm × 10 mm base which connects to a computer chip generating 100 W. The material is silicon with $k = 80$ W/mK. The cooling fluid is water which enters a representative microchannel at $T_{in} = 35°C$. It is required that $\Delta T_{max} = T_{out} - T_{in} = 10°C$.

Find:

(a) Hydraulic and thermal entrance lengths to check if $L_h, L_{th} \ll L = 10$ mm and fully developed flow can be assumed.
(b) The pressure drop across a representative microchannel.
(c) The average heat transfer coefficient, using Table 4.1.

Sketch:	*Assumptions/Postulates:*	*Concepts:*
	• Uniform microchannels with three-sided constant heat flux • Fin-and-channel equivalence • Constant properties • Averaged parameters	• Basic global energy balance • Empirical correlations • Average Po and Nu values

Solution:

- Properties for water at $T_{ref} = 0.5(T_{in} + T_{out}) = 40°C$: $\rho = 992$ kg/m³, $c_p = 4180$ J/(kg · K); $k = 0.63$ W/(m · K); $\mu = 655 \times 10^{-6}$ N · s/m²; Pr = 4.33; and $k_{silicon} = 180$ W/(m · K)
- Entrance lengths, Reynolds number, mass flow rate, and hydraulic diameter: $L_{hyd} = 0.05 \mathrm{Re}_{D_h} D_h$, while $L_{th} = 0.1 \mathrm{Re}_{D_h} \Pr D_h$

Here, $D_h = 4A/P = 4ab/2(a+b) = 87.5$ μm and $\mathrm{Re}_{D_h} = \rho \bar{v} D_h / \mu = \dot{m}_{ch} D_h / (\mu A_{ch})$. From $\dot{Q} = n \dot{m}_{ch} c_p \Delta T$ and with n = 110, i.e., number of channels of width $w = 50$ μm, and spacing $s = 40$ μm we obtain:

$$\dot{m}_{ch} = \frac{\dot{Q}}{n c_p \Delta T} = 1.3 \text{ mL/min}$$

Hence,

$$\text{Re}_{D_h} = \frac{\dot{m}_{ch} D_h}{\mu A_{ch}} = 165$$

(a) Finally,

$$L_{hyd} = 0.72 \text{ mm} << L_{ch} = 100 \text{ mm}$$

and

$$L_{th} = 6.24 \text{ mm} << L_{ch} = 100 \text{ mm}$$

- Pressure drop:
  For steady laminar developing flow in a straight conduit of extent Δx:

  $$\frac{\Delta p}{\Delta x} = 2\mu\bar{v}\left(f_{app} \text{Re}_{D_h}\right)\Big/ D_h^{\ 2} = 2\mu\bar{v}\,\text{Po}\big/ D_h^{\ 2} + K\rho\bar{v}^2\big/2$$

  where mean channel velocity $\bar{v} = \dot{m}_{ch}/\rho A_{ch}$, entrance flow correlation coefficient $K = 4\times\left(f_{app} - f\right)\big/D_h$, f is the friction factor for fully developed flow, and $\text{Po} = f\,\text{Re}$ is the Poiseuille number.
      For $\Delta x \equiv L_{ch} >> L_{hyd}$, $K \approx 0.9$ and $\text{Po} \approx 20$ for the present microchannel geometry.
- In general for a rectangular microchannel (Steinke & Kandlikar, 2005):

  ➢ $K(\kappa \equiv w/h) = 0.6796 + 1.2197\kappa + 3.31\kappa^2 - 9.59\kappa^3 + 8.901\kappa^4 - 2.996\kappa^5$

  and

  ➢ $\text{Po} = f\,\text{Re}_{D_h} = 24\left[1 - 1.355\kappa + 1.9467\kappa^2 - 1.7\kappa^3 + 0.9564\kappa^4 - 0.2537\kappa^5\right]$

(b) Finally, with $\Delta x \equiv L_{ch} = 100 \text{ mm}$ plus inlet effect K:

$$\Delta p = 44 \text{ kPa}$$

- Heat transfer coefficient:

  In light of the constant wall heat flux from three microchannel sides, Table 4.1 provides $\kappa = w/h = 1/7$ and for $\overline{\mathrm{Nu}} = 6.567$.

(c) By definition:

$$\overline{\mathrm{Nu}} = \frac{\bar{h} D_h}{k}$$

so that

$$\bar{h} = k\,\overline{\mathrm{Nu}}/D_h = 47.4 \times 10^3 \ \mathrm{W/(m^2 \cdot K)}$$

4.2.3 Gas Flow in Microchannels

As mentioned in Sect. 3.2.3 for microscale gas flow, especially rarefied gases, the flow adjacent to the solid surface may not be in thermodynamic equilibrium. As discussed, the gas molecules no longer reach the velocity of a moving wall or adhere to a stationary wall or accommodate the actual wall temperature. Thus, a slip condition for the gas velocity and jump condition for the gas temperature have to be considered, both depending on the Knudsen number $\mathrm{Kn} = \lambda_{\mathrm{mfp}}/L_{\mathrm{system}}$. For example, first-order boundary conditions read:

- Velocity slip $v_s = \dfrac{2-\sigma_m}{\sigma_m} \mathrm{Kn}\, D_h \left.\dfrac{\partial u}{\partial n}\right|_{\mathrm{wall}}$ (4.7)

and

- Temperature jump $\Delta T = T_j - T_w = \dfrac{2-\sigma_t}{\sigma_t}\dfrac{2\kappa}{1+\kappa}\dfrac{\mathrm{Kn}\, D_h}{\mathrm{Pr}}\left.\dfrac{\partial T}{\partial n}\right|_{\mathrm{wall}}$ (4.8)

where $\sigma_m \approx 1.0$ is the tangential momentum accommodation coefficient, $\sigma_t \approx 1.0$ is the thermal accommodation coefficient, $\kappa = c_p/c_v$, $\mathrm{Pr} = \alpha/\nu$, $D_h \stackrel{\triangle}{=} L_{\mathrm{system}}$ is the hydraulic diameter, and n is the normal direction exiting the wall. Second-order boundary conditions have been postulated for more complex near-wall transport phenomena (see Kirby, 2010) and are discussed in Chapter 5.

Next, thermal Poiseuille-type flow examples, considering different microconduit geometries and boundary conditions, illustrate the use of Eqs. (4.7) and (4.8).

Example 4.5: Rarefaction Effects of Gas Flow between Parallel Plates

Consider air flow under thermal Poiseuille-type conditions in the Knudsen number range of 0 to 0.1 between symmetrically heated parallel plates $(q_{\text{wall}} = \text{¢})$ a distance $2h$ apart. This microsystem with velocity slip and temperature jump may be regarded as an idealization of a microchannel for computer chip cooling.

Sketch:	*Assumptions/Postulates:*	*Concepts:*

	• Steady laminar, unidirectional, thermally fully developed flow
	 • Constant properties
	 • $-\nabla p = \Delta p / L = \text{¢}$
	    $\vec{v} = [u(y), 0, 0]$
	 • $0 \leq \text{Kn} = \lambda_{\text{mfp}} / D_h \leq 0.1$
		• Reduced momentum and heat transfer equations
		 • First-order slip and jump conditions with
		    $\sigma_m = \sigma_t \approx 1.0$
		 • Decoupled momentum and heat transfer

Solution $(q_{\text{wall}} = \text{constant case})$:

- Based on the postulates, continuity is warranted and the x-momentum equation (Sect. A.5) reduces to:

$$0 = \frac{\Delta p}{L} + \mu \frac{d^2 u}{dy^2} \qquad (\text{E4.5-1})$$

while the heat transfer equation, neglecting axial diffusion and viscous dissipation, reads:

$$u \frac{\partial T}{\partial x} = \alpha \frac{\partial^2 T}{\partial y^2} \qquad (\text{E4.5-2})$$

- Generalized boundary conditions include:

$$u(y = h) = -D_h \, \text{Kn} \frac{du}{dy}\bigg|_{y=h} \quad \text{and} \quad u(y = -h) = -D_h \, \text{Kn} \frac{du}{dy}\bigg|_{y=-h} \qquad (\text{E4.5-3a,b})$$

as well as

$$du/dy\big|_{y=0} = 0 \langle \text{in case of symmetry} \rangle \qquad \text{(E4.5-3c)}$$

$$q_{w,1} \equiv q_1 = k\frac{\partial T}{\partial y}\bigg|_{y=h} = 0 \text{ or } T(y=h) = T_1 - \frac{8\kappa}{1+\kappa}\frac{h}{Pr}\frac{\text{Kn}}{\partial y}\frac{\partial T}{\partial y}\bigg|_{y=h} \qquad \text{(E4.5-4a,b)}$$

and

$$\frac{\partial T}{\partial y}\bigg|_{y=0} = 0 \langle \text{in case of symmetry} \rangle \qquad \text{(E4.5-4c)}$$

$$q_{w,2} \equiv q_2 = k\frac{\partial T}{\partial y}\bigg|_{y=-h} = 0 \text{ or } T(y=-h) = T_2 + \frac{8\kappa}{1+\kappa}\frac{h}{Pr}\frac{\text{Kn}}{\partial y}\frac{\partial T}{\partial y}\bigg|_{y=-h} \qquad \text{(E4.5-4d,e)}$$

where $D_h = 4A/P = 4h$ and $\text{Kn} = \lambda_{\text{mfp}}/4h < 0.1$.

- Integration and invoking the BCs yield the velocity profile in dimensionless form:

$$\hat{u}(\hat{y} \equiv y/h) = \frac{3}{2}\left[1 - \hat{y}^2 + 8 \text{ Kn}\right]/(1+12 \text{ Kn}) \qquad \text{(E4.5-5)}$$

- For the constant wall heat flux case, Eq. (E4.5-2) can be rewritten as:

$$u\frac{dT_m}{dy} = \alpha\frac{\partial^2 T}{\partial y^2}; \; T_m = \frac{1}{\bar{u}A}\int_A uT \, dA \qquad \text{(E4.5-6a,b)}$$

where $dT_m/dx = \cancel{c}$ based on the thermally fully developed flow assumption:

$$\frac{\partial\theta}{\partial x} \equiv \frac{\partial}{\partial x}\left[\frac{T_w(x) - T(x,y)}{T_w(x) - T_m(x)}\right] = 0 \qquad \text{(E4.5-7)}$$

Specifically, from $\theta \equiv (T_w - T)/(T_w - T_m)$ and hence $T = T_w - (T_w - T_m)\theta$ we can conclude:

$$\frac{\partial T}{\partial x} = \frac{dT_w}{dx} - \theta\frac{dT_w}{dx} + \theta\frac{dT_m}{dx} = \begin{cases} dT_m/dx = \cancel{c} & \text{for } q_w = \cancel{c} \\ \theta\, dT_m/dx = \cancel{c} & \text{for } T_w = \cancel{c} \end{cases} \qquad \text{(E4.5-8)}$$

Implications of interest include the following.

- With $\theta = \theta(y)$ only:

$$\left.\frac{\partial \theta}{\partial y}\right|_{y=h} = \cancel{c} \text{ and hence } \frac{q_w/k}{T_w - T_m} = \cancel{c} \qquad \text{(E4.5-9a,b)}$$

- From Newton's law of cooling:

$$q_w = h(T_w - T_m), \frac{h}{k} = \cancel{c}, \text{ and } \frac{dT_m}{dx} = \frac{dT_w}{dx} = \cancel{c} \qquad \text{(E4.5-10a-c)}$$

- Considering only the upper half due to symmetry, Eq. (E4.5-6a) reads:

$$\frac{3}{2(1+12\ \text{Kn})}\left[1-\left(\frac{y}{h}\right)^2 + 8\ \text{Kn}\right]\left(\frac{dT_m}{dx}\right) = \alpha\frac{\partial^2 T}{\partial y^2} \qquad \text{(E4.5-11)}$$

subject to

$$\left.\frac{\partial T}{\partial y}\right|_{y=0} = 0 \text{ and hence } T(y=h) = T_1 - \frac{8\kappa}{1+\kappa}\frac{h}{\text{Pr}}\,\text{Kn}\left.\frac{\partial T}{\partial y}\right|_{y=h} \qquad \text{(E4.5-12a,b)}$$

- Integration and invoking the first BC yield:

$$\frac{dT}{dy} = A\left[\left(\frac{y}{h}\right) - \frac{1}{3}\left(\frac{y}{h}\right)^3 + B\left(\frac{y}{h}\right)\right] \qquad \text{(E4.5-13)}$$

where

$$A \equiv \frac{3h}{2\alpha(1+12\ \text{Kn})}\left(\frac{dT_m}{dx}\right) \text{ and } B = 8\ \text{Kn}$$

- Integrating again and incorporating the temperature jump yield $(T_w = T_1 = T_2 = \text{fct}(x))$:

$$T(y=h) = \frac{Ah}{2}\left(\frac{5}{6} + B\right) + C = T_w - KA\left(\frac{2}{3} + B\right) \qquad \text{(E4.5-14a,b)}$$

where

$$K = \frac{8\kappa}{1+\kappa}\frac{h}{Pr}Kn \quad \text{and} \quad C = T_w(x) - KA\left(\frac{2}{3}+B\right) - \frac{Ah}{2}\left(\frac{5}{6}+B\right)$$

Hence,

$$T(x,y) = T_w(x) + Ah\left[\frac{1}{2}\left(\frac{y}{h}\right)^2 - \frac{1}{12}\left(\frac{y}{h}\right)^4 + \frac{B}{2}\left(\frac{y}{h}\right)^2\right] - KA\left(\frac{2}{3}+B\right) - \frac{Ah}{2}\left(\frac{5}{6}+B\right) \quad \text{(E4.5-15)}$$

- An energy balance for a control volume $h \cdot w \cdot \Delta x$, assuming $w = 1$ (unit plate width), we have:

$$h\bar{u}\rho c_p T_m - h\bar{u}\rho c_p\left(T_m + \frac{dT_m}{dx}\Delta x\right) + 2(h+1)\Delta x q_w = 0 \quad \text{(4.5-16a)}$$

or

$$\frac{dT_m}{dx} = \frac{2(h+1)}{h\bar{u}\rho c_p}q_w = ¢ \quad \text{(4.5-16b)}$$

which implies that $T_m(x)$ is linear.

Graph:

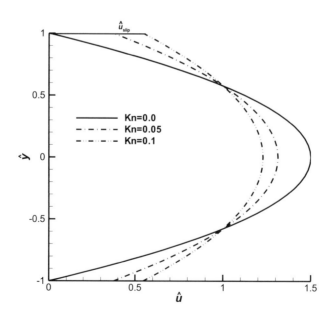

Comments:

- As expected, Kn values in Eq. (E4.5-5) greatly influence the axial velocity profiles.
- For more elaborate thermal case studies see Sadeghi & Saidi (2010).

4.3 Micromixing

Excellent mixing is important for rapid chemical analysis, i.e., bringing reagent and solution together, or for uniform nanodrug delivery when dealing with nanoparticle suspensions, among other applications. In macrofluidics, *turbulent mixing* caused by random fluid and particle velocity fluctuations is the preferred method. In contrast, microscale mixing relies on low-Re-number *chaotic flow due to secondary flows and diffusion,* accomplished by complex geometric features, multistream focusing, microjets, electro/magnetohydrodynamics, or acoustic streaming. Kumar et al. (2011) provide an in-depth review of pressure drops and mixing in microchannels.

Examples of chaotic-flow-causing arrangements which lead to uniform species distributions and/or flow fields include:

- Forced convection through meandering microconduits with designed wall patterns, e.g., herring bones, pins and dimples, etc.
- Use of T-junctions for two-stream contact and mixing.
- Forced micromixing by means of stirring, ultrasonic waves, etc.
- Injection of small fluid substreams in segments.
- Splitting and recombination of several fluid lamellae.

Figure 4.4 lists causes and mechanisms leading to passive or active micromixing, while Figure 4.5 depicts geometric and mechanical elements of passive mixing.

Diffusive Mixing.   In micro- and nanosystem species concentration gradients drive mixing. The mixing rate is determined by the diffusion flux (see Sect. 1.3.3.4):

$$\vec{J} = -\mathcal{D}\nabla c \qquad \left[\frac{\text{kg}}{\text{m}^2 \cdot \text{s}} \right] \tag{4.9}$$

where c is the species concentration and the diffusion coefficient could be expressed by the Stokes-Einstein equation (Sect. 2.5):

$$\mathcal{D} = \frac{k_B T}{3\pi\mu d_p} \qquad \left[\frac{\text{m}^2}{\text{s}} \right] \tag{4.10}$$

where the spherical species molecular or particle diameter is in the range of $0 < d_p \leq 100$ nm.

The diffusion time τ and "mixing length" l_m are correlated via the Fourier number $(0.1 \leq \text{Fo} \leq 1.0)$:

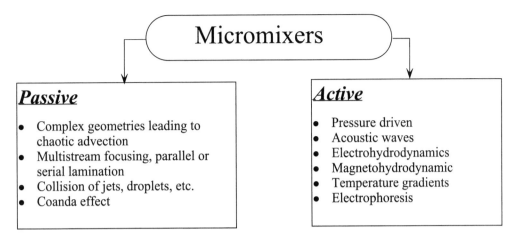

Figure 4.4   Passive and active, disturbance creating micromixers

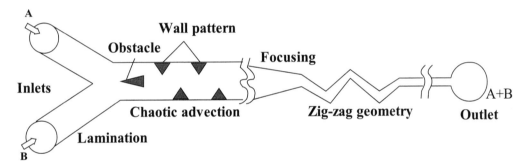

Figure 4.5   Passive mixing elements and processes

$$\text{Fo} = \frac{\mathcal{D}\tau}{l_m^2} \tag{4.11}$$

Actually l_m could be the micro/nano channel width and hence with $l_m \hat{=} L_{\text{system}} \ll 1$, $\tau \to 0$, which implies rapid mixing. One application area of Eq. (4.11) is focusing of streams fostering mixing. For example, a converging microchannel (or dual sheath flow) pushes streams together, causing rapid cross mass transfer because of the large microstream contact areas. Note, concentration-gradient-driven diffusion plus secondary fluid convection is often labeled "dispersion."

Chaotic Mixing.   As documented, molecular diffusion is a series of random events which, when amplified, may lead to chaotic mixing. Thus, transient deterministic fluid flow away from channel walls may exhibit chaotic mixing. Alternatively it is also called chaotic advection because of random species transport by the laminar flow. In simple 2-D flow, chaotic advection can be induced by a transient disturbance, or in 3-D flow even without any perturbation. It is a complex process where, say, flow tracers develop into fractals, i.e., minute self-repeating pattern. This has been observed on all scales from chaotic air flow in the alveoli of the respiratory tract to severe weather patterns. Concerning microchannel flow, chaotic mixing features either trajectories in the flow that become separated by a distance growing exponentially with time or interfacial area between two fluids that enlarges exponentially with time.

Passive Mixing.   As indicated, *passive micromixers* rely on chaotic convection and diffusion, i.e., no external power source is required. Complex geometric channel features, such as patterned wall roughness, serpentine or zigzagging conduits, and staggered obstacles placed in the channel, are causing strong secondary flows, i.e., locally chaotic flow fields. Parallel or serial fluid lamination, Y- and T-junctions, injection/collision of jets and the Coanda effect promote rapid mixing as well. In general, species/solute/ nanoparticle transport during passive mixing can be described by the mass transfer equation:

$$\frac{Dc}{Dt} = \nabla \cdot \boldsymbol{\mathcal{D}} \nabla c \pm S_c \tag{4.12}$$

For example, parallel lamination is a simple mixing method for planar microfluidic systems. The idea is to split a stream into multilamellae structures which are stretched and recombined employing the fluid flow energy. More vigorous mixing is achieved via *sequential lamination,* where a stream is split and rejoined in several (sequential) stages. Specifically, using n such splitting stages, 2^n layers are laminated, which causes $4^{(n-1)}$ faster mixing, while for parallel lamination the mixing time decreases with a factor of n^2. Chapter 10 in Nguyen & Wereley (2006) provides several design examples for lamination mixers. Basic elements and processes of passive micromixers are indicated in Figures 4.4 and 4.5.

Active Mixing.   An external energy source or field can cause all kinds of flow disturbances in microchannels, resulting in near-uniform outputs due to micromixing. The causes of disturbances can be mechanical, electrokinetic, electromagnetic, or thermal (see Figure 4.4). The operational conditions and mixing effectiveness are a function of Reynolds number, Peclet number, and Strouhal number. Specifically,

$$\mathrm{Re} = \frac{\text{inertia forces}}{\text{viscous forces}} = \frac{vD_h}{v} \tag{4.13}$$

$$Pe = \frac{\text{convection species transport}}{\text{diffusion species transport}} = \frac{vl}{\mathcal{D}} \qquad (4.14)$$

and

$$Str = \frac{\text{species residence time}}{\text{species disturbance time scale}} = \frac{D_h / \tau}{v} \qquad (4.15)$$

Actually, at high Peclet numbers (e.g., $Pe = \mathcal{O}(100)$ and low Reynolds numbers (i.e., $Re \leq 1.0$) as well as $l \sim D_h > 1$, active mixing of, say, injected chemical species may not occur. This is known as the low Re and high Pe limit.

Species Mass Transport Equation.   While the Reynolds number appears in the dimensionless momentum equation:

$$\frac{D\tilde{u}}{Dt} = -\tilde{\nabla}\tilde{p} + \frac{1}{Re}\tilde{\nabla}^2\tilde{u} \qquad (4.16)$$

the Peclet number governs the convection-diffusion equation:

$$\frac{D\tilde{c}}{Dt} = \frac{1}{Pe}\tilde{\nabla}^2\tilde{c} \qquad (4.17)$$

Solving Eq. (4.17) in conjunction with Eq. (4.16) and with appropriate BCs yields the mixing patterns.

4.4 Laboratory-on-a-Chip Devices

A LoC is a micro/nanofluidic device that integrates laboratory functions, such as chemical analysis, biocell sorting, DNA sequencing, drug development, environmental monitoring, and medical diagnostics on a single chip. Conventional biochemical labs (with people in huge rooms) handle large volumes of solutions, solvents, reagents and cells/particles for mixing (or separation), heating, concentration, and ultimately product measuring. Now, these main process functions are carried out by LoCs with actuators, microchannels, valves, heaters, sensors, and read-out circuits (Ghallab & Badawy, 2010; among others). For example, electric/magnetic actuators generate forces for fluid pumping and controlling, sensors measure electrical, thermal magnetic, or optical sample properties, while a read-out circuit is used to amplify and condition (electric) output signals. With lab component minimization from the cubic meter to the cubic micrometer scale as well as the integration and automation of the processing steps, lab-on-chips emerged. Lab-on-a-chip and micro-total-analysis systems (μTASs) are subsets of MEMS, employed for biological and/or chemical analyses and discovery (see Figure 3.2). As mentioned, LoCs typically deal with specific biomedical or biochemical processes, such as DNA extraction, cell identification, sorting, and manipulation, but also with pollution detection, analysis, and medical diagnostics. In contrast, μTAS devices integrate the total sequence of lab processes in chemical analysis. Clearly, such microdevices consist of reservoirs connected to microchannels with pumps/actuators, valves, heaters, separators/mixers, and sensors (see also Figures 3.1 and 3.2). Again, fluid or particle flow activation in LoCs or μTAS can be due to capillary (i.e., surface tension), mechanical pressure, electrokinetic, and/or acoustic wave effects. In any case, the key benefits of such a major size reduction from people's labs down to microfluidic systems are:

- lower cost (e.g., much lower capital investment and greatly reduced volume of expensive process materials);
- better performance (i.e., faster and more accurate with controlled multitask processing);
- safer handling of hazardous materials; and
- massive chip parallelization due to compactness.

However, with minimization and hence the very large surface-to-volume ratio, *negative physical and chemical effects* may occur. Examples include entrained air bubbles or impurities in the micro/nanochannels, fouling, high flow resistance, low signal-to-noise ratio, interfering wall materials, etc.

4.4.1 LoC Processing Steps

As indicated with Figure 4.6, sample injection, pretreatment/handling, separation, sensing and detection are typically major processing tasks. Clearly, other product end goals may require sample mixing, heating, and chemical reaction. Different types of LoC semiconductors and their biological and/or chemical analysis applications are reviewed in the book by Ghallab & Badawy (2010), among others.

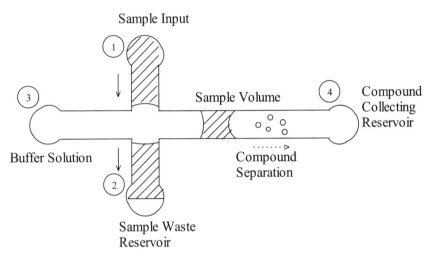

Figure 4.6   Sample supply for LoC processing within a microfluidic system

For example, via electroosmosis the Sample ① is pushed towards Reservoir ②. When appearing at the microchannel intersection, the external electric potential is switched from ①-② to ③-④ causing a small sample volume (pico- to nanoliter) to move towards Reservoir ④. Subject to the external electric field, the tagged compounds with inherently different electrophoretic motilities are naturally separated in the long stretch of the microchannel leading to Reservoir ④. Laser excitation of the fluorescence-tagged compounds allows for measuring a sequence of peaks of light emission over travel time. This method of electrokinetic sample delivery is called "floating injection."

A simpler scheme is the T-junction injection (which is not as controlled), where waste reservoir ② is omitted (see Blas et al. (2008)). In any case, cross-wise sample movements are achieved by switching the four electric potentials, allowing for controlled, valve-less redirection of flow from ①-② to ③-④. These electric fields are applied via external wires situated in the reservoirs or by electrodes attached to the substrate at the four ends of the LoC.

Clearly, in nanochannels for which the width is less than twice the electric double layer (EDL), i.e., Stern plus diffusion layer, say, 1 to 100 nm, the EDLs overlap.

Hence a resulting electroosmotic velocity profile is curved, in contrast to the higher, near-uniform EO flow in microchannels (see Sect. 3.2.4.1).

Continuous-Flow Separation. In a batch operation for compound/particle separation, a tiny sample volume is injected and subsequently tagged molecules/ compounds are separated and identified during each test run (see Figure 4.1). Clearly, the axial flow is aligned with the electrophoretic force field. In contrast, for continuous-flow separation (CFS) in microfluidic systems an angled particle-force field generates different trajectories for different particles/compounds. As for most continuous-flow processes, CFS allows for more flexible, smoother, faster, and higher volume operation.

Sample Injection. Before any injection of a fluid-particle suspension to the LoC, sample pretreatment, e.g., filtration or extraction, may be necessary. For example, relatively large particles which may clog the LoC-microchannels have to be separated via microfilters/membranes, microdialysis filters, or electrophoresis.

Two common sample introduction methods are electrokinetic injection and hydrodynamic injection. Although electroosmosis as a sample driving force is generally preferred, the use of micropumps in hydrodynamics has several advantages. For example, the sample pH and electrolyte concentration as well as special microchannel material with low electrical conductivity (all important in electrokinetic flow) are of no concern in hydrodynamic injection. Furthermore, the absence of an applied electric field avoids interference with LoC electronics for sensing/detecting and measuring.

Constituent Separation. Extraction, isolation, recovery, detection, and labeling of chemical compounds or biological particles/cells in mixtures are major analysis tasks. Chromatography and electrophoresis are two frequently employed separation methods.

4.4.2 LoC Applications

Rapid, reliable, and cost-effective chemical analysis, biological cell screening, and drug development are central in modern chemistry and environmental monitoring as well as medicine and pharmaceutics. Thus, LoC devices are more and more replacing central laboratories requiring skilled workers and large expensive equipment. The PoC tests at home or in the environment concerning pregnancy, blood sugar, and air/water quality are basic examples. They are accomplished by battery-powered "black boxes," i.e., LoCs, performing such chemical, biological, or environmental analyses. Applications include immune or assays to detect cancer cells, viruses, and bacteria for diagnostics and test treatment. LoCs loaded with blood samples can be used to identify infectious diseases, crack cells and extract DNA, and monitor treatment success (or failure). Real-time environmental monitoring of air and water pollution, rather than occasional sampling, is another important LoC application.

Concerning fluidics, the key element in LoC devices is the microchannel with heaters and associated substrate-feed reservoirs, valves, channel bifurcations, etc. Thus, computational and/or experimental microfluidics material given in Chapters 3 and 4 as well as in Chapter 8 are essential for the analysis and design improvement of LoC devices.

4.5 Homework Assignments and Course Projects

While Sect. 4.5.1 lists concept questions derived from the text, Sect. 4.5.2 provides an introduction to course projects in terms of objectives, options, and execution. Clearly, at this stage mini-project topics are restricted to Part B material.

4.5.1 Text-Related Questions and Tasks

4.1   Using Figure 4.1 as a guide correlate different types of "micropumps" with specific applications/devices/systems and provide image and functional descriptions.

4.2   Derive Eqs. (4.1) and (4.2a, b).

4.3   In Example 4.1, derive Eqs. (E4.1-5b), (E4.1-6), and (E4.1-7) and discuss $(J_y B_x)$ -values for which fluid heating may become significant.

4.4   Concerning Example 4.2: Plot a few particle trajectories $(150 \leq R_p \leq 200$ nm$)$ in the microtube for different $d = 1 + \kappa R_M$ values $(0 \leq \kappa \leq 1.0)$.

4.5   Redo Example 4.2:

(a) In cylindrical coordinates for a microtube of radius r_0

(b) In rectangular coordinates for parallel plates a microdistance h apart

4.6   Concerning Example 4.3:

(a) Derive Eqs. (E4.3-5b), (E4.3-7), and (E4.3-9).

(b) Compare the friction factor of Eq. (E4.3-10) $c_f = \mathrm{Re}^*/\Delta p^*$ with the Darcy-Weisbach friction factor $f = 8\tau_{wall}/(\rho v_{avg}^2) = 64/\mathrm{Re}_D$.

(c) Plot $\Delta T^*(D)$ and interpret Eq. (E4.3-11).

4.7   Considering Table 4.1, show that $\mathrm{Po} \equiv f\,\mathrm{Re} = 96$ for flow between parallel plates (and in channels with $b/a \to \infty$).

4.8   Research and tabulate semiempirical Po number and Nu number correlations for steady laminar flow with entrance/exit effects.

4.9   Research, visualize, and discuss the accommodation coefficients σ_m and σ_t (see Eqs. (4.7) and (4.8)).

4.10 Concerning Example 4.5:

• Check Eqs. (E4.5-15) and (E4.5-16a,b).

• How can $T_w(x)$ in Eq. (E4.5-15) be determined?

• Plot $T(x, y)$ for reasonable parameter values.

4.11 Based on a literature search, correlate different types of micromixers with specific applications; illustrate and discuss a few.

4.12 Using scale analysis (see Sect. 1.2), derive the dimensionless groups of Eqs. (4.11) and (4.13) to (4.15).

4.13 Research LoC devices and provide images, descriptions, and applications (see Sect. 4.4).

4.5.2 Set-up for Course Projects (CPs)
General:

- A course project has to be original, i.e., resubmission of selective write-ups of past research projects or ongoing thesis work is verboten!
- Two-person group work is permissible only when dual efforts are apparent, e.g., experimental + computational work.
- Hand in an *overview* (e.g., a preproposal), i.e., project title, system sketch, objectives, solution method, and expected results, before starting the project.
- Follow the format of report writing (see Sect. 7.3 in Chapter 7).

CP Selection Process and CP Options:

- Focus on what interests you most as related to micro/nanofluidics, is within your skill level, and can be accomplished within a set time frame.
- Select an appropriate journal/conference paper, for example, an experimental article on thermal flow (or nanofluid flow) in a microchannel, and perform an analytic/numerical analysis. There are now a half dozen new journals which are directly related to micro/nanofluidics; furthermore, traditional fluid mechanics, fluid-particle dynamics, heat transfer, and biomedical journals feature articles on microfluidics and nanofluidics topics in every issue.
- Perform the micro/nanofluidics part of a microsystem, e.g., a magnetohydrodynamic pump, a lab-on-a-chip, a particle sorting device, a point-of-care fluid-sampling and diagnostics/analysis device, etc.
- Write tutorials, i.e., background information and solved example problems in key micro/nanofluidics areas, such as drug delivery (e.g., in nanomedicine), MEMS or LoC or PoC processes with device-component performance analysis. Alternatively, select new tutorial topics, such as two-phase (liquid-gas) microchannel flow, boiling, magnetohydrodynamics, flow focusing, or particle sorting in microfluidics devices, optofluidics, acoustic-wave-driven micro/nanofluidics, etc.

CP Objectives:

- Deepen your knowledge base in micro/nanofluidics.
- Advance your analytical/computational skills or experimental skills.
- Generate an impressive research report in content and form.

References (Part B)

Blas, M., Delaunay, N., Rocca Dr., J-L., 2008, Electrophoresis, Vol. 29, pp. 20-32.

Chang, H.-C., Yeo, L.Y., 2010, *Electrokinetically Driven Microfluidics and Nanofluidics,* Cambridge University Press, New York.

Davidson, R.C., 2001, *Physics of Non-neutral Plasmas,* World Scientifics, Imperial College Press.

Duan, Z., 2012, International Journal of Thermal Science, Vol. 58, pp. 45-51.

Dutta P., Beskok, A., 2001, Analytical Chemistry, Vol. 73, pp. 1979-1986.

El-Genk, M.S., Yang, I-H, 2008, ASME Journal of Heat Transfer, Vol. 130, pp. 082405-1-13.

Fang, Y., Liou, W.W., 2002, ASME Journal of Heat Transfer, Vol. 124, pp. 238.

Furlani, E.P., 2001, *Permanent Magnet and Electromechanical Devices,* Academic Press, New York.

Furlani, E.P., Ng, K.C., 2006, Physical Review E, Vol.73, 061919.

Ganguly, R., Puri, I.K., 2010, Nanomedicine and Nanobiotechnology, Vol.2, pp. 382-399.

Ghallab, Y.H., Badawy, W., 2010, *Lab-on-a-Chip: Techniques, Circuits, and Biomedical Applications,* Artech House, Boston/London.

Gijs, M.A.M., Lacharme, F., Lehmann, U., 2010, Chemical Reviews, Vol. 110, pp. 1518-1563.

Griffith, D.J., 1999, *Introduction to Electrodynamics,* Prentice-Hall, Upper Saddle River, NJ.

Harkins, W.D., Brown, F.E., 1919, Journal of the American Chemical Society, Vol. 41, pp. 499-524.

Hetsroni, G., Mosyak, A., Pogrebnyak, E., Rozenblit, R., (2011), International Journal of Thermal Science, Vol. 50, pp. 853-868.

Horiuchi, K., Dutta, P., 2004, International Journal of Heat and Mass Transfer, Vol. 47, pp. 3085–3095.

Jang, J., Lee, S.S., 2000, Sensors and Actuators A: Physical, Vol. 80, pp. 84-89.

Kandlikar, S.G., Garimella, S., Li, D., Colin, S., King, M.R., 2006, *Heat Transfer and Fluid Flow in Minichannels and Microchannels,* Elsevier.

Kirby, B.J., 2010, *Micro- and Nano-scale Fluid Mechanics: Transport in Microfluidic Devices*, Cambridge University Press, New York.

Kleinstreuer, C., Li, J., Feng, Y., 2012, *Advances in Numerical Heat Transfer, Volume 4, Nanoparticle Heat Transfer and Fluid Flow*, Minkowycz, W.J., Sparrow, E.M., Abraham, J.P. (eds.), CRC Press, Taylor & Francis Group.

Kumar, V., Paraschivoiu, M., Nigam, K.D.P., 2011, Chemical Engineering Science, Vol. 66, pp. 1329-1373.

Lee, S-J. J., Sundararajan, N., 2010, *Microfabrication for Microfluidics,* Artech House, Boston/London.

Meng, E., 2011, *Biomedical Microsystems,* CRC Press, Taylor & Francis Group, New York, London.

Nguyen, T.K.T., 2012, *Magnetic Nanoparticles from Fabrication to Clinical Applications,* CRC Press, Boca Raton, FL.

Nguyen, N.-T., Wereley, S.T., 2006, *Fundamentals and Applications of Microfluidics,* Artech House, Boston.

Probstein, R.F., 1994, *Physicochemical Hydrodynamics: an Introduction,* Wiley, New York.

Sadeghi, A., Saidi, M.H., 2010, ASME Journal of Heat Transfer, Vol. 732, 072401-1.

Steinke, M.E., Kandlikar, S.G., 2005, Single Phase Liquid Heat Transfer in Microchannels, 3rd International Conference of Microchannels and Minichannels, June 13-15, Toronto, Canada.

Zhang, Z.M., 2007, *Nano/Microscale Heat Transfer,* McGraw-Hill, New York.

Part C: NANOFLUIDICS

Nanofluidics deals with fluid flow and fluid-particle transport around objects or in systems, devices, and conduits where one characteristic length is at least 1 to 800 nm. The advantages of microfluidics, such as small material volumes for fast-processing, high-throughput, disposable low-cost devices, etc., hold also in nanofluidics. However, due to the extralarge surface-area-to-volume ratio, say, $\kappa \leq 10^9 m^{-1}$, the proximity of the walls and their surface characteristics cause formidable scientific and engineering challenges in describing accurately the fluid flow and particle dynamics. In electroosmosis, for example, the EDL (electrical double layer) with an effective thickness in the range of, say, $1 \leq \lambda_D \leq 800 nm$ may greatly affect transport and reactions in nanofluidic structures. When the hydraulic diameter, D_h, of nanoconduits is below, say, 10 nm, fluids may have to be regarded as a collection of molecules rather than continua. Specifically at the nanoscale of, say, 10 to 20 nm one has to deal with greatly amplified surface force/flux/process effects as well as materials, biochemical elements, and/or compounds which may change their mechanical, optical, and electrical properties. In addition, intermolecular forces may have a greater impact than hydrodynamic forces. In fact, intrinsic material properties can be manipulated and controlled via "atom-by-atom build-up," leading to new nanoscale structures and hence material properties. Thus, nanoscale devices feature fluid and solid properties which are governed by a confluence of classical physics and quantum mechanics (see Figure 1.1). Furthermore, at the nanolevel new device manufacturing and production techniques are possible, for example, molecular self-assembly, templating, and fragmentation.

As part of nanotechnology, nanofluidics is of fundamental importance. It encompasses the analysis of fluid flow and heat/mass transfer in nanochannels, the dynamics of nanoparticles in liquids (labeled "nanofluids"), and operation of micro/nanoscale devices, reactors, and systems. Concerning nanofluids, their applications can be found in rather diverse research areas. Examples include consumer products such as cosmetics, sunscreens and cleaners, cooling of microsystems as well as tumor imaging and drug delivery as part of nanomedicine.

Nanoparticles, in general, with their higher surface area and denser surface atom packing as well as diverse shapes (e.g., spheres, tubes, rings, hexagons, wires, ovals, ribbons, etc.), are employed to control large-scale material properties. As a result, lighter and stronger composites, better conducting devices, and stain-resistant fabrics have been marketed.

In summary, whenever possible nanofluidics is described by the system of (reduced) Navier-Stokes equations, augmented by electrostatic or magnetohydrodynamic body forces, as appropriate. However, for flow in nanochannels with $D_h \leq 10\,nm$ the interactions and resulting motion of molecules are numerically evaluated with molecular dynamics (MD) simulation for liquids or a direct simulation Monte Carlo (DSMC) method for rarefied gases.

Of course, with the prevalence of nanomaterials rapidly rising, their possibly detrimental impact on human health and the environment has to be seriously considered.

CHAPTER 5

Fluid Flow and Nanofluid Flow in Nanoconduits

5.1 Introduction

After a chapter overview, differences between macroscale and nanoscale material structures and thermodynamics are briefly discussed. The inherent nanoscale aspects provide further awareness that *nanofluidics* is not just a miniaturized version of macrofluidics but implies fluid and/or particle motion in nanoscale channels, tubes, or pores with new transport phenomena. A natural occurrence of nanofluidics is in ion channels, i.e., pores formed by proteins, which control the movement of ions in and out of cells through cell walls or membranes. In general, nanoporous membranes (or barriers) with pore diameters of 1 to 5 nm can be employed to sort out and analyze molecules, single-stranded DNA polymers, and particles of equivalent mean diameters. An industrial example of nanofluidics is fluid flow in carbon nanotubes with applications in nanomedicine and material processing.

However, while most gaseous and liquid microflows can be described by continuum mechanics theory (or quasi-continuum, adding velocity slip and temperature jump models), nanoflows may exhibit quite anomalous behavior. Specifically, the nanoscale hydraulic diameter and hence the even much higher surface-to-volume ratio in nanofluidic systems turn the fluid into an assembly of molecules and amplify the impact of surface characteristics. As a result, fluid properties such as density and viscosity may fluctuate over interatomic distances, surface roughness effects can be major, and wall boundary conditions, although difficult to define, greatly influence the fluid-wall interaction and hence the flow field in nanoconduits.

5.1.1 Overview

As mentioned, nanofluidics deals with transport phenomena of fluids and particles in channels, devices, and systems of characteristic lengths ranging from 1 nm to 800 nm. For conduits with hydraulic diameters of $D_h < 10$ nm, a fluid has to be regarded as an assembly of molecules, greatly influenced by the channel's surface characteristics. As mentioned, fluid properties such as density and viscosity may fluctuate over interatomic or intermolecular distances. Furthermore, surface roughness and/or electric charge effects significantly influence fluid-wall interaction and hence the flow field in the nanoconduit. As in microfluidics, nanofluidic system components include supply reservoirs, flow conduits, passive or active pumps, actuators, valves, sensors, heaters, etc. New techniques are employed to generate nanostructures, such as nanochannels, nanopores, nanowires, nanodots, and nanoparticles. Examples of novel tools necessary for atom-by-atom construction, nanostructure visualization, or device fabrication include STM (scanning tunneling microscopy), soft lithography, ion-beam lithography, and bottom-up assembly methods. It should be recalled that capillary effects and electrokinetic techniques, i.e., electroosmosis or electrophoresis, as well as magnetohydrodynamics are typically employed in nanofluidics (see Sects. 3.2.4 and 3.2.5). Hence, the nanochannel size (e.g., $D_{\text{hydraulic}}$) in relation to the electroosmotic or magnetohydraulic wall layer characterizes the fluid-particle dynamics.

As mentioned, nanoscale devices and processes can be found in a wide range of applications. For example, deterministic nanofluidic arrays, such as a network of nanopores or nanochannels, allow for fundamental experiments in biophysics, i.e., DNA analysis and cell manipulation. As part of nanomedicine, multifunctional nanoparticles are being used as contrast agents for better magnetic resonance imaging as well as drugs with tailored nanostructures for attachment to cancer cells. Other nanofluidic applications include water purification (e.g., filtration, desalination), and alternative energy sources, such as membrane-based fuel cells. Photovoltaic cells with a special nanoparticle coating convert more efficiently sunlight into electricity. An alternative approach for efficient solar panels is the use of nanostructured glass, having a surface which is water repellent and self-cleaning. Specifically, the panel surface is covered with cones 200 nm at the base and 1 μm high, plus a film, to achieve these superhydrophobic characteristics, which are also observable in nature. A unique *natural* nanofluidic system is the cell membrane, i.e., their trans-membrane protein pathways with pore diameters of 1 to 5 nm. These ion channels achieve particle throughput not (yet) matched by synthetic nanopores, although artificial membranes are more conducive for controlled manipulation. The advantages of nanofluids, i.e., dilute metallic nanoparticle suspensions in liquids, for enhanced microsystem cooling are discussed in Sects. 6.3 and 8.3.

Nanofluidics systems or devices are not just smaller than microsystems. They may differ for three basic reasons, which imply the necessity for a deeper scientific knowledge base and a higher mathematical skill level (see Figure 5.1):

- While the fabrication of microsystems follows miniaturization techniques, i.e., basically those developed in microelectronics, nanosize particles/devices are often "built up" via assemblage of atoms and molecules.
- While material properties on the microscale are the same as those on the macroscale, the same nanomaterial may exhibit different thermal and/or electric conduction, optical behavior, and surface reactions, often due to rearranged lattice atoms or molecules.
- While for the solution of almost all practical microscale transport phenomena the fluid continuum hypothesis holds, flow in nanochannels may require new math modeling considerations and solution techniques, because of the prominent interactions of fluid and wall molecules.

As mentioned, a system's surface-area-to-volume ratio, $\kappa = A_{\mathrm{surf}} / \forall_{\mathrm{syst}} \sim L_{\mathrm{syst}}^{-1}$, can be enormous. For example, a 1-m cube chopped into 1-nm cubes produces a combined surface area of all nanocubes which would cover the state of Delaware. In terms of sphere ratios, a 1-nm particle relates to a soccer ball as the soccer ball relates to planet Earth. As a result, forces which are important in macrofluidics (e.g., gravity) may be insignificant when compared to electromagnetic and molecular near-wall forces, e.g., F_{Lorentz}, F_{Coulomb}, or $F_{\mathrm{v.d.Waals}}$, because of the proximity of nanochannel surfaces. Specifically, a conduit's hydraulic diameter may scale with a liquid's intermolecular distance and Debye length or with the mean-free path of a gas. The attractive or repulsive electrostatic forces, say, F_{Coulomb}, act in the larger part of the EDL of electrolytes, while van der Waals forces are always attractive and act predominantly at distances smaller than 2 nm. The Lorentz force acts on magnetic fluids or particles when subjected to electrostatic and magnetic fields. Also, controlled diffusion trumps convection in channels of height or width $h < 1\,\mu m$, leading to travel times in the nanosecond range (see Sect. 3.2.1). Clearly, in order to solve nanofluidics problems, engineers have to possess a deeper understanding of the basic sciences and meet new applied math challenges (see Figure 5.1).

5.1.2 Nanostructures

The ability to control and manipulate building blocks on the nanoscale, i.e., nanostructures, makes it possible to exploit new physical, biological, and chemical material properties and system applications. For example, Khataee & Mansoori (2011) discussed the following points:

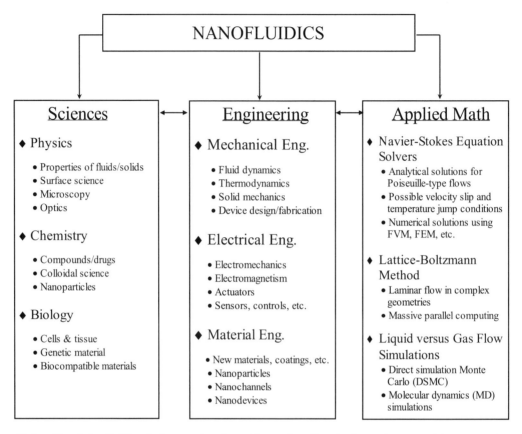

Notes: Nano/microfluidics devices create a favorable cost/benefit ratio due to low material consumption (e.g., solutions, cells, drugs, etc.), faster processing at better performance, less hazard when dealing with toxic materials, and reduced costs. Obviously, mass production of cheap (i.e., disposable) and reliable devices is key to the general acceptance of nanofluidics products and operation.

Figure 5.1   Information and tools needed in nanofluidics

- The quantum mechanical (wavelike) properties of electrons inside matter are influenced by variations on the nanoscale. Thus, via nanoscale design of materials it is possible to vary their micro- and macroscopic properties, such as charge capacity, magnetization, and melting temperature, without changing their chemical composition.
- A key feature of biological entities is the systematic organization of matter on the nanoscale. So placing man-made nanoscale elements inside living cells is envisioned. It would also make it possible to produce new materials

using the self-assembly features of nature. All this leads to a powerful combination of biology with materials science.

- Nanoscale components with their very high surface-to-volume ratio make them ideal for use in composite materials, reacting systems, drug delivery, and chemical energy storage (such as hydrogen and natural gas).
- Macroscopic systems made up of nanostructures can have much higher density than those made up of microstructures. One resulting application is better conductors of electricity spawning new electronic device concepts, smaller and faster circuits, and more sophisticated functions. They greatly reduce power consumption by controlling nanostructure interactions and complexity.

One of the more important nanostructures developed in 1991 are carbon nanotubes (CNTs). They are being used for strengthening composites and as drug carriers to achieve solid tumor targeting, to name just two applications. Originally, a symmetric network of hexagonal graphite was rolled up to form a hollow, single-walled tube, nanometers in diameter and micro to-milimeters in length. Their photochemical, electronic, thermal, and mechanical properties are quite unusual. For example, nanofluids, say, water with a low volume fraction of CNTs, exhibit thermal conductivities twice as high as that of pure water (Kleinstreuer & Feng, 2012), while the tensile strength of a rope of CNTs can be 100 times that of a steel wire, and CNT electrical conductivity is similar to that of copper.

For biomedical applications, e.g., molecular drug delivery and targeting, interesting nanostructures are under development and testing. Examples include liposomes, i.e., nanospheres of 50 to 100 nm, antibody proteins, i.e., a Y-shaped structure of 10 nm in diameter, and as mentioned, the versatile CNTs.

Production, construction, or self-assembly of structured nanoparticles would have been impossible without laser techniques and spectrometry as well as recent scanning probe microscopy. Examples of the latter include STM and atomic force microscopy (AFM).

On a broader, future-oriented scale the benefits of nanostructuring and nanotechnology applications are as follows (Khataee & Mansoori, 2011):

(a) Wear-resistant tires made by combining nanoscale particles of inorganic clays with polymers as well as other nanoparticle-reinforced materials
(b) Greatly improved printing employing nanoparticles that have the best properties of both dyes and pigments as well as advanced ink jet systems
(c) Vastly improved new generation of lasers, magnetic disk heads, nanolayers with selective optical barriers, and systems on a chip made by controlling layer thickness to smaller than a nanometer
(d) Design of advanced chemical, biological, and biomedical detectors

(e) Multifunctional nanoparticles to be used in medicine with vastly improved tumor visualization, drug delivery, and drug targeting capabilities

(f) Chemical-mechanical polishing with nanoparticle slurries, hard coatings, and high hardness cutting tools.

The following selected observations regarding the expected future advances are also worth mentioning at this juncture (Roco et al., 1999):

(i) Mimicking nature, methods of nanotechnology could provide new materials and devices copied from living organisms.

(ii) Photolithographic patterning of matter on the microscale has led to the revolution in microelectronics over the past few decades. With nanotechnology, it will become possible to control matter on every important length scale leading to new systems and devices.

(iii) Micro- and macrosystems constructed of nanoscale components are expected to have entirely new properties that have never before been iden-tified in nature. As a result, by altering and redesigning the structure of materials in the nanoscale range one should be able to systematically and appreciably modify or change selected properties of matter at macro- and microscales. This would include, for example, production of polymers or composites with most desirable properties which nature and existing technologies are incapable of producing.

However, associated with all these positive aspects of building nanostructures atom by atom to control material properties, there are also major challenges to overcome. For example, when building a magnetic memory storage element out of just 12 atoms, IBM researchers encountered several problems. The magnetic data storage bits so far can hold information for only a few hours and at very low temperature, i.e., $T \to 0\,\mathrm{K}$. Also, entering quantum physics, each bit's magnetic field starts interacting and may be unstable, i.e., weakening each bit's ability to hold on to a "one" or a "zero."

5.1.3 Nanothermodynamics

Thermodynamics deals with physical and biochemical phenomena which involve energy transfer (described by heat, work, mass, and temperature changes) as well as total entropy increase (due to irreversibilities, considering the system and its interacting surrounding). In nanothermodynamics, the focus is on state changes of molecules and atoms. For example, considering a monatomic molecule, six coordinates are necessary to describe its state (i.e., three for its location/position and three for its velocity/momentum). Thus, a system of n molecules can be described by a $6n$ set of all position and momentum values. As an aside, this brings to mind the Heisenberg uncertainty principle, i.e., it is impossible to measure a particle's position and momentum simultaneously. Knowing the kinematics and dynamics of the n molecules, the thermodynamic properties and structure of the system (fluid/substance) can be computed via statistical mechanics techniques. For example, the finite macroscopic density is $\rho_n \equiv m_n / \forall$ where m_n is the molecular mass with $n \rightarrow \infty$ and $\forall$ being the associated system's volume. Typical sample droplet populations have $n \geq 10^{23}$ in $\forall = 1 \, \mathrm{cm}^3$.

Laws of Nanothermodynamics. Macroscale as well as nanoscale descriptions of fluid/system states and their property changes due to energy transfer are based on the axiomatic zeroth, first, second, and third laws of thermodynamics (see Sect. 1.2.4). However, nanosystem structure and behavior may differ significantly from macroscale systems, especially in property relations and phase transitions. For example, the pressure in a nanosystem is anisotropic, i.e., being a tensor rather than a scalar, and phases may coexist over bands of temperature and pressure, rather than phase changes occurring at definite T_{sat} and p_{sat} values. Hence, via statistical mechanics techniques working equations of nanothermodynamics have to be formulated (see Khataee & Mansoori, 2011).

The *zeroth law* establishes an absolute temperature scale, where the third law states that the absolute-zero temperature is unattainable. If at all possible, a system/substance/body at zero Kelvin with "zero-point energy" would have zero entropy. On a microscopic level, temperature (and hence other thermodynamic properties) exhibit temporal and spatial fluctuations. It also explains that temperature cannot reach absolute zero because atomic and molecular properties, including energy, fluctuate about their average values. Nevertheless, a definition of the third law of thermodynamics reads: The entropy of a perfect crystal of an element at zero Kelvin is zero.

The familiar *first law* on the macroscopic level states the conservation of mass and energy, barring nuclear reactions. For a closed system ($m = \cancel{c}$), for example, on a differential basis (see Sect. 1.2.4):

$$\delta Q - \delta W = dE \text{ or } \delta \dot{Q} - \delta \dot{W} = \frac{dE}{dt} \approx \left. \frac{\Delta U}{\Delta t} \right|_{\mathrm{system}} \tag{5.1a,b}$$

Here, heat supplied, say, to an ideal piston-cylinder device, generates boundary work, $\delta W = pd\forall$, done by the system onto the surrounding. Equation (5.1) also holds for nanosystems where energy transfer occurs as a result of the collective motion of an assembly of atoms/molecules/particles.

The *second law,* i.e., the total entropy increase principle, indicates both the possible direction a process can proceed and the amount of process irreversibilities and hence waste energy. Specifically, on the macroscopic level:

$$S_{gen} \equiv \Delta S_{total} = \Delta S_{system} + \Delta S_{surrounding} > 0 \qquad (5.2)$$

On the microscopic level, Boltzmann postulated the second law as a statistical statement:

$$S = k_B \ln(N) \qquad (5.3)$$

where $k_B = 1.38054 \times 10^{-23}$ J/K is the Boltzmann constant and N is the probability, or number of distinct ways, of arranging the particles consistently with the overall system properties. Equation (5.3) implies that the entropy of two systems, considered together, is the sum of their separate entropies. Gibbs recognized that this is not necessarily the case and hence defined entropy as:

$$S = -k_B \sum_{i=1}^{N} p_i \ln p_i \qquad (5.4)$$

where n is the system's large number of particles/molecules/elements, p_i is the distribution probability of a set of particles i in the system, and N is the system's number of microscopic possibilities, i.e., energy states, etc. On the nanoscale, we may not deal with "large numbers n and N"; hence, the entropies of inhomogeneous subsystems may not be additive, as implied in Eq. (5.4).

5.2 Liquid Flow in Nanoconduits

5.2.1 Introduction and Overview

To drive a liquid through nanoconduits, say, rectangular channels, membrane pores, or carbon nanotubes, the use of mechanical pumps would not be very suitable because of their relatively large size, high cost, and required pressure drop $\Delta p = Q R_{\text{hydraulic}}$ (see Table 3.4). Instead, nonmechanical "pumps" are employed based on electroosmosis, electrophoresis, magnetohydrodynamics, or capillary effects, i.e., surface tension. For example, capillary pressures (up to 20 bars) are sufficient to fill channels and move finite fluid volumes (known as digital nanofluidics). In electroosmotic processes (see Sect. 3.2.4) positively or negatively charged molecules and particles, e.g., anions or cations in salt water, move due to an applied electric field, causing the bulk fluid to flow (part of continuous nanofluidics).

In any case, the forces between individual atoms or molecules determine all transport phenomena. Hence, ideally one should model nanofluidic processes by accounting for all the individual atomic and molecular interactions. In general, such detailed simulations are cost prohibitive but necessary for flow in nanoconduits with $D_h < 10$ nm. Fortunately, molecular dynamics analysis has shown that the continuum mechanics hypothesis holds for liquid flow in spaces where $\delta_{\text{I.M.}} / L_{\text{system}} < 3 \times 10^{-5}$ (see Sect. 3.2.1). That translates for *liquid flow* in nanoconduits to the requirement of $D_h > 10$ nm.

After a brief summary of solution methods, nanofluidics aspects of velocity slip flow as well as electroosmosis and electrophoresis are summarized, followed by a few examples which rely on the continuum mechanics assumption.

5.2.2 Nontraditional Simulation Methods

With rapid advancements in computer hardware and software, MDS (molecular dynamic simulations) will be the go-to solution approach for fluid flow and fluid-particle dynamics in nanochannels. Nevertheless, for liquid flow in nanochannels where the continuum hypothesis still holds, solutions of the Navier-Stokes equations with possible Navier slip condition for superhydrophobic surfaces are adequate. Alternatively, the lattice Boltzmann method (LBM) can be applied (Succi, 2001).

Molecular Dynamics Simulation. For *liquid* flows in nanochannels with $D_h = 1,\ldots,5$ nm and where $\Delta t \le 10$ ns, molecular dynamics (MD) simulations is suitable. The method is rather simple but computationally limited to very small time steps, where the computing time is proportional to the number of molecules squared. The basic MD simulation steps are:

- Initialize representative molecule positions (making up the flow domain) and velocities.

- Set up Newton's second law of motion with appropriate external forces, e.g., gravity and particle interaction, i.e., intermolecular potential action (see Lennard-Jones potential).
- Time integration to obtain the particle trajectories and analysis of statistical data, yielding system properties such as "fluid" density and viscosity as well as velocity profiles and pressure distributions.
- Taking, for example, the calculation of the density distribution: The flow domain is discretized into finite volumes (or cells). From the particle positions, one knows the number of molecules/atoms in each cell at time t. After sufficient sample calculations the average number of particles per cell volume is known and hence the number density at a "point" in space.

Drawbacks of MD simulations include:

- Prohibitive computing times for realistically large number of atoms/ molecules/ particles (see Example 5.2)
- Lack of accuracy of force fields, i.e., bonding and interacting forces
- Long MD simulation times that may generate cumulative errors in numerical integration

Lennard-Jones Potential for MD Simulations.   Finding a suitable description of the molecule-molecule and molecule-wall interaction forces is essential for a proper simulation of liquid flow in nanoconduits. For example, consider n water molecules in a nanochannel of known initial conditions and geometric dimensions. Say, two molecules with position vectors $\vec{r}_i$ and $\vec{r}_j$ are the distance $r = |\vec{r}_i - \vec{r}_j|$ apart. The forces between molecules i and j are of the same magnitude but opposite direction, i.e., $F_{ij}(r) = -F_{ji}(r)$. The simplest interaction force is derived from the Lennard-Jones potential depicted in Figure 5.2. Specifically,

$$\psi_{ij}(r) = 4\varepsilon \left[a_{ij} (r/d)^{-12} - b_{ij} (r/d)^{-6} \right] \tag{5.5}$$

where ε is the characteristic energy of the molecule, indicating the strength of molecular interaction, d is the diameter of the molecule, and a_{ij} and b_{ij} are coefficients for the attracting/repulsing molecules. Now, the derivative of $\psi_{ij}(r)$ yields the interaction force (see Figure 5.3):

$$F_{ij}(r) = -\frac{d\psi_{ij}}{dr} = \frac{48\varepsilon}{d} \left[a_{ij} \left(\frac{r}{d}\right)^{-13} - \frac{b_{ij}}{2} \left(\frac{r}{d}\right)^{-7} \right] \tag{5.6}$$

Hence, the acceleration of, say, molecule i can be obtained from Newton's second law:

$$m\frac{d^2\vec{r}_j}{dt^2} = \sum_{j=1, j \neq i}^{n} F_{ij} \tag{5.7}$$

Time integration yields the molecular velocity $\vec{v}_j = d\vec{r}_j / dt$ and position/trajectory $\vec{r}_j(t)$ for all n molecules.

Rewriting Eq. (5.5) in a more compact form by summing the attractive and repulsive potentials, the Lennard-Jones (L-J) potential reads:

$$w(r) = A/r^{12} - B/r^6 = 4\varepsilon\left[(\sigma/r)^{12} - (\sigma/r)^6\right] \tag{5.8}$$

Here, ε is the minimum energy which appears as the depth of the *potential well* in Figure 5.2. It should be noted that the minimum total potential energy occurs at $r = \sqrt[6]{2} \cdot \sigma$, where $w(r)_{min} = -\varepsilon$; σ is the finite distance at which the interparticle potential is zero; and r is the distance between the particle molecules. As an example, Figure 5.2 shows the carbon-carbon (C-C) L-J potential versus distance r. For this case, $\sigma = 2.8 \times 10^{-10}$ m and $\varepsilon = 0.0048$ eV. Example 5.2 illuminates Eq. (5.8) with some numerical values for nitrogen being similar to air.

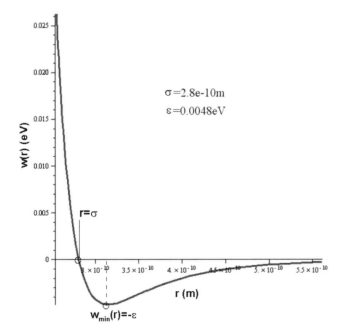

Figure 5.2 Carbon-carbon L-J potential versus r

From Eq. (5.8), the total intermolecular force F versus distance r can be expressed as (see Figure 5.3 for molecular C-C interaction forces):

$$F = -\frac{dw}{dr} = \frac{48\varepsilon}{\sigma}\left[\left(\frac{r}{\sigma}\right)^{-13} - \frac{1}{2}\left(\frac{r}{\sigma}\right)^{-7}\right] \qquad (5.9)$$

Figure 5.4 summarizes both $w(r)$ and $F(r)$, contrasting interacting liquid molecules and gas molecules. Equation (5.9), which is also called the attractive part of the L-J potential, describes the van der Waals force (or dispersion force) as a long-range force. In contrast, the r^{-13}-term, known as Pauli repulsion, has no theoretical justification. The repulsion force should depend *exponentially* on the distance, but the repulsion term of the L-J formula is more convenient due to the ease and efficiency of computing r^{12} as the square of r^6 in Eq. (5.8). Its physical origin is related to the Pauli principle: when the electronic clouds surrounding the atoms start to overlap, the energy of the system increases abruptly. The exponent 12 was chosen exclusively because of ease of computation; thus, the Lennard-Jones potential is an approximation.

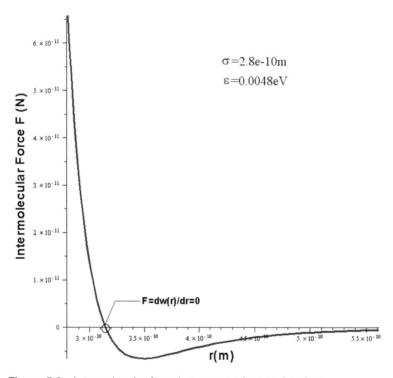

Figure 5.3   Intermolecular force between carbon and carbon

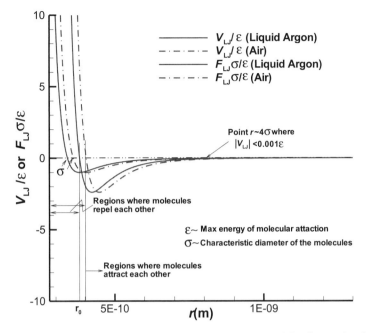

Figure 5.4   Lennard-Jones potential as well as repulsion/attraction forces for liquid argon versus air

---

Example 5.1: What is the number of equations to be solved and how long will it take when describing molecular flow in a nanotube ($R = 30$ nm, $L = 1$ mm; $r_0 = 0.14$ nm of spherical H_2O molecule)

Sketch:	Assumptions:	Concepts:
	• Spherical H_2O molecules consisting of three atoms with three degree of freedom • 66% uniform packing • Uniform flow	• MD based on Newton's second law • Volume balance • $\#_{equations} = \#_{molecules} \times \#_{model\_equations}$ $\qquad \times$ times $\times$ DoF /molecule

Solution:

- $\forall_{nanotube} = 2\pi RL := 1.89 \times 10^{-10}$ m^3

- $\forall_{H_2O\ molecule} = 4\pi r_0^3 / 3 := 1.15 \times 10^{-29}$ m^3

Number of molecules in the nanotube:

- $\dfrac{\forall_{\text{nanotube}}}{\forall_{H_2O \text{ molecule}}} \cdot \dfrac{2}{3} := 1.10 \times 10^{19}$ molecules

Thus, number of equations would be:

- $\#_{\text{equation}} := 1.10 \times 10^{19} \times 3 \times 3 = 10^{19}$ equations

For a computer which can solve 106 equations per second $v_{\text{solution speed}} = 10^{6}/s$. Hence, the required solution time is:

- $\dfrac{\#_{\text{equation}}}{v_{\text{solution speed}}} = 10^{13}s \approx 317,000$ years comment not necessary!

Example 5.2: Use of the Lennard-Jones Potential for Air Molecules

Consider air molecules, represented by nitrogen molecules, to be perfect spheres. Of interest is to check if molecular interaction in air is significant, i.e., $\delta_{\text{I.M.}}^{\text{air}} \approx \sigma_{N_2}$?
    We start with Eq. (5.8):

$$w_{\text{L-J}}(r) = 4\varepsilon \left[(\sigma/r)^{12} - (\sigma/r)^{6} \right]$$

where ε is the minimum attraction energy, r is the distance between a pair of molecules, and σ is the collision diameter. Taking nitrogen values that are similar to air (Bruus, 2008), $\sigma_{N_2} = 0.3667$ nm and $\varepsilon_{N_2}/k_B = 99.8$ K $(k_B = 1.38054 \times 10^{-23}$ J/K$)$ when $p = 101.3$ kPa and $T = 24°C$, we obtain the following:

- At $w_{\text{L-J}}(r = r_0)$ has a minimum, i.e., r_0 is the molecular spacing (see Figures 5.2 and 5.4). Specifically,

$$\frac{\partial w_{\text{L-J}}}{\partial r} := 0 \Rightarrow r_0 = 2^{1/6}\sigma \approx 1.12\sigma \qquad \text{(E5.2-1)}$$

so that

$$w_{\text{L-J}}(r = r_0) = -\varepsilon \qquad \text{(E5.2-2)}$$

- Taking $r = 3\sigma = 1.1$ nm to assure a significant molecular distance, we have:

$$w_{\text{L-J}}(r = 3\sigma) = -0.0055\varepsilon \quad \text{and} \quad \frac{w_{\text{L-J}}(3\sigma)}{k_B} = -0.5 \text{ K} \qquad \text{(E5.2-3)}$$

  where $w_{\text{L-J}}/k_B$ is the interaction energy in Kelvin.
- It is assumed that the intermolecular distance for air is:

$$\delta_{\text{I.M.}}^{\text{air}} \approx \left(\frac{\forall}{N}\right)^{1/3} = \left(k_B T / p\right)^{1/3} = 3.5 \text{ nm} \qquad \text{(E5.2-4)}$$

- Thus, air-molecule interaction can be neglected because $\delta_{\text{I.M.}}^{\text{air}} \gg \sigma_{\text{N}_2}$.
- Comparing molecular energies [K], for air at 24°C, we have the average kinetic energy $E_{\text{kin}}/k_B = 3T/2 = 450$ K, which is much larger than the average interaction energy $w_{\text{L-J}}/k_B = -0.5$ K.

Note: Another application of the L-J potential is given by Travis & Gubbins (2000). They considered simulation of Poiseuille flow through nanoslits.

Nowadays, the Lennard-Jones potential is widely used in molecular dynamics simulation. For convenience, a cut-off distance $r_c = 2.5\sigma$ has been introduced so that the computational L-J potential can be expressed as:

$$w_{\text{comp}}(r) = \begin{cases} w(r) - w(r_c) & r \leq r_c \\ 0 & r > r_c \end{cases} \qquad (5.10)$$

The interaction between liquid molecules and molecules of a solid surface can be described with the L-J force F_{ij} as well. As expected, near any channel wall with its unique molecular structure, the normal distribution of the n molecules varies greatly, resulting in liquid density fluctuations due to the interactions of fluid and solid molecules. Thus, in nanochannels a simple liquid moves in layers of different densities, especially for fluids with high molecular energy ε. This layering effect makes simple fluids, such as water, inhomogeneous. The actual flow is confined in layers smaller than the nanochannel itself. Thus, one can assume an effective viscosity for overall liquid flow, which appears to be higher than the actual fluid viscosity. Additional causes of an elevated "apparent viscosity" assumed in nanofluidics could be due to electroviscous effects, entrapped air bubbles, and other hydraulic resistances.

Clearly, MD simulation of, say, water in nanochannels or nanotubes is very challenging because of the difficulties in modeling the characteristics and influence of bonded and interacting water molecules as well as the interactions with the wall. In order to save computer resources for liquid flow simulations in larger nanochannels, say, $D_h = 10, \dots, 100$ nm, MD simulation resolves the near-wall region while the continuum mechanics approach is used for the bulk motion.

Nanofluidics applications, such as capillary filling in a nanochannel, nanofilters for biochemical separation, water purification, and desalination, are discussed in Abgrall and Nguyen (2009). Particularly useful is the replacement of random anisotropic porous devices/materials, such as membranes, filters, and gels, with controlled customized arrays of pores, i.e., an organized network of nanotubes. These nanopores, or the inverse, an array of nanorods, are integrated in a microsystem with microchannels, actuators, heaters, mixers/separators, and sensors, forming a lab-on-a-chip (LoC) for efficient biochemical/biomedical analyses (see Sect. 4.4).

Lattice Boltzmann Method.   In cases of complex geometries and a nanochannel size range of 2 nm $\leq D_h \leq 10$ nm, the LBM should be considered (Succi, 2001). The LBM is very suitable for massive parallel computing, where the flow domain is discretized with 2-D or 3-D lattices and the probabilistic Boltzmann equation (BE) is numerically solved. The BE, derived from kinetic theory, is a transport equation for a particle (or fluid element) probability density function with a suitable particle collision integral. Starting with an equilibrium distribution function with which the particle/element positions as well as velocity and density fields are initialized on the multinode lattice, two sequential steps are executed at each time level: streaming and colliding. Streaming means that each particle/element moves to the nearest node in the lattice. Collision occurs when particles/elements arrive at a node, interact, and possibly change their velocities and/or directions. Then all quantities are updated at each node and a new distribution function is obtained.

Clearly, further discussion of nontraditional solution methods is beyond the scope of this book; however, some students with experience in using MD simulations or the LBM may employ their special skills when solving suitable course projects (see Sect. 6.5).

5.2.3 Summary of Nanoscale Phenomena and Descriptive Equations

As demonstrated in Part B, it is most convenient and cost-effective when nanofluidics problems can be described by a system of reduced Navier-Stokes equations. The underlying assumption is that the fluid is a quasi-continuum where at the nanochannel walls velocity slip and/or temperature jump may have to be considered. The hydraulic diameter criterion of $D_h \geq 10$ nm holds for nanoconduits of simple geometry and with smooth surfaces. Certain nanofluidics applications are diffusion dominated, using the Stokes-Einstein equation for nanoparticles in the range of $1 \leq d_p \leq 100$ nm, i.e.,

$$D_{AB} = k_B T C_{slip} / \left(3\pi\mu d_p \right) \tag{5.11}$$

where $k_B = 1.38054 \times 10^{-23}$ J/K is the Boltzmann constant and $C_{slip} = \mathcal{O}(1)$.

However, when the species molecules (or nanoparticles) in the carrier liquid are large, diffusion transport is "hindered" and an empirical correlation is appropriate:

$$D_{eff} = D_{AB} \left[1 - 2.104 \left(\frac{d_p}{D_h} \right) + 2.09 \left(\frac{d_p}{D_h} \right)^3 - 0.95 \left(\frac{d_p}{D_h} \right)^5 \right] \tag{5.12}$$

where d_p / D_h is the ratio of nanoparticle-to-hydraulic diameters.

Electrokinetics.   Of the four electrokinetic phenomena mentioned in Sect. 3.2.4, only electroosmosis and electrophoresis are revisited here. As was illustrated in Sect. 3.2.4.1, if a charged nanochannel wall is in contact with an electrolyte and an electric field is applied parallel to the solid-liquid interface, a very thin liquid layer at the wall (i.e., part of the EDL) is set into motion. In turn, due to viscous effects, the moving wall layer drags the liquid bulk with it, resulting in uniform (or plug) flow:

$$u_{EO} = -\frac{\varepsilon_0 \varepsilon_r}{\mu} \zeta E_x = \mu_{EO} E_x \tag{5.13a,b}$$

where $\varepsilon_0 \varepsilon_r = \varepsilon$ is the permittivity, μ is the viscosity, ζ is the zeta potential, $E_x = \Delta V / L$ is the external electric field in the x-direction (with ΔV being the applied net voltage and L being the channel length), while μ_{EO} is the electroosmotic mobility. For example, high near-wall concentrations of positive ions, say, Na$^+$ ions in saltwater on negatively charged glass plates, migrate steadily towards the cathode after an electric field has been externally applied.

Other than changing $\Delta V \sim E_x$, there are also various ways to manipulate the ζ-potential in order to control electroosmotic flow (see Kirby, 2010). However, excess $|\bar{E}|$-values may cause Joule heating.

For a comprehensive mathematical description of ion transport in nanochannels, assuming straight smooth conduits with EDL $< D_h/2$, a set of three equations has to be solved:

(1) The Poisson-Boltzmann equation which describes how the electrostatic potential due to a distribution of charged atoms or molecules varies in space. For example in 1-D:

$$\nabla^2\phi \stackrel{\text{1-D}}{=} \frac{d^2\phi}{dy^2} = -\frac{e}{\varepsilon_0\varepsilon_r}\sum_i n_i z_i \exp\left[-z_i e\phi(y)/k_B T\right] \qquad (5.14\text{a,b})$$

where e is the electron charge, n_i is the EDL ion concentration, and z_i is the valence of ion i.

1.1 Assuming that the electric potential only varies normal to the solid surface and that the ion concentration variation in the normal (i.e., y-) direction at any axial position follows the equilibrium Boltzmann distribution, Eq. (5.14b) reduces to:

$$\nabla^2\phi = -\frac{F}{\varepsilon_0\varepsilon_r}\sum_{i=1}^{2} z_i c_i \qquad (5.15\text{a})$$

$$c_i = c_{i,\infty}\exp\left(-z_i F\phi/RT\right) \qquad (5.15\text{b})$$

and

$$\rho_{\text{electric}} = \sum z_i c_i F \qquad (5.15\text{c})$$

Here $F = 96,485\ \text{C/mol}$ is the Faraday constant, c_i is the ion concentration of the electrolyte, z_i is the valence (where $i = 1$ indicates cations and $i = 2$ anions), and ρ_{electric} is the charge density inside the EDL.

1.2 Assuming a small surface potential, e.g., $z_i\phi_{\text{wall}} < 26$ mV at 25°C, and imposing a linear approximation, Eq. (5.15) reduces to:

$$\frac{d^2\phi}{dy^2} = \kappa^2\phi(y) \qquad (5.16\text{a})$$

and

$$\rho_{\text{electric}} = -\varepsilon E_x d^2\phi/dy^2 \qquad (5.16\text{b})$$

Here $\kappa = \lambda_D^{-1}$ is the Debye-Hückel parameter and λ_D is the Debye length, which is inversely proportional to the square footage of the ionic strength $I_s = \frac{1}{2}\sum c_i z_i^2$.

(2) The Nernst-Planck equation which accounts for the transport of charged atoms/molecules (i.e., ions i) through nanoconduits due to an external

electric potential (see Sect. 3.2.4.4). The total ion flux per unit area is the sum of ion convection, diffusion, and electroosmosis:

$$\vec{j}_i = \vec{v}c_i - \mathcal{D}_i \nabla c_i - \frac{ez_i}{k_B T} \mathcal{D}_i c_i \nabla \phi \qquad (5.17)$$

where $\mathcal{D}_i$ is the ion diffusivity, $e/k_B \hat{=} F/R$, and $-\nabla\phi = \vec{E}$.

(3) The Navier-Stokes equations, including continuity $\nabla \cdot \vec{v} = 0$, and linear momentum:

$$\rho\, D\vec{v}/Dt = \vec{f}_{\text{net pressure}} + \vec{f}_{\text{viscous}} + \vec{f}_{\text{electric}} \qquad (5.18a)$$

For steady-state operation with $\mathrm{Re} < 1$ and constant properties, the balance of forces per unit volume reads:

$$0 = -\nabla p + \mu \nabla^2 \vec{v} - \varepsilon \vec{E} \nabla^2 \phi \qquad (5.18b)$$

Clearly, gravity has been neglected in Eq. (5.18b) but also electroviscous effects due to a streaming potential which, because of EDL interactions, may generate a higher apparent fluid viscosity (Kirby, 2010).

As shown in Sect. 3.2.4.1, the electroosmotic velocity, say, $u_{\text{EO}} = -(\varepsilon/\mu)\zeta E_x$, is generated due to $\vec{f}_{\text{electric}} \equiv \vec{f}_{\text{Coulomb}}$ within a thin wall layer, $1 < \lambda_D < 10$ nm, where ε is the permittivity, ζ is the near-wall electric potential, E_x is the axially applied electric field, and λ_D is the Debye length, i.e., part of the EDL. Clearly, when $\lambda_D \ll D_h/2$ (or $h/2$), u_{EO} can be regarded as a slip velocity. It may have a similar effect as the Navier boundary condition for liquid flow on a rough hydrophobic surface.

Extended Navier Slip Velocity Model for Liquid Flows.   As introduced in Sect. 3.2.3, the basic Navier boundary condition implies that the velocity slip is linearly proportional to the wall velocity gradient and a constant slip length which hypothetically extends linearly from the slip surface into the solid wall (see Figure 3.10):

$$|u| = \delta\, \partial u/\partial x \qquad (5.19)$$

where the slip length δ can be $\delta = 100$ μm for polymeric liquids and $\delta < 100$ nm for flow on a hydrophobic surface. In their review paper, Mathews & Hill (2008) summarize two additional Navier BCs, occasionally necessary to match experimental evidence. One is a nonlinear postulate:

$$|u| = \delta\left(|\partial u/\partial x|\right)^n \qquad (5.20)$$

where $n > 0$ is a parameter to be adjusted to experimental evidence. The other one is a generalized Navier BC, where at higher shear rates the slip length increases rapidly:

$$\delta = \alpha\left(1 - \beta\dot{\gamma}\right)^{-1/2} \qquad (5.21a)$$

where

$$\beta = \dot{\gamma}_{\text{critical}}^{-1} \qquad (5.21b)$$

with $\dot{\gamma}_{\text{critical}}$ being a maximum shear rate which, like α, has to be determined experimentally (see Example 5.3).

Higher Order Slip Flow Model for Rarefied Gases.   New modeling approaches are required when a nanochannel's characteristic length scale (e.g., height, width, or D_h) decreases, the fluid's molecular spacing increases, and/or fluid-wall pairings are superhydrophobic (see Figure. 5.5).

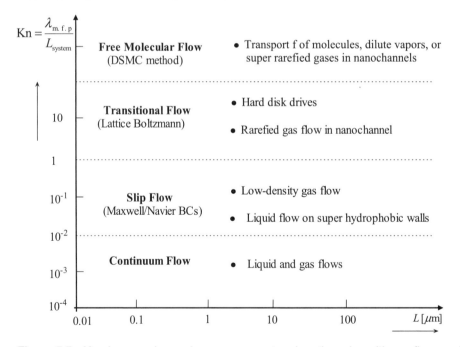

Figure 5.5   Knudsen number regimes versus system length scales with gas flow applications

For example, in the important "slip flow" regime (see Sect. 3.2.3) rarefied gas flow in nanochannels can be simulated via the solution of the (reduced) N-S equation with a first- or second-order boundary condition for the velocity slip. In general, according to Maxwell:

$$u(n=0) = u_{slip} = u_{gas} - u_{wall} = \frac{2-\sigma}{\sigma}\left[Kn(\partial u/\partial n)\big|_{wall} + \frac{1}{2}Kn^2(\partial^2 u/\partial n^2)\big|_{wall}\right] \quad (5.22)$$

where σ is the tangential momentum accommodation factor with $\sigma = 1.0$ for fully diffused and $\sigma = 0$ for fully specular surfaces. Section 5.3 discusses rarefied gas flow in nanochannels and solution methods with applications.

5.2.4 Nanochannel Flow Examples

In the following two quasi-continuum liquid flow applications (i.e., Examples 5.3 and 5.4) different Navier slip BCs are examined. The third example (5.5) deals with reactive species mass transfer in electroosmotic nanochannel flow.

Example 5.3: Slip Flows in a Carbon Nanotube Leading to Flow Rate Augmentation

Consider water flow in a CNT of radius r_0 and length L, assuming Poiseuille-type flow. Analyze the impact of a low-viscosity wall layer h (called a depletion layer) on the volumetric flow rate (see Myers, 2011). Compare results to Navier slip flow with slip length δ.

Sketches:	*Assumptions:*	*Concepts:*
(a) Two-layer flow	• Poiseuille flow conditions • Constant thickness h and properties (μ_h, μ_b)	• Reduced N-S equations • Hagen-Poiseuille (H-P) solution • Interface matching conditions at $r = r_0 - h = r_{bulk} = r_b$
(b) Navier slip model	• Constant slip length δ	• Basic Navier boundary condition

Background:

- Flow rate enhancement $\kappa = Q_{\text{actual}}/Q_{\text{HP}} > 1$
- Hydrophobic surface, i.e., "rough" surface plus vapor/air pockets near wall
- H_2O molecule distribution and orientation at/near the CNT surface varies for $D_{\text{CNT}} < 10$ nm, i.e., lower viscosity (and density) occur in the "depletion layer"

Solution:

- From the reduced N-S equation, we write for the two flow regions:

$$\frac{\mu_b}{r}\frac{d}{dr}\left(r\frac{du_b}{dr}\right) = -\frac{\Delta p}{L} \; ; \text{ in the bulk region } 0 \le r \le r_b = r_0 - h \qquad \text{(E5.3-1a)}$$

and

$$\frac{\mu_h}{r}\frac{d}{dr}\left(r\frac{du_h}{dr}\right) = -\frac{\Delta p}{L} \; ; \text{ in the depletion layer } r_b \le r \le r_0, \text{ i.e., } h = r_0 - r_b \text{ (E5.3-1b)}$$

subject to:

$$\left.\frac{du_b}{dr}\right|_{r=0} = 0 \; ; \; u(r = r_0) = 0 \text{ and at } r = r_b : u_b = u_h; \; \mu_b\frac{du_b}{dr} = \mu_h\frac{du_h}{dr} \qquad \text{(E5.3-2)}$$

Integration yields:

$$u_b = \frac{\Delta p r_0^2}{4\mu_b L}\left[\left(\frac{r_b}{r_0}\right) - \left(\frac{r}{r_0}\right)^2\right] + \frac{\Delta p r_0^2}{4\mu_h L}\left[1 - \left(\frac{r_b}{r_0}\right)^2\right] \qquad \text{(E5.3-3a)}$$

and

$$u_h = \frac{\Delta p r_0^2}{4\mu_h L}\left[1 - \left(\frac{r}{r_0}\right)^2\right] \qquad \text{(E5.3-3b)}$$

Assuming $\hat{u} = u/\left(\Delta p r_0^2/4\mu_b L\right)$, we have:

$$\hat{u}_b = \left[\left(\frac{r_b}{r_0} \right) - \left(\frac{r}{r_0} \right)^2 \right] + \frac{\mu_b}{\mu_h} \left[1 - \left(\frac{r_b}{r_0} \right)^2 \right] \qquad \text{(E5.3-4a)}$$

$$\hat{u}_h = \frac{\mu_b}{\mu_h} \left[1 - \left(\frac{r}{r_0} \right)^2 \right] \qquad \text{(E5.3-4b)}$$

- Flow rate:

$$Q = \int_A u \, dA \; ; \; dA = 2\pi r \, dr \; ; \; A = \pi r_0^2$$

$$Q_{\text{total}} \equiv Q = 2\pi \left[\int_0^{r_b} u_b r \, dr + \int_{r_b}^{r_0} u_h r \, dr \right]$$

$\therefore$

$$Q = \frac{\pi r_b^4 \Delta p}{8 \mu_b L} \left\{ 1 - \frac{2\mu_b}{\mu_h} \left[1 - \left(\frac{r_0}{r_b} \right)^2 \right] \right\} + \frac{\pi r_b^4 \Delta p}{8 \mu_h L} \left[1 - \left(\frac{r_0}{r_b} \right)^2 \right]^2 \qquad \text{(E5.3-5a)}$$

$$\hat{Q} = \frac{Q}{\left(\dfrac{\pi r_0^4 \Delta p}{8 \mu_b L} \right)} = \frac{r_b^4}{r_0^4} \left\{ 1 - \frac{2\mu_b}{\mu_h} \left[1 - \left(\frac{r_0}{r_b} \right)^2 \right] \right\} + \frac{r_b^4}{r_0^4} \frac{\mu_b}{\mu_h} \left[1 - \left(\frac{r_0}{r_b} \right)^2 \right]^2 \qquad \text{(E5.3-5b)}$$

- Hagen-Poiseuille flow solutions:

$$u_{\text{HP}} = \frac{\Delta p r_0^2}{4 \mu L} \left[1 - \left(\frac{r}{r_0} \right)^2 \right] \text{ and } Q_{\text{HP}} = \frac{\pi r_0^4 \Delta p}{8 \mu L} \qquad \text{(E5.3-6a,b)}$$

Thus,

$$\hat{u}_{\text{HP}} = \left[1 - \left(\frac{r}{r_0} \right)^2 \right] \text{ and } \hat{Q}_{\text{HP}} = \frac{Q_{\text{HP}}}{\left(\dfrac{\pi r_0^4 \Delta p}{8 \mu_b L} \right)} = 1 \qquad \text{(E5.3-7a,b)}$$

Hence, the enhancement ratio reads:

$$\kappa \equiv \frac{Q}{Q_{HP}} = \left(\frac{r_b}{r_0}\right)^4 + \frac{\mu_b}{\mu_h}\left[1-\left(\frac{r_b}{r_0}\right)^4\right] \qquad \text{(E5.3-8)}$$

Clearly, $\kappa > 1$ for flow rate increase, which implies $\mu_b \gg \mu_h$, i.e., a thin, depleted water molecule layer at the wall with $\mu_h \ll 1$.

Myers (2011), taking a measured $\kappa = 8$ for a CNT of $r_0 = 20$ nm and $h = 0.7$ nm, calculated:

$$\mu_h = \mu_b\left[\frac{r_0^4 - r_b^4}{\kappa r_0^4 - r_b^4}\right] := 0.018\mu_b \qquad \text{(E5.3-9a,b)}$$

or

$$\frac{\mu_h}{\mu_b} = 0.018, \text{ where, as a reference ratio, } \frac{\mu_{air}}{\mu_{water}} = 0.02$$

Note, when $r_0 \approx h \Rightarrow \kappa \approx 50$.

Expressing κ in terms of h and r_0, we obtain:

$$\kappa = 1 + \frac{4h}{r_0}\left(\frac{\mu_b}{\mu_h} - 1\right)\left[1 - \frac{3h}{2r_0} + \left(\frac{h}{r_0}\right)^2 - \frac{1}{4}\left(\frac{h}{r_0}\right)^3\right] \qquad \text{(E5.3-10)}$$

Hence, Q_{HP} is only valid when $h/r_0 \ll \mu_h/\mu_b$; hence, practically when $r_0 > 3\mu\text{m}$.

- Navier slip-model:
  Invoking $u(r = r_0) = -\delta\,du/dr\big|_{r=r_0}$ instead of $u(r = r_0) = 0$, we obtain:

$$u_{slip} = \frac{\Delta p r_0^2}{4\mu L}\left[1 - \left(\frac{r}{r_0}\right)^2 + \frac{2\delta}{r_0}\right] \qquad \text{(E5.3-11a)}$$

Thus,

$$\hat{u}_{slip} = \frac{u_{slip}}{\dfrac{\Delta p r_0^2}{4\mu L}} = \left[1 - \left(\frac{r}{r_0}\right)^2 + \frac{2\delta}{r_0}\right] \qquad \text{(E5.3-11b)}$$

so that

$$Q_{slip} = Q_{HP}\left[1 + \frac{4\delta}{r_0}\right]$$
(E5.3-12a)

$$\hat{Q}_{slip} = \frac{Q_{slip}}{Q_{HP}}\left[1 + \frac{4\delta}{r_0}\right]$$
(E5.3-12b)

As a result, the enhancement ratio due to slip flow is:

$$\kappa_{slip} = 1 + \frac{4\delta}{r_0}$$
(E5.3-13)

Using the previous values ($r_0 = 20$ nm and $\kappa = \kappa_{slip} = 8$) yields a slip length:

$$\delta = 35 \text{ nm}$$

Combining depletion layer h and slip length δ with the assumed viscosity ratio, we obtain:

$$\delta = h\left(\frac{\mu_b}{\mu_h} - 1\right)\left[1 - \left(\frac{3h}{2r_0}\right) + \left(\frac{h}{r_0}\right)^2 - \frac{1}{4}\left(\frac{h}{r_0}\right)^3\right]$$
(E5.3-14)

Graphs:

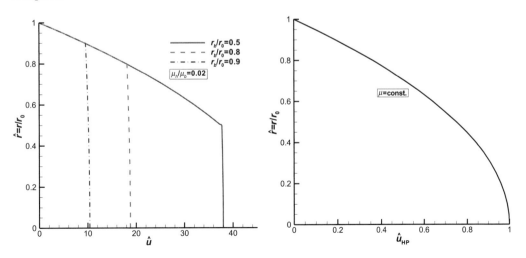

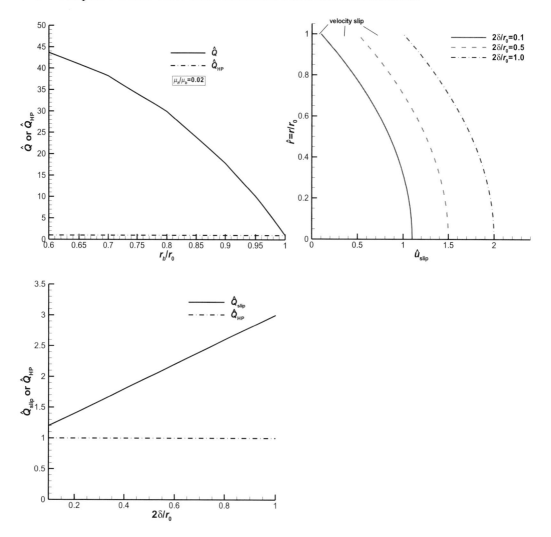

Comments:

(i) For all depletion-layer thicknesses the (dimensionless) bulk velocity is almost uniform because of the dominant additive constant in the $\hat{u}_b(\hat{r})$-equation. For example, with $r_b/r_0 = 0.5$ and $\kappa = 2$: $\hat{u}_b(\hat{r}) = (0.5 - \hat{r}^2) + 1.55$; while for $r_b/r_0 = 19.3/20$ and $\kappa = 8$: $\hat{u}_b(\hat{r}) = (0.985 - \hat{r}^2) + 1.654$.

(ii) The H-P solution provides the familiar parabolic velocity profile $\hat{u}_{HP}(\hat{r})$ as shown in the second graph.

(iii) As expected, the thicker the depletion layer the higher is the throughput because of the larger low-viscosity-flow area.

(iv) The velocity profile with slip at the wall, $\hat{u}_{\text{slip}}(\hat{r})$, greatly depends on the assumed slip length δ but maintains its parabolic character. The actual fluid slip at the wall appears as $\hat{u}_{\text{slip}}(\hat{r}=1.0)$.

(v) The last graph shows that the flow rate gain, expressed as κ_{slip}, is linearly dependent on δ and the increase is modest when compared to the (theoretical) "depletion-layer" approach.

While Example 5.3 deals with the basic Navier BC, in contrast to the "low-viscosity wall-layer" hypothesis, Example 5.4 focuses on the nonlinear and generalized Navier BCs for steady fully developed liquid flow in a nanotube.

Example 5.4: Extended Navier Boundary Conditions

Sketch:	*Assumptions:*	*Concepts:*
	• Poiseuille flow conditions with constant fluid properties	• Cylindrical Poiseuille flow equation for $u(r)$ with extended Navier boundary conditions at $r=r_0$

Solution:

➤ Solving the reduced N-S equation yields (see Example 5.3):

$$\hat{u}(\hat{r}) \equiv \frac{u(r)}{u_{\max}} = 1 - \left(\frac{r}{r_0}\right)^2 + C \qquad \text{(E5.4-1)}$$

where $\hat{r} = r/r_0$, $u_{\max} = \Delta p r_0^2/(4\mu L)$, and C depends on $\hat{u}(\hat{r}=1) = \hat{u}_{\text{slip}}$.

➤ Consider three Navier BCs at $\hat{r}=1$:

• $\hat{u}_{\text{slip}} = -\hat{\delta}\, d\hat{u}/d\hat{r}$ for $\delta = \text{¢}$ (see Eq. (5.19))

• $\hat{u}_{\text{slip}} = -\hat{\delta}\left|\dfrac{d\hat{u}}{d\hat{r}}\right|^n$ for $\delta = \text{¢}$ (see Eq. (5.20))

• $\hat{u}_{\text{slip}} = -\hat{\alpha}\left(1 + \hat{\beta}\dfrac{d\hat{u}}{d\hat{r}}\right)^{-1/2}\dfrac{d\hat{u}}{d\hat{r}}$ for $\delta = \text{¢}$ (see Eq. (5.21)) where $\hat{\delta} = \delta/r_0$, n, $\hat{\alpha} = \alpha/r_0$, and $\hat{\beta} = \beta/r_0$ are all constants greater than zero. The no-slip condition gives the classic Poiseuille profile:

$$\hat{u}(\hat{r})\big|_{\text{Poiseuille flow}} = 1 - \hat{r}^2 \qquad (\text{E5.4-2})$$

The three Navier BCs yield:

- $\hat{u}(\hat{r})\big|_{\text{basic slip}} = 1 - \hat{r}^2 + 2\hat{\delta}$ (E5.4-3)

- $\hat{u}(\hat{r})\big|_{\text{nonlinear slip}} = 1 - \hat{r}^2 + 2^n\hat{\delta}$ (E5.4-4)

- $\hat{u}(\hat{r})\big|_{\text{generalized slip}} = 1 - \hat{r}^2 + 2\hat{\alpha}\big/\sqrt{1 - 2\hat{\beta}}$ with $0 < \hat{\beta} < 0.5$ (E5.4-5)

Graph:

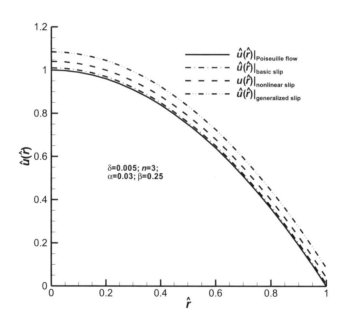

Comments:

- As expected, the no-slip Poiseuille flow profile generates the lowest flow rate, i.e., $Q_{\text{H-P}}$.
- All velocity profiles maintain the basic quadratic form with add-ons for velocity slip.
- Velocity slip corrections from basic to generalized slip increase the through-put.

End Effects in Nanotube Flow. The underlying assumption of "fully developed flow" in micro/nanotubes and channels (see Examples 5.3 and 5.4) is invalid when the conduit is short and reservoirs feed and collect the fluid. Thus, the *entrance and exit flow effects* cause locally a nonlinear pressure drop and hence influence the total volumetric flow rate.

Considering a cylindrical conduit (radius r_0 and length L) as in, say, CNTs or membrane pores, the Hagen-Poiseuille pressure drop for a single CNT with slip length l_s is:

$$\Delta p_{\text{H-P}} = \frac{8\mu L Q_{\text{Slip}}}{\pi r_0^4 + 4\pi r_0^3 l_s} \text{ and } Q_{\text{H-P}} = \frac{\pi r_0^4 \Delta p}{8\mu L} \tag{5.23a,b}$$

where l_s is inversely related to the friction factor, i.e., $f = l_s^{-1}$. The pressure drop due to end effects can be approximated as (Sisan & Lichter, 2011):

$$\Delta p_{\text{end}} \approx \frac{3\mu Q}{r_0^3} \tag{5.24}$$

Hence, the total pressure drop $\Delta p_{\text{total}} = \Delta p_{\text{H-P}} + \Delta p_{\text{end}}$ so that:

$$Q_{\text{total}} = \frac{\Delta p_{\text{total}}}{R_{\text{total}}} \tag{5.25a}$$

where the hydraulic resistance is:

$$R_{\text{total}} = \frac{3\mu}{r_0^3} + \frac{8\mu L}{\pi r_0^4 + 4\pi r_0^3 l_s} \tag{5.25b}$$

Hence,

$$Q_{\text{total}} = \frac{\Delta p_{\text{total}}}{3\mu} r_0^3 \left[\left(1 + \frac{8L}{3\pi r_0} \right) + \frac{2L}{3\pi l_s} \right]^{-1} \tag{5.26a}$$

Clearly, with increasing slip length, say $l_s \gg r_0$:

$$Q_{\text{total}} = \frac{\Delta p_{\text{total}}}{3\mu} r_0^3 \left[1 + \frac{2L}{3\pi l_s} \right]^{-1} \tag{5.26b}$$

where

$$\frac{\Delta p_{total}}{3\mu} r_0^3 = Q_{frictionless} \qquad (5.26c)$$

can be interpreted as the flow rate in a frictionless nanotube.

Graph:

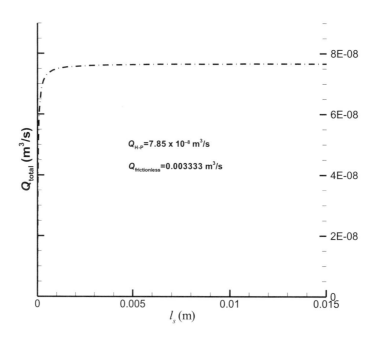

Comments:

- As expected, $Q_{total}(l_s) > Q_{H-P} = Q_{total}(l_s = 0)$.
- For $l_s > 0.1\mu m$, Q_{total} has gained an almost constant value.

Example 5.5: Reactant Conversion in a Nanochannel

Consider pure electroosmotic flow (EOF) in a chemical reactor, represented by a nanochannel of width $2h$ and length L. The inner channel surfaces are coated with a catalyst, causing ideal chemical reaction.

Sketch:	Assumptions:	Concepts:
	• Continuum hypothesis holds • Steady fully developed EOF • Superthin EDL with $\phi_{wall} \approx \zeta = \phi$ • Axial species convection with cross-diffusion	• $u(y) \approx u_{EO} = \phi$ • Reduced mass transfer equation • Constant diffusivity $\mathcal{D}$ • BCs: $c(x=0)=c_0$ and $c(y=\pm h)=0$, implying that species vanish when touching the channel walls

Solution:

• Electrolyte flow solution with $\kappa = \lambda_D^{-1}$ is (see Eq. (3.57b) in Sect. 3.2.4.2):

$$u(y) = u_{EO}\left[1 - \frac{\cosh(\kappa y)}{\cosh(\kappa h)}\right] \text{ for } -h \leq y \leq h \qquad (E5.5\text{-}1)$$

For $\kappa h \gg 1$, i.e., very high electrolyte concentration and hence superthin Debye layer:

$$u(y) \approx u_{EO} = \text{constant} \qquad (E5.5\text{-}2)$$

• Mass transfer (see Eq. (2.79) in Sect. 2.5) is governed here by axial convection and normal diffusion:

$$u\frac{\partial c}{\partial x} = \mathcal{D}\frac{\partial^2 c}{\partial y^2} \qquad (E5.5\text{-}3)$$

subject to:
➤ $c(x=0)=c_0$ inlet reactant concentration
➤ $c(y=\pm h)=0$ superfast chemical reaction causing reactant to vanish
➤ $\partial c/\partial y\big|_{y=0} = 0$ flow symmetry

• Nondimensionalization   with $\hat{c} = c/c_0$, $\hat{u} = u/u_{EO}$, $\hat{x} = x/L$, and $\hat{y} = y/h$ yields:

$$K\frac{\partial \hat{c}}{\partial \hat{x}} = \frac{\partial^2 \hat{c}}{\partial \hat{y}^2} ; K \equiv \frac{u_{EO}h^2}{\mathcal{D}L} \tag{E5.5-4a,b}$$

subject to $\hat{c}(\hat{x}=0)=1$, $\hat{c}(\hat{y}=1)=0$, and $\partial \hat{c}/\partial \hat{y}\big|_{\hat{y}=0}=0$.

- The PDE being linear and separable allows for a product solution

$$\hat{c} = \sum_{n=0}^{\infty} X_n(\hat{x}) \cdot Y(\hat{y}) \tag{E5.5-5}$$

The solution with eigenvalues $\lambda_n = (2n-1)\pi/2$ is:

$$\hat{c}(\hat{x},\hat{y}) = 2\sum_{n=1}^{\infty} \frac{\sin \lambda_n}{\lambda_n} \cos(\lambda_n \hat{y}) \cdot \exp\left(-\frac{\lambda_n^2}{K}\hat{x}\right) \tag{E5.5-6}$$

- The axial reactant conversion is defined as $CON \equiv 1-\hat{c}_{avg}(\hat{x})$, where $\hat{c}_{avg}$ is the cross-sectionally averaged reactant concentration:

$$\hat{c}_{avg}(\hat{x}) = \frac{\int_0^h ucdy}{\int_0^h udy} = \int_0^1 \hat{c}d\hat{y} = 2\sum_{n=1}^{\infty} \frac{\sin \lambda_n}{\lambda_n} \exp\left(-\frac{\lambda_n^2}{K}\hat{x}\right) \tag{E5.5-7}$$

Graph:

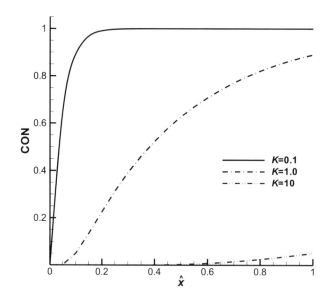

Comments:

- At low K-values, where $K \sim h^2$ and $h \ll 1$ or $K \sim (u_{EO}, \mathcal{D}^{-1})$, the initial substrate amount c_0 is quickly converted, i.e., $CON(\hat{x} = 0.2) = 1.0$.
- When $K \geq 10$, e.g., at high u_{EO}-values, the reactant is swept through the nanochannel without hardly any conversion.

5.3 Rarefied Gas Flow in Nanochannels

5.3.1 Overview

Compared to water where there are about $n_{water} = 10^{10}$ H$_2$O molecules in 1 μm^3, gases are quite dilute, e.g., $n_{air} = 10^7 \, \mu$m^{-3}, and rarefied gases even more so. In fact, when the system characteristic length is less than the mean-free path, say, $\lambda_{mfp} \geq 100$ nm, collisions of the gas molecules occur more often with the nanochannel walls than with each other. Diffusion, typically the dominant transport mechanism in nanochannels, relies on the Knudsen diffusion coefficient rather than the Stokes-Einstein equation (Probstein, 1994):

$$D_{Kn} = \frac{1}{3} u_{molecule} \times L_{system} \tag{5.27}$$

where $u_{molecule} = \sqrt{2RT}$ with R being the gas constant and L_{system} the channel height or width or hydraulic diameter $D_{hydraulic} = 4A/P$.

Knudsen Number Regimes.   As discussed, when convection is dominant and the Knudsen number is in the range of $0.01 < \text{Kn} = \lambda_{mfp}/L_s \leq 0.1$, velocity slip and possible temperature jump boundary conditions have to be invoked when solving the Navier-Stokes equations (see Sect. 3.2.3). The need for imposing such modified BCs arises when the fluid flow adjacent to the wall is not in thermodynamic equilibrium. Specifically, collisions between fluid molecules and the solid surface are not as superfrequent as in normal gases, which allows in the statistical average for a tangential fluid velocity on the solid surface. For (rarefied) gas flows in nanochannels the Knudsen number may greatly increase as the hydraulic diameter decreases. This gas flow description moves from the continuum via the slip flow regime to the transition or even the free molecular regime where $\text{Kn} \equiv \lambda/D_h > 10$ (see Sects. 3.2.3 and 4.2.3). The scenarios can be illustrated with simple micro-to-nano Couette flow (see Figure 5.6).

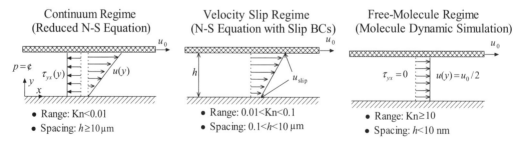

Figure 5.6   Gas flow regimes in Couette macro-to-nano conduits

In the "free molecular regime," i.e., considering a very low gas density or channel wall spacing where $h \leq \lambda_{mfp}$, molecules may travel from one wall to the other unperturbed. Hence, statistically all locations within the flow field are equivalent, i.e., no gradients exist. Furthermore, the gas being far from an equilibrium state, the concept of viscosity and hence the Stokes macroscale hypothesis, $\tau_{yx} = -\mu du/dy$ in this case, fail.

Clearly, nanofluidics problems falling into the continuum and velocity slip regimes, i.e., $0 < \text{Kn} < 0.1$, can be readily handled by solving the (reduced) system of N-S equations with appropriate boundary conditions, as illustrated in Parts A and B. For solving transport phenomena in the transitional regime, a statistical method, say, direct simulation Monte Carlo, is appropriate while for $\text{Kn} > 10$ MD simulations are necessary.

Direct Simulation Monte Carlo.   In the transitional Knudsen number region, i.e., $0.1 \leq \text{Kn} \leq 3$, the direct simulation Monte Carlo (DSMC) method is appropriate. DSMC is a powerful numerical solution approach for high-Knudsen-number gas flows in complex microconduits. However, based on the number of necessary packages of molecules to be considered, the computational effort can be prohibitive, and the reliability of input parameters, slow convergence, as well as the level of statistical noise can be taxing.

The basic DSMC steps are:

- Selecting m-packages, i.e., a large number of actual molecules, representing "simulation molecules"
- Discretizing flow domain into cells of size $\lambda_{mfp}/3$ tracking representative packages of molecules forming cells, which are moved over time interval $\Delta t < \Delta t_{collision}$
- Initialization of cells of molecules and inlet/outlet boundary conditions
- Molecule indexing and cell/particle tracking
- Simulation of particle-particle and particle-wall collisions employing a probabilistic procedure. Note, molecular motion and intermolecular collision are decoupled, recalling that $\Delta t_{motion} < \Delta t_{collision}$, and hence a probabilistic treatment is required.
- Sampling of macroscopic properties of the flow field given within each cell, i.e., at the cell centers. For unsteady flows they are obtained by ensemble averaging of numerous independent calculations.

5.3.2 Nanochannel Flow Examples

As stressed numerous times, steady, laminar, and fully developed are the common assumptions for gas flow in long nanochannels at low Knudsen numbers (i.e., $0 < \text{Kn} < 0.1$). An exception is illustrated with Example 5.6, while Example 5.7

assumes again Poiseuille-type flow. More realistic applications are discussed in Chapter 6 and suggested as course projects in Sect. 6.5.

Example 5.6: Empirical Second-Order Velocity Slip of a Rarefied Gas between Two Parallel Plates a Distance h Apart

Consider a half-channel of developing flow where we define $\text{Kn} = \lambda/2h$ and take

$$u_{\text{slip}} = -2c_1 \text{Kn} \frac{\partial u}{\partial n}\bigg|_{\text{wall}} - 4c_2 \text{Kn}^2 \frac{\partial^2 u}{\partial n^2}\bigg|_{\text{wall}} \tag{E5.6-1}$$

Arkilic et al. (1997) determined experimentally the constants to be $c_1 = 1.466$ and $c_2 = 0.9756$.

Sketch:	Assumptions:	Concepts:
$y = h$ $n \downarrow$ $u(y,z)$ y x z $y = h/2$ u_{slip}	• Steady laminar pseudo-1-D flow, i.e., $u = \bar{u}(z) \cdot \text{fct}(y)$ • Constant properties	• Integral form of the momentum equation (see Sect. 1.3.2.2)

Solution:

The 1-D integral form of the axial momentum equation (see Sect. 1.3.2.2) reads (see Graph I):

$$-A\,dp - 2A\frac{\tau_w}{h}dz = d\left[\int_A (\rho u^2)\,dA\right] \tag{E5.6-2}$$

Graph I:

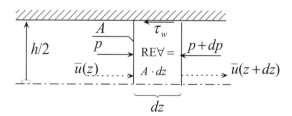

Integration, expressing $\tau_w \sim \Delta p$, assuming symmetry, and invoking the velocity slip condition (E5.6-1) yield:

$$u(y,z) = \bar{u}(z)\frac{\left(\dfrac{y}{h}\right) - \left(\dfrac{y}{h}\right)^2 + 2c_1\mathrm{Kn} + 8c_2\mathrm{Kn}^2}{1/6 + 2c_1\mathrm{Kn} + 8c_2\mathrm{Kn}^2} \qquad (E5.6\text{-}3)$$

where $u_{slip} = u(y = h, z)$, i.e.,

$$u_{slip} = \bar{u}(z)\frac{2c_1\mathrm{Kn} + 8c_2\mathrm{Kn}^2}{1/6 + 2c_1\mathrm{Kn} + 8c_2\mathrm{Kn}^2} \qquad (E5.6\text{-}4)$$

and

$$\hat{u}(\hat{y}) = u(y/h)/\bar{u}(z) \qquad (E5.6\text{-}5)$$

Graph II:

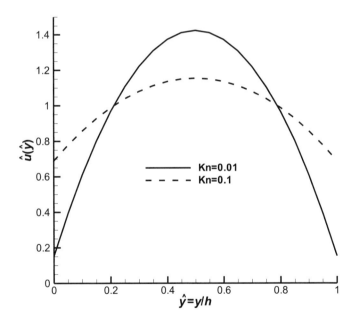

Comment:

With any given $\bar{u}(z)$ the parabolic axial velocity profile greatly depends on the Knudsen number, as encapsulated in the empirical u_{slip}.

Example 5.7: Heating of Nanochannel Gas Flow in the Slip Regime

Consider thermal Poiseuille flow between parallel plates ($2h$ apart and at the wall constant $q_1 = q_2 = q_w$) with velocity slip and temperature jump. Find expressions for the temperature profile and Nusselt number.

Sketch:	*Assumptions:*	*Concepts:*
	• Decoupled Poiseuille flow, i.e., constant properties, etc. • First-order Maxwell slip model • Thermally fully developed flow	• Velocity profile from Example 3.6 • Here, $\dfrac{\partial T}{\partial x} = \dfrac{dT_m}{dx} = \cancel{c};$ $T_m \triangleq$ mean temperature

Solution:

• Nondimensionalization

$$\hat{u} = \frac{u}{u_m}; \quad \hat{y} = \frac{y}{h}; \quad \hat{x} = \frac{\alpha x}{u_m h^2}; \quad \Theta = \frac{T - T_w}{q_w h / k},$$

where $u_m = (1/A)\int u dA$, $\alpha = k/\rho c_p$, $T_w = T(y = h)$, $q_w = -k\,\partial T/\partial y\big|_{y=h}$.

• Gas flow field with velocity slip $u_{slip} = \lambda_{mfp}\,\partial u/\partial y\big|_{y=\pm h}$ and symmetry $\partial u/\partial y\big|_{y=0} = 0$ reads:

$$\hat{u} = \frac{3}{2}\left(\frac{1 - \hat{y}^2 + 4\mathrm{Kn}}{1 + 6\mathrm{Kn}}\right); \quad \text{where } \mathrm{Kn} \equiv \frac{\lambda_{mfp}}{h} \qquad \text{(E5.7-1a,b)}$$

• Reduced heat transfer equation:

$$u\,dT_m/dx = \alpha\,\partial^2 T/\partial y^2 \qquad \text{(E5.7-2)}$$

Subject to constant-wall-heat-flux or temperature jump conditions:

$$q_1 = -k\frac{\partial T}{\partial y}\bigg|_{y=h} = q_2 = -k\frac{\partial T}{\partial y}\bigg|_{y=-h} = q_w, \; T(x, y = \pm h) = T_w + \kappa\frac{\partial T}{\partial y}\bigg|_{y=\pm h}$$

where $\kappa = \left(2\gamma/\left(1+\gamma\right)\right)\cdot\lambda_{\mathrm{mfp}}/\mathrm{Pr}\ \left(\gamma \equiv c_{p}/c_{v},\ \mathrm{Pr} = \alpha/\nu\right)$.

- In nondimensional form with $\Theta = \Theta\left(\hat{y}\right)$ only, so that Eq. (E5.7-2) reads:

$$\frac{d^{2}\Theta}{d\hat{y}^{2}} = \frac{3}{2}\left(\frac{1-\hat{y}^{2}+4\mathrm{Kn}}{1+6\mathrm{Kn}}\right)\frac{dT_{m}}{d\hat{x}}\frac{1}{hq_{w}/k} \tag{E5.7-3}$$

- Integration and using symmetry, i.e., $\Theta'(0)=0$, yield:

$$\frac{d\Theta}{d\hat{y}} = \left[\frac{3\hat{y}}{2}\left(\frac{1-\hat{y}^{2}/3+4\mathrm{Kn}}{1+6\mathrm{Kn}}\right)\right]\frac{dT_{m}}{d\hat{x}}\frac{k}{hq_{w}} \tag{E5.7-4}$$

Assuming $\left(dT_{m}/d\hat{x}\right)\cdot k/hq_{w} = -1$, a second integration, and using $\Theta(1) = 2\mathrm{Kn}$ provide:

$$\Theta\left(\hat{y}\right) = \left[\frac{3}{2}\frac{1}{\left(1+6\mathrm{Kn}\right)}\frac{dT_{m}}{d\hat{x}}\frac{k}{hq_{w}}\left(\hat{y}^{2}/2-\hat{y}^{4}/12+2\hat{y}^{2}\mathrm{Kn}-5/12-2\mathrm{Kn}\right)\right] \tag{E5.7-5}$$
$$+2\mathrm{Kn}$$

Graph:

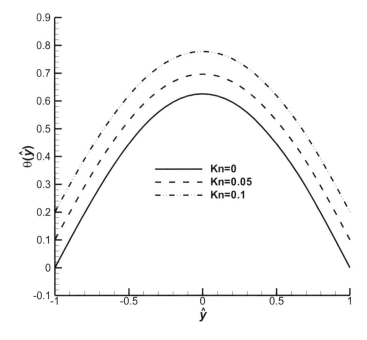

Comment:

As the Knudsen number is increased into the velocity slip regime, the symmetric fourth-order temperature profile is shifted to larger values because of the temperature jump effect.

5.4 Homework Assignments and Course Projects

As in Section 4.5, the homework assignments consist of concept questions (Sect. 5.4.1) and suggested course projects (see Sect. 5.4.2) with a focus on Chapter 5 material.

5.4.1 Text-Related Questions and Tasks

5.1 Discuss the main physical and mathematical differences between microfluidics and nanofluidics. Provide illustrations and contrasting sample applications.

5.2 Show that, indeed, (i) the surface area of n nanocubes obtained from a 1-m cube covers the state of Delaware and (ii) a 1-nm sphere relates to a soccer ball as the soccer ball relates to planet Earth.

5.3 Set up basic flow in a nanotube and compare convection times with diffusion times for typical scenarios.

5.4 Research nanofluidics applications derived from nature as well as engineered areas, i.e., provide images and discuss characteristics and functions.

5.5 Section 5.1 lists numerous useful nanofluidics systems. What are potentially hazardous nanotechnology products, applications, or devices where nanofluidics plays a role?

5.6 Write a technical brief discussing Eqs. (5.3) and (5.4).

5.7 Based on an updated literature review, contrast various "pumps" for liquid flow in nanoconduits, i.e., provide images, descriptions, and applications.

5.8 Particle-particle and particle-wall interactions are important in many macro- to nanoscale systems. Focusing on liquid flow in nanoconduits, analyze attractive/repulsive forces, their physical and mathematical description, as well as practical applications in MD simulations and other solution techniques.

5.9 Write a research report comparing features and distinct applications of "nontraditional simulation methods," i.e. DSMC, MDS, and LBM (see Sects. 5.2 and 5.3.1)

5.10 Examples 5.3 and 5.4 illustrate the use of different slip-velocity models for liquid flow in nanoconduits. Provide a critical analysis, based on experimental evidence, detailing when each model is most applicable.

5.11 Entrance and exit effects in micro- and nanoscale flow systems are often important because the developing flow regions may be a substantial part of the entire conduit length. (a) Visualize and discuss $\Delta p_{total} = \Delta p_{H\text{-}P} + \Delta p_{end}$. (b) Using Eq. (5.24) and the associated analysis of CNT flow, develop a criterion for $L_{end}/L_{tube} \leq 10\%$.

5.12 Write a technical brief on "rarefied gas flow," focusing on images, physics and math descriptions, and nanoconduit flow applications.

5.13 Concerning Example 5.5, the wall BC $c(y = \pm h) = 0$ is a major simplification. Alternative BCs are:

$dc/dy|_{y=\pm h} = 0$, implying that the species cannot diffuse though the solid wall;

$c(y = \pm h) = c_{wall}(K)$, where K is the absorption coefficient; or

$c(y \to \pm \infty) = 0$, assuming that the species never reaches the walls.

Redo the problem considering a more realistic set-up, including an appropriate wall BC.

5.14 Concerning Example 5.6, evaluate the percentages in flow rate change when using the second-order velocity slip condition (E5.6-1) for different Kn numbers. Graph $Q(\text{Kn})$ for the two velocity slip cases.

5.15 Concerning Example 5.7: (a) Provide the missing steps in the derivation, analyze/justify the assumptions, and plot additional temperature distributions with comments. (b) Set up the thermal nanochannel flow problem for the case of $q_1 \neq q_2$. Suggest and detail an appropriate solution method. Sketch anticipated temperature profiles.

5.4.2 Set-Up for Course Projects

General:

- A course project has to be original, i.e., resubmission of selective write-ups of past research projects or ongoing thesis work is verboten!
- Two-person group work is permissible only when dual efforts are apparent, e.g., experimental + computational work.
- Hand in an *overview* (like a preproposal), i.e., project title, system sketch, objectives, solution method, and expected results, before starting the project.
- Follow the format of report writing (see Sect. 7.3 in Chapter 7).

CP Selection Process and CP Options:

- Focus on what interests you most as related to micro/nanofluidics, is within your skill level, and can be accomplished within a set time frame
- Select an appropriate journal/conference paper, for example, an experimental article on thermal flow (or nanofluid flow) in a microchannel, and perform an analytic/numerical analysis. There are now a half dozen new journals which are directly related to micro/nanofluidics; furthermore, traditional fluid mechanics, fluid-particle dynamics, heat transfer, and biomedical journals feature articles on microfluidics and nanofluidics topics in every issue.

- Perform the micro/nanofluidics part of a microsystem, e.g., a magnetohydro-dynamic pump, a lab-on-a-chip, a particle sorting device, a point-of-care fluid-sampling and diagnostics/analysis device, etc.
- Write tutorials, i.e., background information and solved example problems in key micro/nanofluidics areas, such as drug delivery (e.g., in nanomedicine), MEMS or LoC or PoC processes with device component performance analysis. Alternatively, select new tutorial topics, such as two-phase (liquid-gas) microchannel flow, boiling, magnetohydrodynamics, flow focusing, or particle sorting in microfluidics devices, optofluidics, acoustic-wave-driven micro/nanofluidics, etc.

CP Objectives:

- Deepen your knowledge base in micro/nanofluidics.
- Advance your analytical/computational skills or experimental skills.
- Generate an impressive research report in content and form.

CHAPTER **6**

Applications in Nanofluidics

6.1 Introduction

"Applications in nanofluidics" implies insightful results from analytic or numerical solutions of somewhat idealized fluid flow in nanochannels or fluid-nanoparticle transport phenomena in microchannels. They play essential roles in nanotechnology and contribute to the proper design and operation of microsystems. While Chapter 5 dealt with basic liquid flow and gas flow in nanochannels, Chapter 6 focuses on rarefied gas flow in nanotubes as well as convection of *nanofluids,* i.e., dilute suspensions of nanoparticles $(d_p < 50 \text{ nm})$ in liquids. Specifically, Sect. 6.2 introduces metal nanoparticles used for enhanced cooling when added to liquids, as well as nanodrugs with multifunctional capabilities, e.g., tumor detection, visualization, and cancer treatment. Typically, multistep fabrication methods based on fundamental physical and/or chemical processes are employed for the two very different nanoparticle applications. Then in Sect. 6.3, after discussing viscosity and thermal conductivity models for nanofluids, metal-oxide nanoparticle suspensions are applied to improve cooling. Applications of nanodrugs are given in Sect. 6.4, discussing transport mechanisms of nanodrug delivery to cancer cells.

6.2 Nanoparticle Fabrication

There are two basic nanoparticle fabrication methods, i.e., the traditional "top-down" and the novel "bottom-up." The traditional one is a miniaturization process, while the latter assembles nanoparticles and nanostructures from atoms and molecules. Especially fabrication of nanoparticles for drug delivery from the molecular level upwards allows for built-in multifunctionality, e.g., drug encapsulation, surface coating for visualization as well as attachment of ligands for nanoparticle anchoring to cancer cells. In contrast, production of metal-oxide nanofluids is a one-step, but more typically a two-step, chemical and physical process. The overall goal for effective micro-system cooling is to generate a stable suspension of uniformly dispersed spherical, monosized nanoparticles.

6.2.1 Metals and Metal Oxides for Cooling

It has been shown that small-volume fractions of metallic nanoparticles in liquids, say, 1% to 5%, measurably increase the effective thermal conductivity of the mixture, and hence heat transfer performance, when compared to the pure carrier fluid (see Kleinstreuer & Feng, 2012). While this thermal enhancement is generally accepted for more rod-like nanoparticles, particularly carbon nanotubes forming aggregates or even meshes, some experimental results with certain nanofluids are still controversial.

Production of Nanofluids.   In a typical two-step process making oxide nanoparticles or nanotubes, first a dry powder is produced by chemical/physical means, such as chemical vapor deposition or inert gas condensation. Then the powder is dispersed in a liquid, where nanoparticle aggregation poses a major problem. Nanoparticle dispersion and stability can be achieved via sonication, adding surfactants, or pH control. For metallic nanoparticles a one-step process is recommended. Specifically, synthesis and dispersion of nanoparticles into the liquid take place simultaneously. As summarized by Choi (2009), nanoscale vapor from metallic source material, e.g., aluminum, copper, or silver, can be directly dispersed into low-vapor-pressure liquids, such as water, ethylene glycol, or oil. Alternative physical top-down processes include wet grinding with bead mills, or optical laser ablation of particles suspended in liquids.

6.2.2 Drug Carriers in Nanomedicine

Due to the very low Stokes numbers of nanofluid flow, nanoparticles (NPs) largely follow the fluid stream, e.g., inhaled nanodroplets in air penetrating deep into nasal cavities or injected nanoparticles cruising in nutrient-supply arteries towards tumors (Kleinstreuer, 2006). Of great importance in nanomedicine are drug carriers, often multifunctional nanoparticles, which have encapsulated typically poorly soluble, unstable, highly aggressive drugs. They are called *nanocarriers* and provide extended blood circulation time and reduce side effects. Once they reach a tumor site, the therapeutic drug is released in response to a specific stimulation, say, an induced change in local blood temperature or pH value. Silica or polymer-based NPs, liposomes, and gold nanoparticles are common examples of "smart drug delivery systems" (Kleinstreuer et al., 2012).

Nanocarrier Fabrication.   Silica nanoparticles (SNPs) are mesoporous structures which are ideal for encapsulation and surface attachment of all kinds of cargo, e.g., drugs, proteins, and/or imaging agents. In a typical NP synthesis, known as the sol-gel process (Brinker & Scherer, 1990), the silicate source is mixed with a surfactant in a hot aqueous solution $(pH = 11)$ to form micellar templates of, say, 2 nm which grow into quasi-spherical SNPs with diameters ranging from 50 to 200 nm. By removing the surfactants with acidic alcohol, hexagonal pores of about 2 nm remain in the SNPs.

In comparison to inorganic NPs, *polymers* have several advantages, such as optical transparency, biocompatibility, processability, and versatility. For example, *liposomes* are spherical vesicles, containing an aqueous solution, formed by a bilayer membrane made out of phospholipids. They can be prepared by disrupting biological membranes, e.g., via sonication, and can encapsulate both hydrophilic and hydrophobic agents, protecting the cargo during circulation in the blood stream.

Of the metallic drug-loaded particles, *gold nanoparticles* (AuNPs) have emerged as a desirable drug delivery system. Gold is basically inert, nontoxic, and biocompatible. AuNPs are becoming popular because of controlled size fabrication $(1 < d_p < 100 \text{ nm})$, featuring a high surface-to-volume ratio which allows for appending a high density of ligands for targeting of and anchoring to cancer cells. For example, a 2-nm AuNP can accommodate about 100 ligands. Furthermore, the core gold particle can be coated with a multifunctional monolayer, protective of loaded drugs, containing visualization agents, etc. One fabrication method is the reduction of an $AuCl_4^-$ salt with sodium-borohydride $(NaBH_4)$ in the presence of the desired ligands, where the particle core size can be controlled by varying the ligand-gold stoichiometry. Alternatively, the $HAuCl_4$ compound is reduced with sodium citrate in water. In a postprocessing step a mixed monolayer is added.

6.3 Forced Convection Cooling with Nanofluids

Depending on the nanoparticle size and concentration, nanofluid flow can be modeled with the Euler-Euler approach for spherical particles when $d_p < 100$ nm, or in the Euler-Lagrange frame when $d_p > 100$ nm, or by regarding the suspension as a two-fluid mixture (see Sect. 2.3). In case the nanoparticles are nonspherical, employing the "equivalent-spherical-diameter" approach with associated drag and lift coefficients may be suitable.

As alluded to in Sect. 2.4, high rates of heat transfer in mechanical, chemical, and biomedical *microsystems* require heat exchangers which are very small, light, and efficient. *Microchannels* made out of glass, silicon, or polymers form the basic elements of such microsystems as well as built-in heat exchangers. Improving the thermal performance of compact devices requires better coolants than conventional fluids such as oil, water, or ethylene glycol. One solution to microscale cooling problems is the addition of solid nanoparticles to the fluid. The resulting *nanofluids* may significantly change the mixture's properties, most notably its thermal conductivity and dynamic viscosity.

Nanoparticles considered for microsystem cooling range from metals and metal-oxides to carbon nanotubes with inner diameters of 1 to 50 nm. Indeed, prevailing experimental evidence indicates a greater enhancement of *nanofluid thermal conductivity,* k_{nf}, than predicted by the "effective medium" theory of Maxwell (1891). Such an increase of k_{nf} over $k_{base-fluid}$ varies with nanoparticle volume fraction and characteristics, e.g., size, shape, material, surface charge/coating, and degree of particle aggregation, as well as with the type of base fluid, its temperature, conductivity, pH value, and additives. Nevertheless, while in the past enhanced k_{nf} values have been reported when employing the transient hot-wire method, some data based on recent nonintrusive (optical) techniques could not confirm such high k_{nf} values, or even an increase over the values obtained with Maxwell's theory (Kleinstreuer & Feng, 2012). Nevertheless, nanofluid flow dynamics is an ongoing research area because of its broader importance in both nanotechnology and nanomedicine.

For a better understanding of the underlying physics of k_{nf} enhancement, six major sources of improved heat transfer should be considered:

- micromixing because of Brownian motion of the nanoparticles affecting the surrounding fluid;
- higher pathway conduction of clustered nanoparticles or connected carbon nanotubes;
- liquid-molecule layering around nanoparticles causing lower heat resistance;
- larger heat conduction in the case of certain metallic nanoparticles;
- thermal wave impact and/or thermophoresis;
- interactions between the nanoparticles and shear layers at the conduit walls.

6.3.1 Nanofluid Properties

The basic nanofluid properties are a function of nanoparticle volume fraction φ and mixture temperature T. Such nanofluids are assumed to be dilute suspensions, i.e., the monosized noninteracting nanoparticles are well dispersed. So, the nanofluid density and heat capacity can be expressed as:

$$\rho_{nf} = \varphi \rho_p + (1-\varphi)\rho_{bf} \tag{6.1}$$

$$(\rho c_p)_{nf} = \varphi(\rho c_p)_p + (1-\varphi)(\rho c_p)_{bf} \tag{6.2}$$

where the subscripts nf, bf, p indicate nanofluid, base fluid, and particle, respectively.

The temperature-dependent properties of, say, water are assumed to be (Feng & Kleinstreuer, 2010), in international units:

$$\rho_{water} = 1000 \cdot \left(1 - \frac{(\tilde{T}+15.7914)}{508929.2 \cdot (\tilde{T}-205.0204)} \cdot (\tilde{T}-277.1363)^2\right)\left[\frac{kg}{m^3}\right] \tag{6.3a}$$

$$c_{p,water} = 9616.873445 - 48.7364833 \cdot \tilde{T} + 0.1444662 \cdot \tilde{T}^2 - 0.000141414 \cdot \tilde{T}^3 \left[\frac{J}{m^3 K}\right] \tag{6.3b}$$

$$\mu_{water} = 0.02165 - 0.0001208 \cdot \tilde{T} + 1.7184e - 7 \cdot \tilde{T}^2 \left[\frac{kg}{m \cdot s}\right] \tag{6.3c}$$

$$k_{water} = -1.1245 + 0.009734 \cdot \tilde{T} - 0.00001315 \cdot \tilde{T}^2 \left[\frac{W}{m \cdot K}\right] \tag{6.3d}$$

where $\tilde{T} = T/(1[K])$ is the nondimensional temperature.

Effective Dynamic Viscosity.   Most of the reported data for nanofluid viscosities have been discussed in terms of formulations proposed by Einstein (1906), Brinkman (1952), Batchelor (1977), and Graham (1981), to name a few. The conventional viscosity models of nanofluids are summarized in Table 2.2. It turns out that none of the models mentioned can predict the viscosity of nanofluids very well for a wide range of nanoparticle volume fraction.

Masoumi et al. (2009) proposed a viscosity model by including the effect of Brownian motion of the nanoparticles on the viscosity of nanofluids in terms of $\mu_{nf} = \mu_{bf} + \mu_{app}$, where μ_{app} is the apparent viscosity. Specifically, temperature, nanoparticle diameter, volume fraction, nanoparticle density as well as the base fluid physical properties were all considered:

$$\mu_{nf} = \mu_{bf} + \frac{\rho_p v_B d_p^2}{72 C \delta} \tag{6.4}$$

where $v_B = \sqrt[3]{18\kappa_b T / \pi \rho_p d_p} / d_p$ is the Brownian velocity, κ_b is the Boltzmann constant, $\delta = \sqrt[3]{\pi / 6\varphi} d_p$ is the distance between nanoparticles, and $C = \text{fct}(\mu_{bf}, d_p, \varphi)$ is a correction factor.

The correlation factor C was calculated by using experimental data for water-based nanofluids consisting of 13-nm and 28-nm Al_2O_3 nanoparticles as well as the 36-nm Al_2O_3-water nanofluid by Nguyen et al. (2008).

Kleinstreuer et al. (2012) illustrated nanofluid flow in a microchannel by comparing the Brinkman equation with the effective nanofluid viscosity model postulated by Masoumi et al. (2009) as given in Figure 6.1. As already noticed by Nguyen et al. (2008), the conventional Brinkman model (see Table 2.2) underpredicts the nanofluid viscosity. Furthermore, the functional dependence $\mu_{nf}(\varphi)$ is highly nonlinear in the model by Masoumi et al. (2009), especially for $\varphi > 3\%$ and $d_p < 40$ nm (see Figure 6.1a). Such a viscosity enhancement of nanofluids may increase the pressure drop and, hence, the requirement of pumping power. As expected, the temperature influence on μ_{nf} (see Figure 6.1b) is much less dramatic, being expressed as $\mu_{bf}(T)$.

Effective Thermal Conductivity.   For the thermal performance analysis of nanofluid flow in microconduits, several thermal conductivity models have been employed (see Kleinstreuer & Feng, 2012; among many others). In this paper, three different models were applied and compared, i.e., the conventional Maxwell model, the correlation by Patel et al. (2010), as well as the newly developed Feng-Kleinstreuer (F-K) model (Kleinstreuer & Feng, 2012). Maxwell (1891) derived a "static" thermal conductivity model for conventional fluids containing at that time micrometer/millimeter particles. It was assumed that the effective thermal conductivity of the mixture is a function of the thermal conductivity of the suspensions and the base fluid as well as the volume fraction of the suspensions:

$$k_{static} = \left(1 + \frac{3\left(\dfrac{k_p}{k_{bf}} - 1\right)\varphi}{\left(\dfrac{k_p}{k_{bf}} + 2\right) - \left(\dfrac{k_p}{k_{bf}} - 1\right)\varphi} \right) k_{bf} \tag{6.5}$$

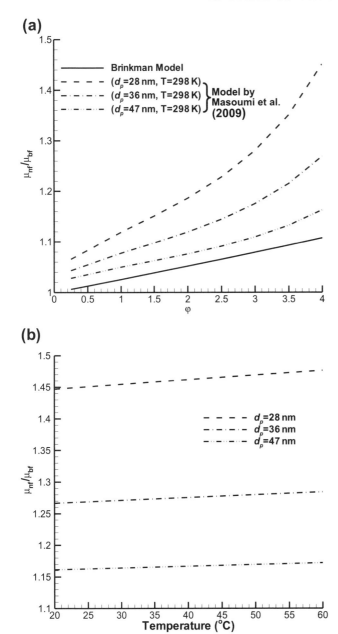

Figure 6.1 Dynamic viscosity models for nanofluids: (a) viscosity versus volume fraction; (b) viscosity change with temperature ($\varphi = 4\%$; model by Masoumi et al., 2009)

The emergence of nanofluids as a new field of nanoscale heat transfer, with applications to microsystem cooling, is directly related to miniaturization trends and nanotechnology. The unexpected high thermal conductivity of nanofluids documented in many experiments showed that the effective thermal conductivity of nanofluids depends not only on the nanostructures of the suspensions but also on the dynamics of nanoparticles in liquid. Thus, it was postulated that the thermal conductivity of nanofluids consists of a static part (k_{static}) after Maxwell (1891) and a micromixing part (k_{mm}), i.e., enhancement due to Brownian motion of nanoparticles, so that the F-K model can be expressed as:

$$k_{nf} = k_{static} + k_{mm} \tag{6.6}$$

The static part is Eq. (6.5), while the micromixing part, based on sound physics, is given by:

$$k_{mm} = 49{,}500 \cdot \frac{\kappa_B \tau_p}{2m_p} \cdot C_c \cdot \left(\rho c_p\right)_{nf} \cdot \varphi^2 \cdot \left(T \ln T - T\right)$$

$$\cdot \frac{\exp\left(-\zeta \omega_n \tau_p\right) \sinh\left(\sqrt{\dfrac{\left(3\pi\mu_{bf}d_p\right)^2}{4m_p^2} - \dfrac{K_{p-p}}{m_p}\dfrac{m_p}{3\pi\mu_{bf}d_p}}\right)}{\tau_p \sqrt{\dfrac{\left(3\pi\mu_{bf}d_p\right)^2}{4m_p^2} - \dfrac{K_{p-p}}{m_p}}} \tag{6.7}$$

Here, C_c is equal to 38 for metal-oxide nanofluids which can be derived theoretically (which also holds for the number 49,500), instead of being obtained via a curve-fitting technique (Feng, 2009). The damping coefficient ζ, natural frequency ω_n, and characteristic time interval τ_p can be expressed as:

$$\zeta = \frac{3\pi d_p \mu_{bf}}{2m_p \omega_n} \tag{6.8}$$

$$\omega_n = \sqrt{K_{p-p}/m_p} \tag{6.9}$$

$$\tau_p = \frac{m_p}{3\pi\mu_{bf}d_p} \tag{6.10}$$

Specifically, for metal-oxide nanofluids, the magnitude of particle-particle interaction intensity K_{p-p} is determined for different particle diameters as:

$$K_{p-p} = \rho_p \cdot \sqrt{d_p \cdot 10^{-9}} \cdot \left(\frac{32.1724 \cdot 273 \text{ K}}{T} - 19.4849 \right) \text{ for } 20 \text{ nm} < d_p \leq 50 \text{ nm} \quad (6.11)$$

$$K_{p-p} = \rho_p \cdot \sqrt{d_p \cdot 10^{-9}} \cdot \left(\frac{24.6402 \cdot 273 \text{ K}}{T} - 18.7592 \right) \text{ for } d_p > 50 \text{ nm} \quad (6.12)$$

In light of experimental evidence, the F-K model is suitable for several types of metal-oxide nanoparticles in water with volume fractions up to 5% and mixture temperatures below 350 K.

   In contrast, Patel et al. (2010) provided a correlation for the effective thermal conductivity of nanofluids, based on a regression analysis of several experimental data sets:

$$k_{nf} = k_{bf} \left(1 + 0.135 \times \left(\frac{k_p}{k_{bf}} \right)^{0.273} \times \varphi^{0.467} \times \left(\frac{T - 273}{20} \right)^{0.547} \times \left(\frac{100}{d_p} \right)^{0.234} \right) \quad (6.13)$$

where T is the temperature of nanofluids in Kelvin; d_p is the average nanoparticle diameter in nanometers. Apparently, the correlation is valid for suspensions of spherical nanoparticles of 10 to 150 nm diameter, a thermal conductivity range of 20 to 400 W/mK; base fluids having thermal conductivities of 0.1 to 0.7 W/mK, particle volume fractions of 0.1 to 3%, and suspension temperatures from 20°C to 50°C. Clearly, the k_{nf} correlation by Patel et al. (2010), having a much simpler form than the F-K model, is easier to use but lacks physical insight and a broad range of applications. Figures 6.2 and 6.3 provide model comparisons with some recent benchmark experimental data sets. Overall, the F-K model generates a better matching in trend and precision for different volume fractions and temperatures, although, the measured k_{nf} increase with nanoparticle volume fraction above 3% by Mintsa et al. (2009) is surprisingly low (see Maxwell model, Eq. (6.5)).

6.3.2 Thermal Nanofluid Flow

To illustrate the impact of the key nanofluid property on heat transfer and entropy production, different microchannels and k_{nf} models of alumina-water have been considered. Specifically, a trapezoidal channel with different base angle α, all with $D_h = 187$ μm, as well as the correlation by Patel et al. (2010) and the F-K model for k_{nf} have been selected (see insert in Figure 6.4).

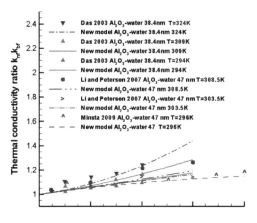

Figure 6.2   Comparison of F-K model with experimental data sets at different volume fraction

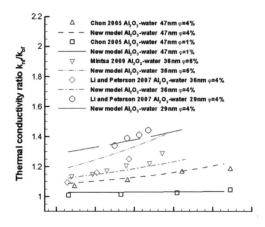

Figure 6.3   Comparison of F-K model with experimental data sets at different temperature

6.3.3 Friction Factor and Pressure Drop Results

Figure 6.4 compares the pressure gradients at different Reynolds numbers for water and different nanofluid parameters. While the conventional Brinkman viscosity model with a 1% nanoparticle volume fraction generates almost the same result as for pure water, the pressure gradient increases when employing the more realistic model by Masoumi et al. (2009), especially when the particle size is small and/or the volume fraction is large.

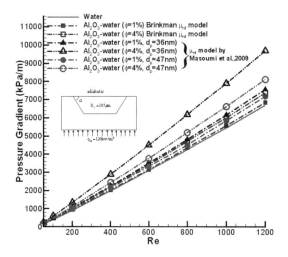

Figure 6.4   Pressure gradient versus Reynolds number for nanofluid flow using two different viscosity models

Of practical interest is the power requirement necessary to generate different pressure drops across the microchannel length. Figure 6.5 depicts the pumping power change with different base angles at the same Reynolds number for different mixtures. A larger base angle can help to decrease the required pumping power. In order to achieve the same Reynolds number, a larger pumping power is required for Al_2O_3−water nanofluids, i.e., a 18% increase in pumping power for 1% Al_2O_3-water nanofluids with d_p = 36 nm over pure water. For a 2% ZnO-EG nanofluid, the required pumping power increases, when compared with pure EG, by 8% when using the Brinkman model.

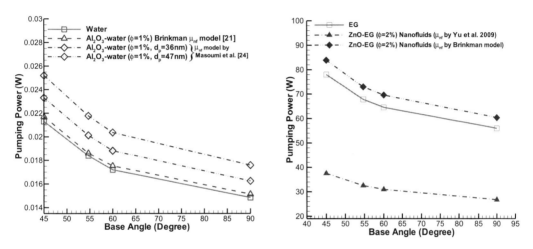

Figure 6.5   Pumping power versus base angle at Re = 600 (a) water and Al_2O_3 −water; (b) EG and ZnO-EG

6.3.4 Convective Heat Transfer

In order to compare the thermal performance of nanofluid flow in microchannels, the local heat transfer coefficient as well as the thermal resistance for different pumping powers were compared for pure water flow and nanofluid flow with different volume fractions, employing the viscosity model by Masoumi et al.(2009) and the k_{nf} model by Kleinstreuer and Feng (2012).

The local heat transfer coefficient developing along the microchannel at the same inlet Reynolds number (Re = 600) and heat flux for different fluids is shown in Figure 6.6. As expected, the local heat transfer coefficient increases when using nanofluids, where smaller nanoparticles yield elevated heat transfer coefficients. It should be noted that the Reynolds number of nanofluids depends on the kinematic viscosity of the nanofluids, which increases with an increase in nanoparticle volume fraction and smaller nanoparticles, implying that the inlet velocity should be increased to keep the Reynolds number constant. Figure 6.6 also depicts the higher heat transfer coefficient in the thermal entrance region. It was found that the thermal entrance length computed follows the correlation by McHale and Garimella (2010). A smaller base angle shows better heat transfer coefficients for all the cases especially for water and Al_2O_3-water nanofluids. Interestingly, the thermal resistance, i.e., $R_{thermal} = (T_{w,ave} - T_{in})/q$, decreases when employing nanofluids, especially for nanofluids with larger volume fractions and smaller nanoparticles. The reason is that the average wall temperature is lower due to the higher thermal performance of nanofluids without a significant increase in pumping power. For a typical case, the average enhancement of thermal performance for Al_2O_3-water with a volume fraction of 4% is about 14% for 36-nm nanoparticles and 8% for 47-nm nanoparticles.

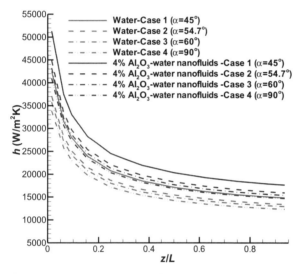

Figure 6.6   Local heat transfer coefficient developing along the microchannel walls for different geometries at the same operational conditions

6.4 Nanodrug Delivery

Multifunctional nanodrug delivery (NDD) is an intense research area encompassing several engineering branches as well as molecular pharmaceutics. Computational fluid-particle dynamics techniques can be realistic and accurate tools to achieve efficient NDD. Indeed, major improvements in drug delivery are necessary because presently most of the drugs have poor solubility, causing aggregation or degrade rapidly, require high dosage, and are highly toxic leading to *severe side effects due to nonspecific delivery.* Basically, a therapeutic substance is introduced into the body with the goals of high efficacy and safety via controlled drug release location, timing, and dosage. This may be achieved via various systemic drug delivery routes, e.g., gastrointestinal, (parental) injection, transdermal, and pulmonary. Major NDD applications range from controlling severe pain to combating malignant tumors. Ideally, drug delivery systems should allow for diagnostic measurements and controlled drug release as well as visualization of treatment efficacy. Of the various system components, the physico-biochemical properties of the drug carrier (i.e., the formulation), a suitable targeting methodology, and the delivery device are most critical for successful treatment of a specific disease. Present drug carriers for targeting include hydrogels, polymeric micelles, nanoparticles (NPs) or multifunctional nanostructures, porous micro/nanoparticles, drug-encapsulating dendrimers or carbon nanotubes, and radioactive microspheres (see Kleinstreuer et al., 2013; among many others). Common drug delivery devices include needles, syringes, micropumps, catheters, implants, and inhalers. Targeting mechanisms are traditionally grouped into *active* and *passive* methods. Examples include "image-guided drug delivery" with affinity-based cancer drugs, the use of external magnetic fields to steer ferric (nano-) particles towards tumors, or just injection of drug carriers into the blood stream in the hopes that sufficient NP circulation time will bring them close to the tumor site so that enhanced local vessel permeability as well as receptor connections will achieve attachment to cancer cells only. Presently, applications of most of the multifunctional nanoparticles mentioned are still in the experimental, animal-testing, or clinical trial stage.

6.4.1 Types of Drug-Loaded Nanoparticles

Polymeric NPs are either nanospheres or nanocapsules where their matrix or shell is made of PLGA (poly-D, L-lactide co-glycolide) or PLA (polylactide). The active components are absorbed on their surfaces or entrapped/dissolved inside the polymeric matrix. *Ceramic (or silica) NPs,* being porous, unaffected by pH changes and less than 50 nm in mean diameter, are novel nanodrug carriers. *Micelles and liposomes* are typically spherical NPs made from solid lipids forming the particle surface, i.e., the hydrophilic heads stabilized by surfactants, and multiple hydrophobic tails. Their advantages for drug delivery include small nanosize, biodegradable, easy surface manipulation, and multiroute options for injection with targeted gene delivery applications. *Magnetic nanoparticles (MNPs)* are useful as enhanced contrast agents in MRI (magnetic resonance imaging), or may function as drug carriers which can be directed towards tumors via a locally applied magnetic field. They are basically magnetic iron-oxide particles (Fe_3O_4 or Fe_2O_3) with synthetic or natural polymer coating of their surfaces to avoid particle agglomeration, where the nanodrugs are either surface attached or encapsulated. For example, surface-bound drugs can be released from the drug carriers by changing the physiological conditions locally (e.g., temperature and/or pH) when they are near/at the target site so that the nanodrugs can diffuse directly into the diseased cells. As discussed, *metal-based nanoparticles* are very useful because they are very small $\left(d_p < 50 \text{ nm}\right)$, of stable shapes, with possibly negative charge, high surface reactivity, and biocompatible. *Dendrimers* are polymeric macromolecules which branch out tree-like from a central core, forming an inner and an outer shell. They are monodisperse, highly symmetric, i.e., quasi-spherical, water-soluble compounds which can be employed for encapsulation of hydrophobic nanodrugs. *Hydrogels* are hydrophilic, containing over 99% water, made of a network of synthetic or natural polymer chains. They have the ability to sense changes in pH, temperature, or metabolite concentrations; hence, when functioning as nanodrug carriers they can release their pay-load upon detecting such changes in blood properties near a tumor. *Aptamers* are DNA or RNA oligonucleotide sequences that selectively bind to target cells with high affinity and selectivity. Thus, aptamer-based drugs are most suitable for the prevention and treatment of chronic and acute diseases. The use of *carbon nanotubes (CNTs)* as drug carriers holds great promise for cancer diagnosis and therapy. CNTs are made up of thin sheets (e.g., 0.3 nm thick) of benzene-ring carbons rolled into single- or multiwall tubes. This novel structure belongs to the family of fullerenes, distinctly different from the carbon forms of graphite and diamonds. CNTs are ultralight with a high aspect ratio (i.e., $d_p = O(10 \text{ nm})$ and $L = O(100 \text{ μm})$), water solubility, of high mechanical strength and electrical/thermal conductivities. Due to their ultrahigh surface areas, nanodrugs, peptides, and/or DNA can be readily attached to their walls and tips.

6.4.2 Mechanisms of Nanodrug Targeting

As indicated, the conventional "targeting" mechanism relies on free nanoparticle convection in the blood stream, i.e., possible circulation towards the diseased tissue. Then, in "*passive* targeting" nanodrugs are released with possible migration into diseased cells. In "*active* targeting" biochemical docking is followed by nanodrug release and diffusion into the cancer cells. Clearly, the particle injection point, e.g., upstream or intratumor delivery, a patient's local vascular system and flow waveform, tumor location and extent, as well as the differences in normal and diseased tissues play key roles in the success of these delivery mechanisms. For passive and active targeting the enhanced permeability and retention (EPR) effect comes into play because tumor vasculature is leaky and hence circulating (nano-) particles accumulate preferentially in cancerous regions and also in inflamed tissue (Figure 6.7a). However, the EPR effect is less dominant at sites of inflammation because of lower drug retention due to lymphatic drainage, washing out the drugs. Once the particles equipped with ligands are very near a tumor, success of active targeting depends on the number of ligand-receptor sites, their level of affinity, and binding strength. Specifically, particle-cell coupling can be achieved by molecular recognition via the ligand-receptor, antigen-antibody interactions, or by the use of aptamers, which are DNA strands that selectively bind their target with high affinity and specificity. As an example, Figure 6.8 depicts a multifunctional semiporous silica (or polymer) particle, covered with ligands as well as a hydrophilic coating and loaded with nanodrugs plus ferric nanocrystals, in case of magnetic steering towards the diseased site.

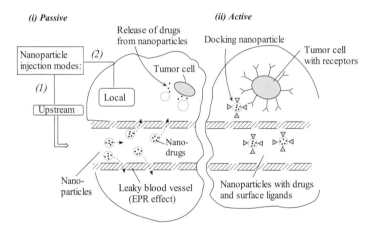

(a) Free transport of drug-loaded nanoparticles towards tumor cells:
   (i) passive biochemical targeting and (ii) active biochemical targeting

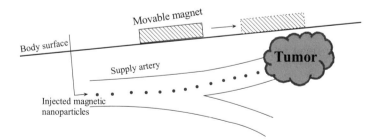

(b) Forced convection of drug-loaded nano/microparticles with ion crystals
   for magnetic steering

Figure 6.7   Schematics of drug-particle transport and targeting mechanisms

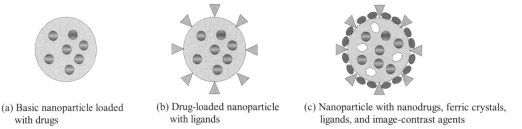

(a) Basic nanoparticle loaded with drugs

(b) Drug-loaded nanoparticle with ligands

(c) Nanoparticle with nanodrugs, ferric crystals, ligands, and image-contrast agents

Figure 6.8   Schematics of multifunctional nanoparticle development from (a) to (c)

The second unique targeting mechanism, i.e., forced convection of magnetic NPs with a translating external magnet, is restricted to cases where the tumor resides near the body surface and the blood perfusion is not too high. Typically, the drug is placed on the ferric nanoparticles which are injected near the tumor (Figure 6.7b). An external magnet assures high NP concentrations at the tumor site and hence rapid nanodrug diffusion into the cancer cells (Pankhurst et al. 2003; Cregg et al. 2012; among others).

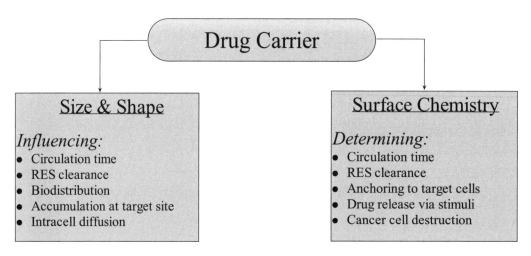

Figure 6.9   Characteristics of drug carriers

In summary, advances in new drug formulations and multifunctional nanoparticles as drug carriers have measurably improved the pharmacokinetics and biodistribution of drugs, typically suffering from poor water solubility, low stability, and unwanted toxicity. Still, for such nanoparticles to avoid clearance from circulation or destruction by the immune system, achieving high cancer cell selectivity and sufficient target concentrations is a prevailing challenge. Specifically, drug carriers, while cruising in the blood stream, interact with the mononuclear phagocyte system

(MPS). It is the body's primary mechanism of clearing foreign particles, employing macrophages and filtration. Hence, NPs have to shield themselves from such clearance mechanisms to reach and attach to the predetermined tissue site. Figure 6.9 summarizes the characteristics of NPs to come closer in achieving these two goals. Clearly, all parameters are dependent on the "circulation time," which correlates with the probability of finding the target cells and the associated biodistribution.

Drug delivery routes and devices have been reviewed by Jain (2008) and mathematical models for simulating drug release, i.e., species mass transfer from carriers are discussed by Arifin et al. (2006). These topics are, of course, part of *optimal targeted drug delivery* (TDD), as described in Sect. 8.2. Clearly, optimal TDD has the potential to overcome the major problems associated with the NP transport mechanisms discussed so far.

6.5 Homework Assignments and Course Projects

6.5.1 Text-Related Questions and Tasks

6.1 Section 6.2 briefly discussed nanoparticle fabrication for applications in microsystem cooling as well as in nanomedicine. Write and update a literature review of nanoparticle manufacturing with an application of your choice.

6.2 Section 6.3 introduced nanofluids as *dilute* suspensions for microsystem cooling. In other industries producing food, medicine, cosmetics, cleaners, composites, etc., nanofluids may be dense mixtures which require appropriate property models. Select a specific end product and research the characteristics and property models of dense nanofluids needed for product development.

6.3 Considering Eq. (6.4), derive the Brownian velocity v_B and distance between nanoparticles δ with that of c in the Graham model. Plot $\mu_{nf}(\varphi, d_p)$ as proposed by Graham (1981) and Masoumi et al. (2009). Comment!

6.4 An important parameter in the F-K model for k_{nf} is the "particle-particle interaction intensity" K_{p-p} as given in Eqs. (6.11) and (6.12). Write a technical brief on particle-particle interaction and clustering in nanofluids, i.e., images, physics, models, and applications.

6.5 Almost all naturally occurring and most man-made nanoparticles are nonspherical and polydisperse. Discuss math models and solution techniques suitable to handle such characteristics for proper nanofluid flow simulations.

6.6 Research the use of magnetic nanoparticles (MNPs) in nanomedicine for:

(a) tumor visualization

(b) tumor treatment

Specifically, provide MNP images, fabrication steps, and applications for case (a) and case (b).

6.7 A major difficulty for non-MNPs is the secure delivery from the body entry point to the target. Discuss the physics and math for multifunctional NP delivery routes with the goal of minimizing the "circulation time."

6.8 Contrast organic vs. synthetic nanodrugs (NDs), i.e., tabulate examples for each category, list their characteristics, and discuss the pros and cons of their applications.

6.9 Elaborate on "reticulo-endothelial system" (RES) in conjunction with nanodrug delivery (NDD). Specifically, develop a compartmental model and discuss applications.

6.10 Develop a math model for "enhance permeability and retention" (EPR), i.e., the leaky blood vessel effect (see Figure 6.7a).

6.5.2 Set-Up for Course Projects

General:

- A course project has to be original, i.e., resubmission of selective write-ups of past research projects or ongoing thesis work is verboten!
- Two-person group work is permissible only when dual efforts are apparent, e.g., experimental + computational work.
- Hand in an *overview* (e.g., a preproposal), i.e., project title, system sketch, objectives, solution method, and expected results, before starting the project.
- Follow the format of report writing (see Sect. 7.3 in Chapter 7).

CP Selection Process and CP Options:

- Focus on what interests you most as related to micro/nanofluidics, is within your skill level, and can be accomplished within a set time frame
- Select an appropriate journal/conference paper, for example, an experimental article on thermal flow (or nanofluid flow) in a microchannel, and perform an analytic/numerical analysis. There are now a half dozen new journals which are directly related to micro/nanofluidics; furthermore, traditional fluid mechanics, fluid-particle dynamics, heat transfer, and biomedical journals feature articles on microfluidics and nanofluidics topics in every issue.
- Perform the micro/nanofluidics part of a microsystem, e.g., a magnetohydro-dynamic pump, a lab-on-a-chip, a particle sorting device, a point-of-care fluid-sampling and diagnostics/analysis device, etc.
- Write tutorials, i.e., background information and solved example problems in key micro/nanofluidics areas, such as drug delivery (e.g., in nanomedicine), MEMS or LoC or PoC processes with device component performance analysis. Alternatively, select new tutorial topics, such as two-phase (liquid-gas) microchannel flow, boiling, magnetohydrodynamics, flow focusing, or particle sorting in microfluidics devices, optofluidics, acoustic-wave-driven micro/nanofluidics, etc.

CP Objectives:

- Deepen your knowledge base in micro/nanofluidics.
- Advance your analytical/computational skills or experimental skills.
- Generate an impressive research report in content and form.

References (Part C)

Abgrall, P., Nguyen, N-T., 2009, *Nanofluidics,* Artech House, Norwood, MA.

Arifin D.Y., Lee L.Y., Wang C-H., 2006, Advanced Drug Delivery Reviews, Vol. 58, pp. 1274-1325.

Arkilic, E.B., Schmidt, M.A., Breuer, K.S., 1997, Journal of Microelectromechanical Systems, Vol. 6, pp. 167-178.

Batchelor, G.K., 1977, Journal of Fluid Mechanics, Vol. 128, pp. 240.

Brinker, C.J., Scherer, G.W., 1990, *Sol-Gel Science: The Physics and Chemistry of Sol-Gel Processing,* Academic Press, San Diego, CA.

Brinkman, H.C., 1952, Journal of Chemistry Physics, Vol. 20, pp. 571-581.

Bruus, H., 2008, *Theoretical Microfluidics,* Oxford University Press, Oxford.

Choi, S.U.S., 2009, ASME Journal of Heat Transfer, Vol. 131, pp. 033106-1-9.

Chon, C.H., Kihm, K.D., Lee, S.P., Choi, S.U.S., 2005, Applied Physics Letters, Vol. 87, p. 153107.

Cregg P.J., Murphy K., Mardinoglu A., 2012, Applied Mathematical Modeling, Vol. 36, pp. 1-34.

Das, S.K., Putra, N., Theisen, P., Roetzel, W., 2003, ASME Journal of Heat Transfer, Vol. 125, pp. 567-574.

Einstein, A., 1906, Annalen der Physik, Vol.19, pp. 289-306.

Feng, Y., 2009, MS Thesis, NC State University, Raleigh, NC.

Feng, Y., Kleinstreuer, C., 2010, International Journal of Heat and Mass Transfer, Vol. 53, pp. 4619-4628.

Graham, A.L., 1981, Applied Scientific Research, Vol. 37, pp. 275.

Harkins, W.D., Brown, F.E., 1919, Journal of the American Chemical Society, Vol. 41, pp. 499-524.

Jain K.K., 2008, *Methods in Molecular Biology,* Vol. 437, pp. 1-50.

Khataee, A., Mansoori, G.A., 2011, *Nanostructured Titanium Dioxide Materials: Properties, Preparation and Applications,* World Science, Hackensack, NJ.

Kirby, B., 2010, *Micro- and Nanoscale Fluid Mechanics: Transport in Microfluidic Devices,* Cambridge University Press.

Kleinstreuer, C., 2006, *Biofluid Dynamics: Principles and Selected Applications,* CRC Press, Boca Raton, FL; Taylor & Francis Group, London, New York.

Kleinstreuer, C., Feng, Y., 2012, Journal of Heat Transfer, Vol. 134, 051002-1-11.

Kleinstreuer, C., Basciano, C.A., Childress, E.M., Kennedy, A.S., 2012, ASME Journal of Biomechanical Engineering, Vol. 134, pp. 051004-1-10.

Kleinstreuer, C., Li, J., Feng, Y., 2012, *Advances in Numerical Heat Transfer, Volume 4, Nanoparticle Heat Transfer and Fluid Flow*, Minkowycz, W.J., Sparrow, E.M. and Abraham, J.P. (eds.), CRC Press, Boca Raton, FL.

Kleinstreuer, C., Childress, E.M., Kennedy, A.S., 2013, Chapter 10 in *Transport in Biological Media,* Becker, S., Kuznetsov, A. (eds.), Elsevier, London, New York.

Li, C.H., Peterson, G.P., 2007, International Journal of Heat and Mass Transfer, Vol. 50, pp. 4668-4677.

Masoumi, N., Sohrabi, N., Behzadmehr, A., 2009, Journal of Physics D: Applied Physics, Vol. 42, pp. 055501-1-6.

Mathews, M.T., Hills, J.M., 2008, International Journal of Nanotechnology, Vol. 5, pp. 218-242.

Maxwell, J.C., 1891, *A Treatise on Electricity and Magnetism,* 3rd ed., Clarendon Press, Oxford, UK.

McHale, J.P., Garimella, S.V., 2010, International Journal of Heat and Mass Transfer, Vol. 53, pp. 365-375.

Mintsa, H.A., Roy, G., Nguyen, C.T., Doucet, D., 2009, International Journal of Thermal Sciences, Vol. 48, pp. 363-371.

Myers, T.G., 2011, Microfluidics and Nanofluidics, Vol. 10, pp. 1141-1145.

Nguyen, C.T., Desgranges, F., Galanis, N., et al., 2008, International Journal of Thermal Science, Vol. 47, pp. 103-111.

Pankhurst Q.A., Connolly J., Jones S.K., Dobson J., 2003, Journal of Physics D: Applied Physics, Vol. 36, pp. 167-181.

Pennathur, S., Santiago, J.G., 2005a, Analytical Chemistry, Vol. 77, pp. 6772-6781.

Pennathur, S., Santiago, J.G., 2005b, Analytical Chemistry, Vol. 77, pp. 6782-6789.

Patel, H.E., Sundararajan, T., Das, S.K., 2010, Journal of Nanoparticle Research, Vol. 12, pp. 1015-1031.

Probstein, R.F., 1994, *Physicochemical Hydrodynamics: an Introduction,* Wiley, New York.

Roco, M.C., Williams, S., Alivisatos, P., (eds), 1999, *Nanotechnology Research Directions: IWGN Workshop Report — Vision for Nanotechnology R&D in the Next Decade,* WTEC, Loyola College in Maryland.

Sisan, T.B., Lichter, S., 2011, Microfluid Nanofluid, Vol. 11, pp. 787-791.

Succi, S., 2001, *The Lattice Boltzmann Equation for Fluid Dynamics and Beyond* (*Numerical Mathematics and Scientific Computation*), Clarendon Press, Oxford.

Travis, K.P., Gubbins, K.E., 2000, Journal of Chemical Physics, Vol.112, pp.1984-1994.

Part D: COMPUTER SIMULATIONS OF FLUID-PARTICLE MIXTURE FLOWS

Most micro/nanofluidics systems deal with discrete or continuous flow of fluid-particle suspensions for biochemical analyses. Once the standard assumption of Poiseuille-type flow is not appropriate any more, the use of mathematical software (e.g., MATLAB, MAPLE, Mathematica, etc.) or engineering software (e.g., COMSOL, Fluent, CFX, STAR-CCM, etc.) becomes necessary to solve more realistic problems. While some type of math software is generally available and well mastered by students, the high licensing fees, the unavailability of engineering workstations, and some difficulties in using commercial software still bars students from solving more advanced microfluidics and nanofluidics problems. Furthermore, once the continuum mechanics assumption becomes invalid, most likely for fluid flow in nanoconduits, the direct simulation Monte Carlo method or molecular dynamics simulations will be necessary.

Thus, Chapters 7 and 8 provide an introduction to setting up advanced fluidics problems and illustrate a few numerical applications, stressing both new physical insight and computer model validation.

CHAPTER 7

Modeling and Simulation Aspects

7.1 Introduction

Analytic solutions of micro/nanofluidics problems employing greatly reduced forms of the system of Navier-Stokes equations (see Sect. 1.3.3) provide both physical insights into basic transport phenomena as well as means of computer model validation. Figure 7.1 depicts essential system modeling and problem solution steps, with possible feedback if the actual results do not match expectations or measurements. The first step, i.e., system and associated problem recognition as well as data collection, can involve the most challenging tasks. Clearly, lacking a clear understanding of the problem at hand and not having reliable data sets for model input and validation may generate incomplete or erroneous results. For example, proper identification of the system geometry, flow regime, type of fluid, dominant transport phenomena, and form of fluid-particle and possible fluid-structure interactions is quite challenging. The next two steps, i.e., math model development and solution technique, are closely connected. The more detailed (and hence realistic) the mathematical system description is, the more complex will be the solution method and ultimately the required computer resources.

The results, in terms of a new or deeper scientific understanding and novel engineering design, have to be checked against the initial goals and real-world evidence. If necessary, the system or problem description has to be refined. The modern professional approach of tackling macro-to-nanoscale problems develops in two phases, i.e., focusing on first the fundamentals and then applied reseach aspects. Thus, the modern engineer fulfills a dual role:

(1) Starting out as a *scientist* one has to understand the basic transport and conversion phenomena to generate physical and/or biochemical insight, an advanced knowledge base, a refined math model, a new theory, etc.

(2) Based on that, one works as an *engineer* to apply the newly gained understanding or novel theory to improve existing techniques or devices or for new system/device design and subsequent testing.

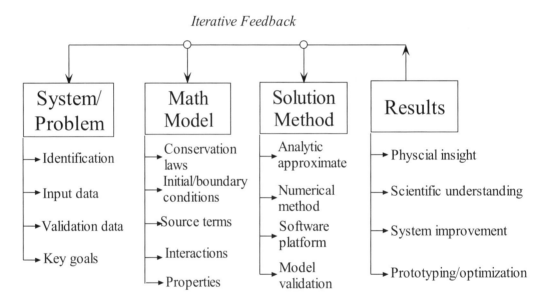

Figure 7.1   Key aspects of mathematical modeling

7.2 Mathematical Modeling

Models can be classified as:

- Verbal or sketches, i.e., a first-stage descriptive theory or concept
- Physical, in terms of a laboratory set-up and measurements
- Mathematical, i.e., equations describing a theory
- Computational, which could be:
    - ➤ Deterministic or stochastic
    - ➤ Analytical or numerical

However, most models carry problem-specific labels, e.g., "nanofluid flow model," without identifying any characteristics. In the case of a mathematical model, it represents a real process by describing key transport phenomena in terms of partial differential equations (PDEs) and associated initial and/or boundary conditions. If, typically, an analytic solution is not available, the flow domain has to be discretized with a mesh (or grid or lattice) and algebraic forms of the PDEs are solved with a numerical program, known as the code, on a computer. Figure 7.2 extends Figure 7.1, summarizing key elements of math modeling and computer simulation.

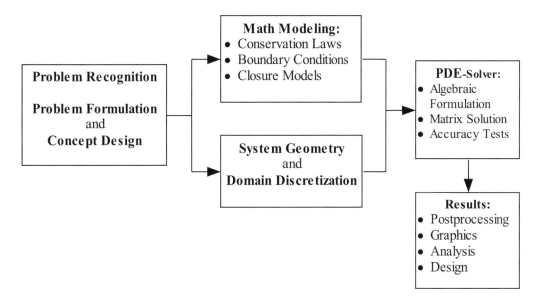

Figure 7.2 Key elements of math modeling and computer simulation

The ultimate goal of new product development is profit maximization. For that reason alone, "*concept design*" should be executed at an early stage of any

Research & Development effort because, if the concept of an innovative product or device is not well thought out, the rest of product development could be marred. Traditionally, with an emphasis on detailed design, prototype testing, and manufacturing preparation, concept design is often overlooked. Ways of designing a concept range from simple hand-drawn sketches to complex multicolor 3-D CAD images for (global) information exchange, digital prototyping, and 3-D printing. That can lead to physical models via rapid prototyping for laboratory testing as well as computer simulations and system/device prototyping. Clearly, the more advanced the development stage is, the more it is removed from the freedom of making conceptual changes which lead to improvements.

In summary, concept design can be sophisticated but in any case should occur at a very early stage of product development. Executed correctly and coherently, concept design can greatly save time and cost, leads to better prototype designs, and may eliminate frustration because of failed products at later stages.

Clearly, when a problem at hand is very complex, it is advisable to formulate a *simplified math model* of the given system. Benchmark analytic, empirical, or numerical solutions of such reduced problems often reveal (after a parametric sensitivity analysis) what the effects are of basic transport phenomena, geometric features, and operational conditions. The Examples in Parts A to C illustrate the usefulness of simplified math modeling results for physical insight and computer model validation. Parts A to C also show when to employ which conservation laws and how the given system geometry and characteristics determine the necessary boundary conditions. Closure models assure that the number of unknowns is matched by the number of available equations. Challenging examples include multiphase flow and turbulence as well as molecular dynamics. The remaining boxes of Figure 7.2 are discussed in the next section.

7.3 Computer Simulation

Numerical microfluidics and nanofluidics simulations are an integral application part of computational fluid dynamics (CFD). After problem formulation and math modeling has been accomplished (see Figure 7.2), the next very important step is mesh (or grid) generation. A given flow geometry (or system) is subdivided into 2-D or 3-D cells (or elements, or control volumes); that is, a discretization of the computational domain with a structured, unstructured, or hybrid mesh has to be accomplished (see Figure 7.3). At each cell corner and/or at each cell center all dependent variables are computed, i.e., velocity components, pressure, temperature, and/or species concentration, based on algebraic forms of the PDEs. Boundary conditions, e.g., no velocity slip, are specified on each edge of a 2-D system or on each face of a 3-D geometry. The same holds for the inlet/outlet planes of the flow domain. Values of material properties, such as fluid density and viscosity, have to be prescribed as well.

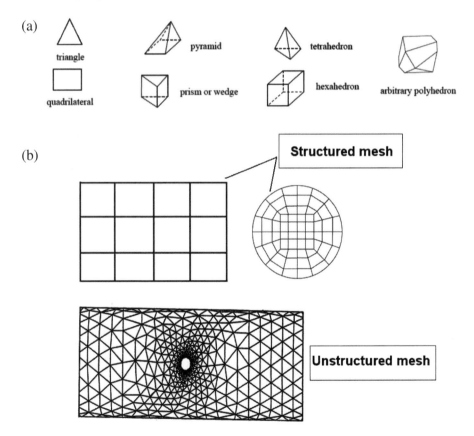

Figure 7.3 (a) Types of 2-D and 3-D elements (or cells) and (b) meshes accommodating simple (structured) or complex (unstructured) geometries

Starting with initial guesses for all dependent variables, the discretized forms of the governing equations are solved interactively, usually at each cell center. Various numerical methods (e.g., finite difference, finite volume, finite element) are available to discretize the transport phenomena, typically described by coupled PDEs (see Chapter 2), into a set of algebraic equations. A suitable matrix solution algorithm and associated convergence parameters are then activated to obtain the solution. Clearly, a numerical solution is reached when, after hundreds (or thousands) of *iterations,* the mass and momentum residuals have decreased to, say, 10^{-4} or less.

Simulation Accuracy.   *Computer model validation,* a very important task, has a numerical part and a comparison part. *Numerical validation* includes a refined mesh so that the results are just independent of the mesh density, while mass and momentum residuals should be less than 10^{-6}, or at least 10^{-4}. The *comparison part* involves matching of simulation results with exact (analytical) solutions and/or benchmark experimental data sets. If there is no confidence in the accuracy (and realism) of the simulation results, there is no reason to talk or write about predictive model capability, new physical insight, or quantitative data for system and device improvements. In this context, the article by Roache (2009) on computer model validation vs. numerical code verification is of interest.

Nevertheless, computer simulation results are never exact for usually three reasons:

- Incomplete mathematical description
- Simplified geometry/flow data
- Numerical errors

Potentially the most severe error sources are shortcomings in modeling when the physics of complex transport phenomena or property functions are mathematically not fully described. This may very well be the case when modeling complex transport phenomena such as turbulence, multiphase flows, biophysical processes, and fluid-structure interactions. Input data as well as critical and/or boundary conditions can be erroneous when the system geometry, inlet/outlet flows, and fluid properties are only assumed and not measured for realistic scenarios. Intrinsic to all numerical methods is the discretization error, which depends mainly on the degree of flow domain resolution with a fine mesh as well as the type and accuracy of the numerical method employed. Obviously, the finer the mesh, the better are the system geometry and the local flow field represented. Simply put, a mesh which fills the flow region, bounded by walls or free surfaces, is the underlying grid for the discrete representation of the governing equations, typically PDEs. Indicative of a mesh (or grid) is the grid spacing h, i.e., the characteristic length scale of a 2-D or 3-D mesh element or cell. In fact, the order of numerical code accuracy is directly related to the power of h. Most numerical algorithms are second-order approximations of the governing equations. It is important to implement smooth regional transitions from coarse to fine meshes and to avoid mesh element (or cell) distortion.

As shown in Figure 7.3b, there are basically structured and unstructured meshes (Thompson et al., 1999; Durbin & Medic, 2007; among others). Structured meshes feature identifiable grid lines which can be numbered sequentially, say, in 3-D: *i, j, k,* forming grid points x (i, j, k), y (i, j, k), and z (i, j, k). Cells (or elements) of structured grids may be quite distorted when geometrically complex flow domains are meshed, i.e., discretized. However, straight-wall geometries can be easily accommodated and boundary conditions can be readily enforced. In contrast, unstructured meshes, typically triangles for 2-D and tetrahedral elements for 3-D flow domains, map faithfully any physical space to the computational space (see Figure 7.3).

Equation Discretization.   Along with flow domain discretization, i.e., meshing, it is required to discretize the governing equations, typically the Navier-Stokes equations, subject to given initial/boundary conditions. Most codes use the finite-volume method (FVM), employing unstructured meshes. The mesh decomposes the fluid flow domain into connected cells. To each cell the flow variables are assigned; typically, velocity and pressure are stored at the center of each cell, i.e., the individual control volume. Specifically, integral forms of the conservation equations (see RTT in Sect. 1.3.2) are solved via an iteration method of successive approximations (see Durbin & Medic, 2007). For example, the linear momentum equation integrated over a cell of "control volume" $\forall$ with the "control surface" S and individual (inter) faces reads:

$$\frac{\partial}{\partial t}\left(\forall \rho \vec{v}\right)_{\text{cell}} = -\sum_{f_i}\left(\rho \vec{v}\vec{v}\cdot \hat{n}S\right)_i - \sum_{f_i}\left(p\hat{n}S\right)_i + \sum_{f_i}\left(\mu \nabla \vec{v}\cdot \hat{n}S\right)_i \qquad (7.1)$$

The LHS is the time rate of change of momentum inside the cell (i.e., control volume) due to net momentum efflux across the cell (inter) faces f_{ij} as well as all the pressure and viscous forces acting on the cell faces f_i. Clearly, the equations of each cell are coupled with all neighboring cells because the interface values needed are interpolated between the cell centers and so is the velocity gradient in the viscous force term generated with adjacent center velocity values. Furthermore, the local time derivative (see LHS of (Eq. 7.1)) is also replaced by a second-order (Euler) approximation. Discretization of the nonlinear convection term, $\rho \vec{v}\vec{v}$, is most crucial, requiring usually second-order (or higher) upwinding, such as the numerical QUICK scheme (see Patankar, 1980; or Hoffman, 2001).

Focusing on numerical solutions of fluid dynamics problems, standard *commercial software codes,* known as Navier-Stokes equation solvers, include CFD-ACE +, CFX, COMSOL, FLOW-3-D, FLUENT, and Star CCM. Except Flow-3D, these codes are based on the finite-volume method (Patankar, 1980; Versteeg &

Malaslaekera, 1996) and feature their preferred mesh generators, i.e., CFD-GEOM, ICEM-CFD, Flow-VU, and Gambit, respectively. CFX and FLUENT belong to the ANSYS family of codes which, like COMSOL, can tackle a variety of multiphysics problems. Most software vendors offer mesh-size restricted versions for affordable student use. Linear PDEs and some PDE solvers are available free of charge via the Internet (see FlexPDE, Calculix, etc.). Basically free open-source codes contain the Navier-Stokes equation solver, several modules, and library functions to solve most CFD problems. An internationally used code is OpenFOAM, accessible at www.openfoam.com. Solid mechanics problems can also be tackled with commercial software, typically based on the finite-element method, such as Abaqus, Adina, ANSYS, I-Deas, and Nastran. Multi-Physics codes (e.g., ANSYS and COMSOL) allow for the solution of fluid-structure interaction (FSI) problems.

Boundary Conditions. The most common boundary conditions are prescribed velocity inlet, no-slip wall, and zero-gage-pressure outlet conditions. Alternatively, the inlet pressure is prescribed and a zero-velocity gradient, i.e., via a long exit conduit, is assumed at the extended outlet, implying fully developed flow. Across a line/plane of symmetry, the gradients of all field variables are zero. For systems with repetitive geometries, such as turbine blades or heat exchanger tubes, periodic boundary conditions are imposed. Concerning initial conditions (ICs), the pressure and velocity at all discrete points in the flow field have to be known initially (i.e., at time $t=0$). The ICs should not affect the final results, just the convergence path via the number of iterations (in steady mode) and time steps required to reach the solution. As the initial assumption for the pressure and velocity fields are very important for fast convergence, in case of complicated transient flow problems CFD codes are typically run in the steady mode for a few iterations to obtain better ICs. Figure 7.4 provides a flow-chart of flow problem set-up and execution, employing computational fluid dynamics techniques.

Mathematically, *parabolic PDEs,* e.g., Prandtl's boundary-layer equation, have solutions which march forward towards the open end, guided by the wall or edge boundary conditions. In contrast, for *elliptic PDEs,* e.g., the Laplace heat conduction equation, the domain-surrounding boundary conditions greatly determine the solution. If the magnitude of any field variable on a boundary (or inlet/outlet) is assigned, it is called a *Dirichlet condition,* while enforcing the gradient of a dependent variable; it is labeled a *Neumann condition* (or Neumann problem).

7.3.1 Result Interpretation

In a postprocessing step, dependent variables or dimensionless groups are graphed for data interpretation or visualized in videos for physical insight and conference presentation. For proper result interpretation, i.e., new physical insight and useful applications, it is important that the flood of data sets is turned into smart 2-D or 3-D color figures. Graphing the results in terms of dimensionless groups and visualizing complex flow fields with streamlines or secondary velocity vector plots is most helpful. In any case, after successful computer simulation a comprehensive research report should be written, which is the point of departure for conference papers, journal articles, theses, and dissertations (see Figure 7.5).

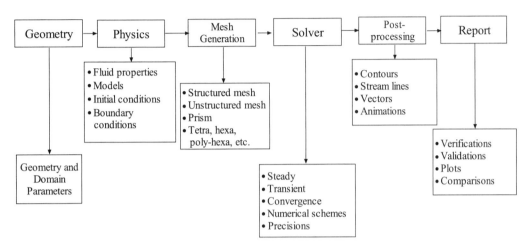

Figure 7.4 Flow chart of a basic CFD process

7.3.2 Computational Design Aspects

After gaining sufficient physical insight into the mechanics of fluid flow, fluid-particle dynamics, and/or fluid-structure interactions, the next major task is new (or improved) engineering system design. That is accomplished via *virtual prototyping,* i.e., on the computer, rather than employing physical models in the laboratory or wind tunnel.

In general, computational design analysis and simulation, supported by experimental testing, assist in evaluating the performance of a system described by a set of dimensions, material properties, (thermodynamic) loads, and operational requirements. Once established, validated computer simulations can also be used very efficiently in system parametric studies. A parametric analysis can be either deterministic or "probabilistic," the latter when taking measured uncertainties into account. In the deterministic case, all parameters vary continuously within the expected design/operational space, resulting in continuous responses, i.e., system performances. However, actual device dimensions, material properties, loads, and output are not nessecarily deterministic parameters, which pose the question of "design robustness." Again, computational analysis can swiftly assess how significant fluctuations in system parameter values may influence the system performance, including potential device failure.

- **ABSTRACT**

  Summarize project objectives, approach taken, and new results.

- **INTRODUCTION**

  Problem statement with project goals, justified by a thorough literature review, system sketch, assumptions, concepts/approach, and application of anticipated results.

- **THEORY AND (NUMERICAL) SOLUTION METHOD**

  List basic equations, postulates, and reduced set of equations (equal the number of unknowns) with initial/boundary conditions; provide details of solution method, e.g., for numerical solutions the following is needed: information/discussion concerning the computer program used/developed, platform, and mesh generation; numerical validation tests, i.e., mesh independence of the results, as well as 10^{-6} residuals for both mass and momentum conservation.

- **RESULTS AND DISCUSSION**

  For numerical solutions, *model validation* with experimental data sets, parametric sensitivity analyses, i.e., changes in boundary conditions, fluid properties, etc.; supply graphs with discussion of *basic research results*; graphs with explanations of *design applications*, and future work.

- **CONCLUSION**

  Similar to the Abstract; but add project result *limitations* and give future work goals.

- **REFERENCES**

  Follow "author (year)" format, e.g., Author (1999); nomenclature with definitions (if necessary) and Appendices (e.g., program listing, lengthy derivations, copies of key journal articles or relevant book pages, etc.)

Figure 7.5   Elements of a research report

Taking the methodology for a smart inhaler system, i.e., optimal drug-aerosol targeting, as an example, the underlying idea was first tested via computer simulations of controlled air-particle streams (see Kleinstreuer & Zhang, 2003 and Kleinstreuer et al., 2008). Another example is uniform nanoparticle transport and mixing for microcooling devices or nanomedicine systems (see Kleinstreuer & Li, 2008a, 2008b). Still, for inventions to be acceptable, laboratory proof of concept has to follow any convincing computer simulation results, while for medical devices clinical outcome is the ultimate litmus test.

For the much more frequent case of device/system improvement, the transport phenomena have to be first fully understood via CFD analysis. Then, geometric and operational changes can be tested via computer experiments. Examples include automobile or aircraft design for drag reduction and/or (downward) lift modification, heat exchangers with improved mixing and reduced size/weight, fuel injection systems generating higher combustion efficiencies, bio-MEMS for targeted drug delivery, medical implants to improve the quality of life, green-energy converters of high efficiency and reliability, etc.

CHAPTER 8

Computational Case Studies

8.1 Introduction

Using Figure 7.3 and the associated discussions as a guide, several internal laminar fluid flow and fluid-particle transport studies are outlined. First, transient laminar flow and steady particle suspension flow in single bifurcations are solved as demonstration projects (Sect. 8.2). Such results could serve as validations of more complex computer simulation programs for which experimental observations are lacking. Considering more complex systems, Sects. 8.3 and 8.4 then deal with optimal microsphere targeting of solid tumors and thermal nanofluid flow for microsystem cooling, respectively. Suggestions for math modeling (see Chapter 7) and numerical course projects are given in Sect. 8.5.

8.2 Model Validation and Physical Insight

As emphasized in Chapter 7, computer model validations via numerical accuracy tests and (if possible) comparisons to benchmark experimental data sets are a basic requirement. They verify the code's accuracy and instill confidence in the predictive capability of the model. In order to illustrate the steps for a comprehensive accuracy test, leading to full CFD model validation, experimental systems were simulated and their results were duplicated for two quite different bifurcating flows.

8.2.1 Transient Laminar Flow in a Single Bifurcation

Following Lieber and Zhao (1998), transient laminar flow in a symmetric bifurcation was analyzed. This simple system could represent part of a tubular network, lung airways, or an LoC (see Sect. 4.4). Figure 8.1 depicts the bifurcation geometry with inlet diameter $D_{in} = 3.81$ cm, daughter-tube diameter $D_{out} = 0.714 D_{in}$, and branch angle $2\alpha = 70°$. The CAD package SolidWorks 2010 was used to generate the 3-D surface model. The fluid was a water-glycerin mixture with $\mu = 0.0714$ Pa·s and $\rho = 1133$ kg/m³. For the transient flow, the Womersly number was kept constant, i.e.,

$$\text{Wo} = \frac{D}{2}\sqrt{2\pi f / v} = 4.3 \tag{8.1}$$

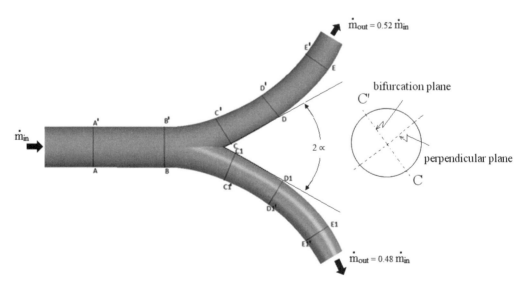

Figure 8.1   Bifurcation geometry

Three Reynolds numbers, i.e., $\mathrm{Re} = (D/2)u_\ell/v = 700,\ 1278,\ 2077$, were considered. The peculiar inlet-and-outlet profile in terms of the centerline velocity u_ℓ at $\mathrm{Re}_{max} = 2077$, as chosen by Lieber and Zhao (1998), is shown in Figure 8.2. A time-varying parabolic inlet velocity profile was implemented.

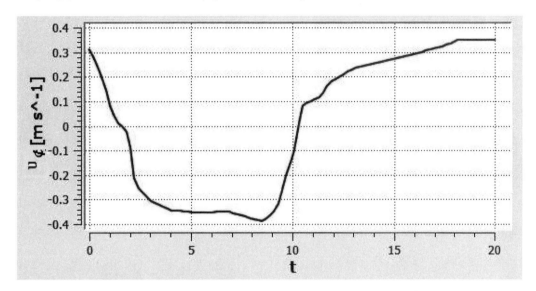

Figure 8.2   Inlet velocity varying with time

Solution Steps.   The engineering software CFX 13.0 (ANSYS, Inc., Canonsburg, PA) was employed to solve the describing N-S equations (see Eqs. (1.80) and (1.82)). Part of the necessary (unstructured) mesh is depicted in Figure 8.3. The final mesh size of 411,850 tetrahedral elements was constructed based on the mesh independence criterion of changes in $u_{average}$ being less than 0.05%. The second numerical accuracy test gave mass and momentum residuals of less than 10^{-6}. The no-slip velocity boundary condition as well as the measured 52:48 mass flow rate split between the daughter branches was enforced.

Validation Results.   Of the various possibilities for velocity profile comparisons, the fluid inlet phase at $t = 14$ s with $\mathrm{Re} = 1036$ was selected in terms of two midplane profiles right after the bifurcation, i.e., in cross section C'-C where the two planes are perpendicular to each other (see Figure 8.1). The skewed velocity profile in the bifurcation plane (Figure 8.4a) reveals the strong inertial effect, forming somewhat of a boundary layer. In the perpendicular plane, however, the velocity profile is M-shaped due to secondary flow effects (Figure 8.4b). Discrepancies between measured data points and computational results may have been caused by experimental uncertainties and slight geometric differences in the physical and meshed models. It should be recalled that the finite-volume code employed fully conserves mass and momentum.

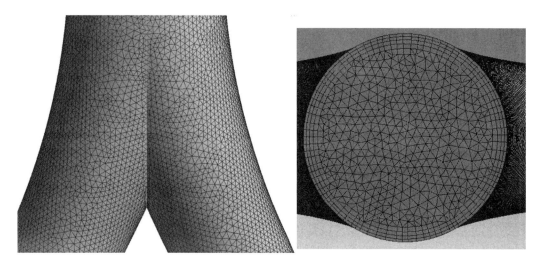

Figure 8.3   Unstructured mesh at bifurcation and tube inlet

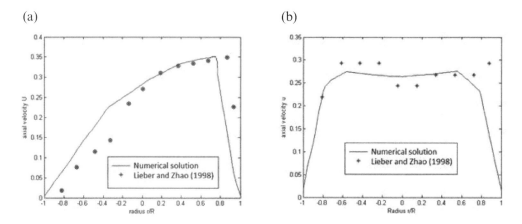

Figure 8.4   Comparison of velocity profile at (a) bifurcation plane and (b) perpendicular plane

Physical Insight.   The successful data comparison instills confidence that the computer simulation model is a predictive tool to analyze flow pattern and gain additional physical insight. Using the representative inlet pulse of Figure 8.2 with $Re_{max} = 2077$, Figures 8.5a-c depict the axial midplane velocity profiles at three time levels, i.e., $t = 1$ s, 5 s, 10 s. In the beginning at $t = 1$ s, when $Re = 605$, the flow field is smooth and rather branch-symmetric with skewed velocity profiles towards the inner

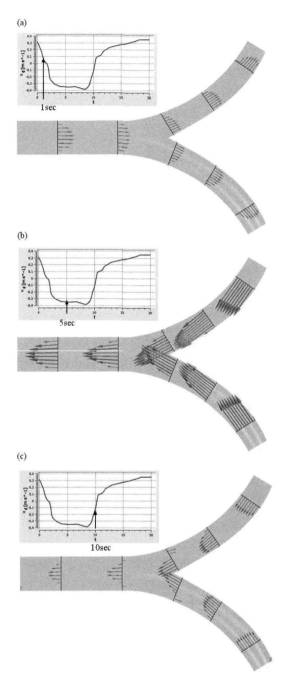

Figure 8.5 Velocity profile at different times: (a) $t = 1$ s; (b) $t = 5$ s; and (c) $t = 10$ s

walls of the daughter tubes (Figure 8.5a) due to inertia effects. At $t = 5$ s($\text{Re} = 2068$), strong reverse flow generates almost uniform profiles upstream in the daughter branches with merging streams at the point of confluence and a quickly developing parabolic outlet profile in the parent tube (Figure 8.5b). At $t = 10$ s ($\text{Re} = 605$), the reverse flow is decelerating, causing flow separation and hence significant recirculation regions around the bifurcation point (Figure 8.5c). A special effect of curved tubes on fluid kinematics is the appearance of Dean vortices.

Figure 8.6 shows the secondary (arrows) and axial (contour) flow fields in cross section $C_1' - C_1$ (see Figure 8.1) at $t = 20$ s where two symmetric counter-rotating vortices appear. Specifically, high-momentum midplane fluid flow impacts the outer curved wall where part of it separates symmetrically and is turned around to the inner (less curved) wall from which it proceeds back towards the outer wall.

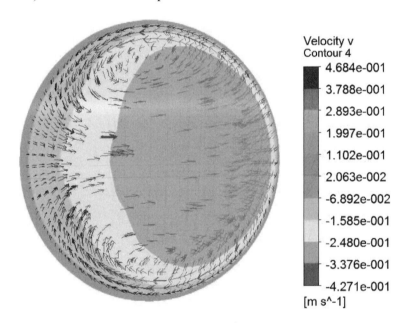

Velocity v
Contour 4

■	4.684e-001
	3.788e-001
	2.893e-001
	1.997e-001
	1.102e-001
	2.063e-002
	-6.892e-002
	-1.585e-001
	-2.480e-001
	-3.376e-001
■	-4.271e-001

[m s^-1]

Figure 8.6   Velocity contour in cross section $C_1' - C_1$ at $t = 20$ s

Summary.   This single bifurcation system exhibits a variety of basic flow phenomena:

- stagnation point flow at the bifurcation point where $\vec{v} = 0$ and $p = p_{max}$;
- recirculation zones due to flow separation;
- inertia (or axial momentum) effects forming boundary-layer type flow on the inner walls of the daughter tubes;
- merging streams during reverse flow;

- axial flow developments when secondary flows convert the parabolic velocity profile in the parent tube into M-shaped profiles in the curved daughter tubes; and
- additional tubular curvature effects causing Dean vortices in the daughter tubes.

Clearly, the causes of these unique flow patterns are due to the 3-D geometry with bifurcation and wall curvature and cyclic inlet/outlet flows. Although the bifurcation is symmetric, a maintained 52:48 flow-rate split generates slightly different velocity profiles in the two daughter tubes. While the outflow period ended at $t \approx 1.5$ s, most of the bulk flow continued due to the inertia effect; in contrast, near-wall flow followed the change in direction of the pressure gradient, aided by secondary flows in the daughter tubes.

8.2.2 Fluid-Particle Dynamics in a Bifurcation

Again, similar to the project discussed in Sect. 8.2.1, an experimental study was selected for computer model validation and extended physical insight (Bushi et al., 2005). Specifically, an asymmetric bifurcation (Figure 8.7) with different steady laminar flow rates through the daughter tubes and two spherical particle sizes, i.e., d_p values of 0.6 mm and 1.6 mm, were considered. The primary goal was to compare

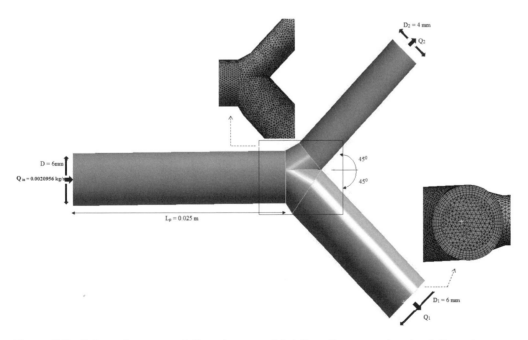

Figure 8.7   Schematic representation of asymmetric bifurcation geometry plus 3-D meshes

the computational results with experimental data in terms of particle number ratio N_1/N_2 leaving the daughter branches as a function of flow rate ratio Q_1/Q_2 and particle diameter d_p, i.e., $N_1/N_2 = \text{fct}(Q_1/Q_2, d_p)$. Rigid walls, constant fluid properties, and one-way fluid-particle coupling were assumed. A Poiseuille inlet velocity profile was enforced with Q_{in} being held constant.

Thus, the equations describing the fluid-particle system are the reduced N-S equations (Sect. 5.5) and Newton's second law of motion. Specifically, the individual particle trajectories are obtained from:

$$m_p \frac{d\vec{v}}{dt} = \vec{F}_D + \vec{F}_p \qquad (8.2)$$

where

$$m_p = \rho_p \frac{\pi}{6} d_p^3 \qquad (8.3)$$

$$\vec{F}_D = \frac{\pi}{8} \rho_f d_p^2 C_D (\vec{v}_f - \vec{v}_p)|\vec{v}_f - \vec{v}_p| \qquad (8.4)$$

$$C_D = \frac{24}{\text{Re}_p} + 3.6\,\text{Re}_p^{-0.313} \qquad (8.5)$$

and

$$\vec{F}_p = -\frac{\pi}{6} d_p^3 \nabla p \qquad (8.6)$$

The drag coefficient is the Schiller-Naumann correlation and the pressure gradient for $\vec{F}_p$ has to be obtained from the momentum equation.

Solution Steps. An unstructured mesh with 411,850 tetrahedral elements ($h \approx 300\ \mu m$) was necessary to achieved grid independence of the results. Selecting one-way fluid-particle coupling, ANSYS-CFX was employed to obtain the steady-state solutions. The convergence criterion was 10^{-6} for the rms values of mass and momentum residuals. While the inlet flow rate was kept constant ($\text{Re}_{in} = 500$), the four measured outlet flow rate ratios of the daughter branches (Bushi et al., 2005) were enforced to obtain the simulation results, considering 10,000 particles with d_p of either 0.6 mm or 1.6 mm.

Validation Results.   Table 8.1 summarizes the numerical values of number ratio of particles, i.e., N_1/N_2, which exited each branch for four different branch flow rate ratios Q_1/Q_2. A comparison with measured data sets is given in Figure 8.8.

Table 8.1   Numerical Particle Partitioning Results

	Q_1/Q_2	N_1/N_2		Q_1/Q_2	N_1/N_2
Particle	9.95E-1	4619/5013	Particle	1.008	4646/5054
diameter	1.5188	5638/4045	diameter	1.5322	5676/4009
(0.6mm)	2.3252	6606/3128	(1.6mm)	2.4462	6777/2964
	3.9919	7705/2067		4.2741	7834/1927

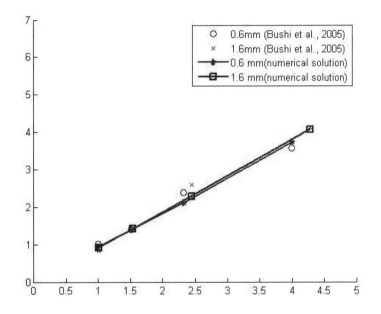

Figure 8.8   Comparison with experimental data sets

As the flow rate through the larger tube increases, proportionally more spheres are swept through daughter branch ①. The particle diameter hardly influences the linear $N_1/N_2(Q_1/Q_2)$ function.

Physical Insight.   Assuming the case of 1.6-mm particles first, the midplane velocity distributions are examined for the four flow rate ratios (see Figure 8.9a). When the branch flow rates are the same, i.e., actually $Q_1/Q_2 = 1.008$, the higher average velocity (recall $Q = v_{avg}A$) in the smaller tube ② carries more particles, so that $N_2 > N_1$ (see Figure 8.9a). This scenario quickly changes when $Q_1/Q_2 > 1$ and hence $N_1/N_2 > 1$.

(a)

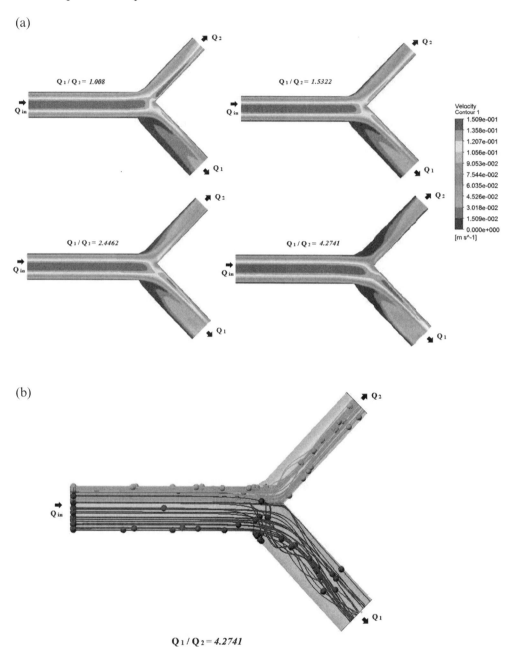

(b)

$Q_1 / Q_2 = 4.2741$

Figure 8.9 (a) Axial midplane velocity contours for the outlet flow ratios given in Table 8.1 for the 1.6-mm particle diameter case and (b) particle trajectories for $Q_1/Q_2 = 4.2741$.

In all four cases major reverse flow can be observed in the larger daughter tube ① as well as strong shear layers at/near the inner walls of both branches. Figure 8.9b depicts the particle trajectories in the bifurcation where 10,000 particles were injected from the inlet. Following basically the steady streamlines, most particles transport through the large branch, i.e., $N_1 = 7834$, and only $N_2 = 1927$ particles passed through branch ②, while some particles $(N_d = 239)$ were deposited on the walls.

8.3 Solid Tumor Targeting with Microspheres

As mentioned in Sect. 6.4, for *direct* tumor targeting it is necessary to employ microspheres as drug carriers (or treatment agents) in a laminar flow environment to assure *deterministic* particle trajectories. So, in Sect. 8.3.1 the methodology for direct (or optimal) particle delivery to solid tumors is outlined. As a computational case study, this new approach is discussed for optimal delivery of radioactive microspheres to liver tumors (Kleinstreuer et al., 2012).

8.3.1 Direct Targeting Methodology

In order to water a flower from a distance with a garden hose or to convert a boundary-value problem into an initial-value problem, the precise initial water jet, i.e., velocity vector and position, have to be known. Similarly, in 100% targeting of a solid tumor connected downstream to an arterial system, the position of a catheter releasing the microspheres has to be accurately predetermined. This is accomplished via "backtracking." Specifically, tens of thousands of particles are randomly released from the artery plane of the best axial location of the catheter's tip. During, say, 10 intervals of the local blood pulse, those particles which have hit the target (e.g., a single tumor or several tumor supply vessels) are backtracked to record the departure points in the initial arterial plane. This procedure generates 10 particle release maps (PRMs) which can be combined to a single PRM indicating several optimal targeting zones, possibly during the entire pulse (see Sect. 8.3.2).

Clearly, the computer simulation model predicting such PRMs for subject-specific cases has to be realistic, accurate, and comprehensive. Secondly, a deployed, telemetrically controlled smart microcatheter (SMC) system is needed to automatically synchronize microsphere supply from an external reservoir to the SMC and, most importantly, assure optimal radial SMC positioning according to the PRM to fully target the predetermined site(s). Figure 8.10 contrasts *schematically* nondirectional tumor targeting with a randomly positioned catheter (Figure 8.10a) and optimal targeting with an assumed SMC (Figure 8.10b).

The next section describes the computational steps for optimal tumor targeting with radioactive microspheres, using a well-positioned catheter. An experimental proof-of-concept of optimal targeting is given in Richards et al. (2012) and a computational, subject-specific analysis can be found in Childress et al. (2012).

8.3.2 Optimal Liver Tumor Targeting Study

Cancer of the liver, i.e., hepatic cellular carcinoma (HCC) and hepatic metastases, is worldwide a leading cause of death. Surgery, transplantation, chemotherapy, radiation, and/or ablation are treatment options to be considered on a case-by-case basis. An alternative technology is radioembolization (RE), also known more precisely as

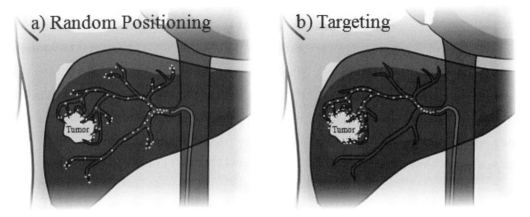

Figure 8.10   Nondirectional versus optimal tumor targeting with microspheres using a conventional catheter (a) versus an optimally positioned smart microcatheter (b)

selective internal radiation therapy (SIRT). In SIRT, millions of 20 to 60 μm spheres, loaded with the β-emitter yttrium 90, are injected into the blood stream via a catheter well placed near the tumor. Ideally, these ^{90}Y-microspheres do both partially embolize the very small tumor supply vessels and deliver locally intense radiation to destroy the tumor cells. Clearly, the goal is to target only the tumor via the typically enhanced blood supply to the tumor. This would minimize any damage to the surrounding (healthy) liver tissue and avoid microsphere migration to other organs. Similarly, drug-eluting beads (i.e., chemo-embolization) along or with ^{90}Y-microspheres could be directly delivered to the tumor with an optimally positioned catheter in the hepatic artery.

So, given a patient's hepatic artery system in terms of geometry, local blood waveform, pressure distributions, and tumor location, the problem solution discussed is how to find the unique, axial and radial catheter position. This, in turn, will allow for delivery of all radioactive microspheres (or chemo-drugs) at right dosages during suitable time intervals to predetermined tumor sites or tumor supply vessels.

Theory.   The computational domain of a representative hepatic artery system, with adjustable catheter, measured blood inlet waveform, and pressure distribution at the four outlets, is shown in Figure 8.11. Selectively, a tumor is assumed to be located at any of the four outlets D1 to D4.

The Cauchy equation and modified Quemada model for non-Newtonian blood flow as well as the coupled microsphere trajectory equation with drag, pressure gradient, and gravity forces are listed elsewhere (see Part A and Kleinstreuer et al., 2012). A discussion of the numerical solution method, including mesh generation (6,458,247 elements) and model validations, is also provided by Kleinstreuer et al. (2012).

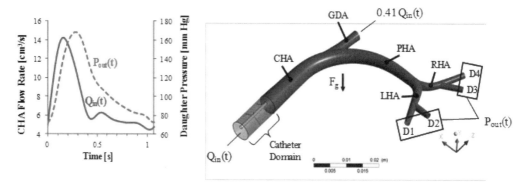

Figure 8.11   Inlet/outlet conditions and a representative hepatic artery system with common hepatic artery (CHA), proper hepatic artery (PHA), gastro-duodenal artery (GDA), right hepatic artery (RHA), left hepatic artery (LHA), and four daughter vessels (D1–D4)

Results and Discussion. First, PRMs with catheter present at three orientations were produced (see Figure 8.12). Specifically, 10,000 SIR-Spheres® ($d_p \approx 32$ μm, $\rho = 1600$ kg/m³, from Sirtex Medical, Sydney, Australia) were released via a randomized distribution across the upstream injection plane, typically at 10 equal intervals during the local blood cycle. Via backtracking, the individual PRMs were constructed, including a composite PRM, as outlined in Sect. 8.3.1.

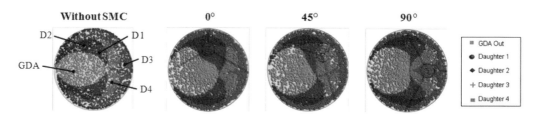

Figure 8.12   Velocity contours and particle release maps (PRMs) without and with the presence of the SMC for three catheter orientations. The PRMs revealed a measurable effect of the catheter presence on the particle trajectories.

For a basic targeting exercise, daughter vessel D1 connecting to a tumor was selected. The PRMs to fully target the five outlets, i.e., GDA and D1 to D4, during the 10 intervals are shown in Figure 8.13. Clearly, optimal targeting of D1 can be achieved during 8 of the 10 intervals.

Finally, Figure 8.14 depicts continuous targeting of a solid tumor situated at the outlet of daughter vessel D1 during intervals 2 to 9. A video based on an experimental proof-of-concept of this new methodology can be viewed via the website http://www.mae.ncsu.edu/cmpl/.

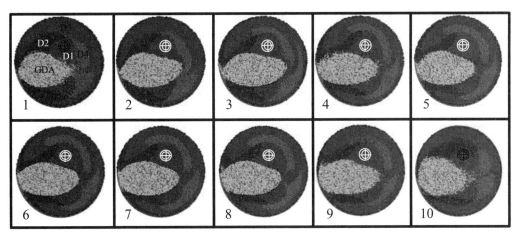

Figure 8.13 Particle release maps for targeting D1 with the basic catheter configuration. The white crosshairs indicate targeted injection to D1, while the two black crosshairs indicate nontargeted injection. Clearly, D1 targeting can be fully achieved during intervals 2–9.

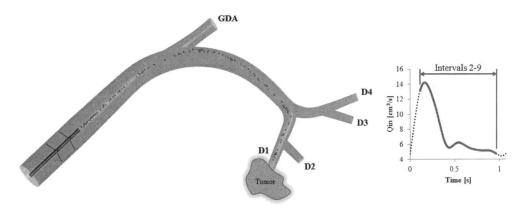

Figure 8.14 Optimal targeting of a liver tumor located at outlet D1

No doubt, subject-specific optimal drug targeting is most appealing because of the potentially high treatment efficacy, significant reduction in detrimental side effects, and wide range of controlled applications. However, realistic and accurate computer simulations necessary to generate optimal subject-specific data sets are presently still resources taxing and time consuming. Furthermore smart microcatheters and microprocessor-controlled treatment techniques are still in the prototyping-and-testing stage. A present-day solution would be a combination of a basic subject-specific computational analysis for the best conventional catheter positioning, to be verified via the use of ejected biodegradable microspheres to assure near-optimal targeting.

8.4 Homework Assignments and Course Projects

Clearly, homework assignments and course projects for Part D are especially challenging and taxing because of the inherent subject complexities. Nevertheless, the growing availability of powerful engineering software and workstations as well as the more advanced computational skills of undergraduate and graduate students alike make it possible to either retrace solved CFD problem solutions for just the learning experience or come up with new project results.

Starting with the material discussed in Chapter 7, Sect. 8.4.1 provides a few suggestions for math modeling using Figure 7.1 and Figure 7.2 as a guide. Section 8.4.2 then reiterates the course project essentials.

8.4.1 Mathematical Modeling

8.1 Contrast with sample applications the differences between (a) math modeling and computer simulation and (b) engineers and scientists.

8.2 Provide examples of micro/nanoscale models which fall into the four bulleted categories listed in Sect. 7.2.

8.3 As discussed in Sect. 7.2, "concept design" is an important early task (see Figure 7.2).

   (a) Develop a flowchart for key steps in concept design.

   (b) Carry out problem 8.3a for a microfluidics device.

   (c) Carry out problem 8.3a for a nanofluidics device.

8.4 Develop a math model (see appropriate boxes in Figures 7.1 and 7.2) for: (i) a heat exchanger cooling of a high-heat-flux microsystem; (ii) a two-stream mixer as part of an LoC; (iii) a microdevice for particle sorting and identification, e.g., cell, pathogens, toxicants, pollutants, etc.; (iv) a micropump for liquid micro/nanochannel flow, based on pressure drop, electroosmosis, electrophoresis, surface tension, or magnetohydrodynamics; (v) a micropump for gas flow in a microchannel or nanoconduit; (vi) a microdevice/system of your choice.

8.5 Consider 2-D heat conduction in a rectangular plate with a hole (see Figure 7.3b). (a) Using Matlab, create structured and unstructured meshes with various degrees of refinement. (b) Review three discretization methods using the transient 2-D heat conduction equation (see Sect. 1.3.3.4) as an example.

8.6 What is the difference between *model validation* and *result verification*? Develop a working flowchart for computer model validation.

8.7 Almost all mathematical modeling tasks focus on an isolated part or a segment of the complete system because it is too costly and difficult to simulate the entire body. Examples include a few bifurcations of the lung airways or the cardiovascular system, the compressor of a jet engine, a river near a power plant, a microchannel within a MEMS, etc. (a) Develop a methodology for modeling (or techniques for describing) inlet/outlet pressures in fluidics subsystems. (b) Select a subsystem of your choice and demonstrate the use of part (a).

8.8 Develop a flowchart for "computational design," i.e., *virtual prototyping;* then, illustrate its application with an example.

8.9 Compare virtual prototyping with the use of a physical model (e.g., in a wind tunnel) for measurements and design, considering (a) a large device (say, an automobile) and (b) a micro/nanosystem (say, an ink-jet printer head).

8.10 Set up the modeling equations and inlet/outlet boundary conditions for (a) the project described in Sect. 8.2.1 and (b) the project described in Sect. 8.2.2.

8.11 Concerning Figure 8.4b, explain with supporting sketches the evolution of the M-profile.

8.12 Review the physics and math of Dean vortices.

8.13 Set up the reduced N-S equations needed for the solution of the Sect. 8.2.2 problem.

8.14 Is Eq. (8.5) a good choice for the case study discussed?

8.15 Derive and discuss Eq. (8.6)!

8.16 What is the influence of F_p versus F_D in Eq. (8.2)? What would be the impact of $F_{gravity}$ in this case?

8.17 Concerning Figure 8.8, why has the particle diameter no visible influence on N_1/N_2 as a function of Q_1/Q_2?

8.18 Set up the mathematical description of the Sect. 8.3.1 problem for nanoparticles, i.e., $1 \text{ nm} < d_p < 100 \text{ nm}$!

8.19 A key requirement for the problem solution in Sect. 8.3.2 was that the therapeutic particles had to be microspheres! However, most chemo-drugs of the future are multifunctional *nanoparticles.* Provide a math model for optimal targeting with an SMC using spherical nanoparticles.

8.20 As a major course project (or thesis), develop a realistic and accurate computer simulation model to solve the project sketched in problem 8.19.

8.4.2 Set-Up for Course Projects
General:

- A course project has to be original, i.e., resubmission of selective write-ups of past research projects or ongoing thesis work is verboten!
- Two-person group work is permissible only when dual efforts are apparent, e.g., experimental + computational work.

- Hand in an *overview* (e.g., a preproposal), i.e., project title, system sketch, objectives, solution method and expected results, before starting the project.
- Follow the format of report writing (see Sect. 7.3 in Chapter 7).

CP Selection Process and CP Options:

- Focus on what interests you most as related to micro/nanofluidics, is within your skill level, and can be accomplished within a set time frame
- Select an appropriate journal/conference paper, for example, an experimental article on thermal flow (or nanofluid flow) in a microchannel, and perform an analytic/numerical analysis. There are now a half dozen new journals which are directly related to micro/nanofluidics; furthermore, traditional fluid mechanics, fluid-particle dynamics, heat transfer, and bio-medical journals feature articles on microfluidics and nanofluidics topics in every issue.
- Perform the micro/nanofluidics part of a microsystem, e.g., a magnetohydro-dynamic pump, a lab-on-a-chip, a particle-sorting device, a point-of-care fluid-sampling and diagnostics/analysis device, etc.
- Write tutorials, i.e., background information and solved example problems in key micro/nanofluidics areas, such as drug delivery (e.g., in nanomedicine), MEMS or LoC or PoC processes with device component performance analysis. Alternatively, select new tutorial topics, such as two-phase (liquid-gas) microchannel flow, boiling, magnetohydrodynamics, flow focusing, or particle sorting in microfluidics devices, optofluidics, acoustic-wave-driven micro/nanofluidics, etc.

CP Objectives:

- Deepen your knowledge base in micro/nanofluidics.
- Advance your analytical/computational skills or experimental skills.
- Generate an impressive research report in content and form.

References (Part D)

Bushi, D., Grad, Y., Einav, S., et al., 2005, Stroke, Vol. 36, pp. 2696-2700.

Childress, E.M., Kleinstreuer, C., Kennedy, A.S., 2012, ASME J. Biomechanical Engineering, Vol. 134 pp. 051005-1-10.

Durbin, P.A., Medic, G., 2007, *Fluid Dynamics with a Computational Perspective,* Cambridge University Press, New York.

Hoffman, J.D., 2001, *Numerical Methods for Engineers and Scientists,* 2nd ed, Marcel Dekker, New York.

Kleinstreuer, C., Li, J., 2008a, *Encyclopedia of Micro and Nanofluidics,* Li, D. (ed.), pp. 1314–1325, Springer, Heidelberg, DE.

Kleinstreuer, C., Li, J., 2008b, ASME Journal of Heat Transfer, Vol. 130, pp. 025501-1-3.

Kleinstreuer, C., Zhang, Z., 2003, International Journal of Multiphase Flow, Vol. 29, pp.271-289.

Kleinstreuer, C., Zhang, Z., Donohue, J.F., 2008a, Annual Review of Biomedical Engineering, Vol. 10, pp. 195-220.

Kleinstreuer, C., Zhang, Z., Li, Z., Roberts, W.L., Rojas, C., 2008b, International Journal of Heat and Mass Transfer, Vol. 51, pp. 5578-5589.

Kleinstreuer, C., Basciano, C.A., Childress, E.M., Kennedy, A.S., 2012, ASME Journal of Biomechanical Engineering, Vol. 134 , pp. 051004-1-10

Lieber, B.B., Zhao, Y., 1998, Annals of Biomedical Engineering, Vol. 26, pp. 821-830.

Patankar, S.V., 1980, *Numerical Heat Transfer and Fluid Flow,* McGraw-Hill, New York.

Richards, A.L., Kleinstreuer, C., Kennedy, A.S., Childress, E., Buckner, G.D., 2012, Biomedical Engineering, Vol. 59, pp.198-204.

Roache, P.J., 2009, Journal of Fluids Engineering, Vol. 131, pp. 034503-1-4.

Thompson, J.F., Soni, B.K., Weatherill, N.P., 1999, *Handbook of Grid Generation,* CRC Press.

Versteeg, H.K., Malalasekera, W., 1996, *An Introduction to Computational Fluid Dynamics: The Finite Volume Method,* Prentice Hall, Upper Saddle River, NJ.

APPENDICES

A. Math Tools and Equations

B. Property Data and Charts

APPENDIX A

Review of Tensor Calculus, Differential Operations, Integral Transformations, and ODE Solutions plus removable Equation Sheets

A.1 TENSOR CALCULUS

Here we restrict our review to tensor manipulations as needed in the text. Further information and solved examples can be found in Aris (1989), Schey (1973), and Appendix A of Bird et al. (2002).

A.1.1 Definitions

<u>Recall:</u> Tensors of rank n have 3^n components. For example:

- A tensor of rank "zero" is a *scalar* which has only one component, i.e., its magnitude (e.g., pressure).
- A tensor of rank "one" is a *vector* which has in general three components, i.e., three magnitudes and three directions (e.g., velocity).
- A tensor of rank "two" is usually labeled a *tensor* which has nine components, e.g., stress.

Coordinate Systems

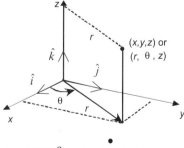

$x = r\cos\theta$ (x, y) or (r, θ)

$y = r\sin\theta$

$z = z$

Vector Products

Dot product

$\vec{u} \cdot \vec{v} = \vec{v} \cdot \vec{u} = |\vec{u}||\vec{v}|\cos\alpha \rightarrow$ scalar

Cross product

$\vec{u} \times \vec{v} = -\vec{v} \times \vec{u} = \vec{\omega} \rightarrow$ vector

Dyadic product:

$\vec{u}\vec{v} = \vec{\vec{a}}$

$\rightarrow$ tensor

397

Clearly, the type of vector product may result in a scalar (see dot product) or a tensor of rank two with nine components (see dyadic product). This is further illustrated when using the *del operator*, which has the characteristics of a vector (see A.1.2).

A.1.2 Operations with ∇

- By definition:

(i) In rectangular coordinates: $\nabla \equiv \hat{i}\,\dfrac{\partial}{\partial x} + \hat{j}\,\dfrac{\partial}{\partial y} + \hat{k}\,\dfrac{\partial}{\partial z}$ (A.1.1a)

(ii) In cylindrical coordinates: $\nabla \equiv \hat{e}_r\,\dfrac{\partial}{\partial r} + \hat{e}_\theta\,\dfrac{1}{r}\dfrac{\partial}{\partial \theta} + \hat{e}_z\,\dfrac{\partial}{\partial z}$ (A.1.1b)

- When operating on a scalar, say, the pressure, ∇p generates a vector, i.e., the *pressure gradient*:

$$\nabla p = (\hat{i}\,\frac{\partial}{\partial x} + \hat{j}\,\frac{\partial}{\partial y} + \hat{k}\,\frac{\partial}{\partial z})p = \frac{\partial p}{\partial x}\hat{i} + \frac{\partial p}{\partial y}\hat{j} + \frac{\partial p}{\partial z}\hat{k}$$ (A.1.2)

- When operating on a vector, it can produce:
  a scalar in case of a dot product, e.g., $\nabla \cdot \vec{v}$;
  a vector in case of a cross product, e.g., $\nabla \times \vec{v}$;
  a tensor in case of a dyadic product, e.g., $\nabla \vec{v}$.

<u>Note:</u> $\nabla \cdot \nabla = \nabla^2$ is the Laplacian operator (see Eqs. (A.1.8a,b))

Specifically,
(a) $\nabla \cdot \vec{v} \equiv \operatorname{div} \vec{v}$ is the divergence of the velocity field:

$$\nabla \cdot \vec{v} = \left(\hat{i}\,\frac{\partial}{\partial x} + \hat{j}\,\frac{\partial}{\partial y} + \hat{k}\,\frac{\partial}{\partial z} \right) \cdot \left(u\hat{i} + v\hat{j} + w\hat{k} \right)$$
$$= \left(\hat{i} \cdot \hat{i} \right)\frac{\partial u}{\partial x} + \left(\hat{i} \cdot \hat{j} \right)\frac{\partial v}{\partial x} + \left(\hat{i} \cdot \hat{k} \right)\frac{\partial w}{\partial x} +$$
$$\left(\hat{j} \cdot \hat{i} \right)\frac{\partial u}{\partial y} + \left(\hat{j} \cdot \hat{j} \right)\frac{\partial v}{\partial y} + \left(\hat{j} \cdot \hat{k} \right)\frac{\partial w}{\partial y} +$$
$$\left(\hat{k} \cdot \hat{i} \right)\frac{\partial u}{\partial z} + \left(\hat{k} \cdot \hat{j} \right)\frac{\partial v}{\partial z} + \left(\hat{k} \cdot \hat{k} \right)\frac{\partial w}{\partial z} = \frac{\partial u}{\partial x} + \frac{\partial v}{\partial y} + \frac{\partial w}{\partial z}$$ (A.1.3)

because $\hat{i} \cdot \hat{i} = |i||i| \cos \alpha; \alpha = 0 \Rightarrow \hat{i} \cdot \hat{i} = 1$, while $\hat{i} \cdot \hat{j} = 0 \langle \alpha = 90° \rangle$.

In general, $\hat{\delta}_i \cdot \hat{\delta}_j = \delta_{ij} := \begin{cases} 1 & \text{if } i = j \\ 0 & \text{if } i \neq j \end{cases}$, known as the *Kronecker delta*

(b) $\nabla \times \vec{v} \equiv \text{curl } \vec{v}$ is the curl (or rotation) of the velocity field:

$$\nabla \times \vec{v} = \left(\hat{i} \frac{\partial}{\partial x} + \hat{j} \frac{\partial}{\partial y} + \hat{k} \frac{\partial}{\partial z} \right) \times \left(u\hat{i} + v\hat{j} + w\hat{k} \right)$$

$$\left(\hat{i} \times \hat{i} \right) \frac{\partial u}{\partial x} + \left(\hat{i} \times \hat{j} \right) \frac{\partial v}{\partial x} + \left(\hat{i} \times \hat{k} \right) \frac{\partial w}{\partial x} +$$

$$\left(\hat{j} \times \hat{i} \right) \frac{\partial u}{\partial y} \ldots \text{ etc.} \tag{A.1.4}$$

Recalling the results for the cross products between unit (or base) vectors:

$$\begin{array}{lll} \hat{i} \times \hat{j} = \hat{k} & \hat{j} \times \hat{i} = -\hat{k} & \hat{i} \times \hat{i} = 0 \\ \hat{j} \times \hat{k} = \hat{i} & \hat{k} \times \hat{j} = -\hat{i} & \hat{j} \times \hat{j} = 0 \\ \hat{k} \times \hat{i} = \hat{j} & \hat{i} \times \hat{k} = -\hat{j} & \hat{k} \times \hat{k} = 0 \end{array}$$

we obtain:

$$\nabla \times \vec{v} = \hat{i} \left(\frac{\partial w}{\partial y} - \frac{\partial v}{\partial z} \right) + \hat{j} \left(\frac{\partial u}{\partial z} - \frac{\partial w}{\partial x} \right) + \hat{k} \left(\frac{\partial v}{\partial x} - \frac{\partial u}{\partial y} \right)$$

$$\nabla \times \vec{v} = \text{curl } \vec{v} = \begin{vmatrix} \hat{i} & \hat{j} & \hat{k} \\ \dfrac{\partial}{\partial x} & \dfrac{\partial}{\partial y} & \dfrac{\partial}{\partial z} \\ u & v & w \end{vmatrix} = \vec{\zeta} \left(\text{vorticity vector} \right) \tag{A.1.5}$$

or

(c) $\nabla \vec{v} \equiv \text{grad } \vec{v}$ is the dyadic product, or gradient, of the velocity field:

$$\nabla \vec{v} = \left(\hat{i} \frac{\partial}{\partial x} + \hat{j} \frac{\partial}{\partial y} + \hat{k} \frac{\partial}{\partial z} \right) \left(u\hat{i} + v\hat{j} + w\hat{k} \right)$$

$$\left(\hat{i}\hat{i} \right) \frac{\partial u}{\partial x} + \left(\hat{i}\hat{j} \right) \frac{\partial v}{\partial x} + \left(\hat{i}\hat{k} \right) \frac{\partial w}{\partial x} +$$

$$\left(\hat{j}\hat{i} \right) \frac{\partial u}{\partial y} + \ldots \text{ etc.} \tag{A.1.6}$$

Now, the unit vector dyadic product $\hat{\delta}_i\hat{\delta}_j$ indicates the *location* (i.e., $\hat{\delta}_i$ is the unit normal to the particular surface) and *direction* (i.e., $\hat{\delta}_j$ gives the direction) of a tensor. Thus,

$$\nabla\vec{v} = \text{grad } \vec{v} = \begin{vmatrix} \dfrac{\partial u}{\partial x} & \dfrac{\partial v}{\partial x} & \dfrac{\partial w}{\partial x} \\[2mm] \dfrac{\partial u}{\partial y} & \dfrac{\partial v}{\partial y} & \dfrac{\partial w}{\partial y} \\[2mm] \dfrac{\partial u}{\partial z} & \dfrac{\partial v}{\partial z} & \dfrac{\partial w}{\partial z} \end{vmatrix} \sim \vec{\vec{\tau}} \; \langle \text{stress tensor} \rangle \tag{A.1.7}$$

Notes:

- The use of $\nabla\cdot\vec{v}$, $\nabla\times\vec{v}$ and $\nabla\vec{v}$ plus illustrations were introduced in Chapter 1.
- The dot product *reduces* the rank of a tensor, e.g., $\nabla\cdot\vec{v}\to$ scalar and $\nabla\cdot\vec{\vec{\tau}}\to$ vector.
- The dyadic (or gradient) product increases the rank, e.g., $\nabla p\to$ vector and $\nabla\vec{v}\to$ tensor.
- The divergence of a scalar gradient is $\nabla\cdot\nabla s = \nabla^2 s = \Sigma(\partial^2 s/\partial x^2)$, where ∇^2 *is the Laplacian operator* producing a scalar field:

$$\text{Rectangular coordinates: } \nabla^2 \equiv \frac{\partial^2}{\partial x^2} + \frac{\partial^2}{\partial y^2} + \frac{\partial^2}{\partial z^2} \tag{A.1.8a}$$

$$\text{Cylindrical coordinates: } \nabla^2 \equiv \frac{1}{r}\frac{\partial}{\partial r}\left(r\frac{\partial}{\partial r}\right) + \frac{1}{r^2}\frac{\partial^2}{\partial\theta^2} + \frac{\partial^2}{\partial z^2} \tag{A.1.8b}$$

- The transpose of a second-order tensor, $\vec{\vec{a}}$, with components a_{ij} is denoted by $\vec{\vec{a}}^{\,\text{tr}}$ and is defined by

$$\left[a^{\text{tr}}\right]_{ij} = a_{ji} \tag{A.1.9a}$$

For example,

$$\vec{\vec{a}} \equiv \vec{v}\vec{w} = \begin{pmatrix} v_1 w_1 & v_1 w_2 & v_1 w_3 \\ v_2 w_1 & v_2 w_2 & v_2 w_3 \\ v_3 w_1 & v_3 w_2 & v_3 w_3 \end{pmatrix} \tag{A.1.9b}$$

whereas

$$\vec{a}^{\,\text{tr}} \equiv (\vec{v}\vec{w})^{\text{tr}} = \begin{pmatrix} v_1w_1 & v_2w_1 & v_3w_1 \\ v_1w_2 & v_2w_2 & v_3w_2 \\ v_1w_3 & v_2w_3 & v_3w_3 \end{pmatrix} \qquad (A.1.9c)$$

Sample Problem Solutions:

To illustrate a few tensor manipulations, the following sample problems are solved. Given the components of a symmetric tensor $\vec{\vec{\tau}}$, i.e., $\tau_{ij} = \tau_{ji}$:

$$\tau_{xx} = 3 \qquad\qquad \tau_{xy} = 2, \qquad\qquad \tau_{xz} = -1$$
$$\tau_{yy} = 2, \qquad\qquad \tau_{yz} = 1$$
$$\tau_{zz} = 0$$

and the components of a vector $\vec{v}$, e.g., $v_x = 5, v_y = 3, v_z = 7$; evaluate:

(a) $\vec{\vec{\tau}} \cdot \vec{v}$; (b) $\vec{v} \cdot \vec{\vec{\tau}}$; (c) $\vec{v}\vec{v}$; and (d) $\vec{\vec{\tau}} \cdot \vec{\vec{\delta}}$

where $\vec{\vec{\delta}}$ is the unit tensor, i.e.,

$$\delta_{ij} = \begin{pmatrix} 1 & & \varphi \\ & 1 & \\ \varphi & & 1 \end{pmatrix}$$

Solution:

A good preliminary exercise is to write down the vectors and tensors in component form:

(a) $\vec{\vec{\tau}} \cdot \vec{v} = \Sigma_i \vec{\delta}_i \left\{ \Sigma_j \tau_{ij} v_j \right\} = \left(\vec{\delta}_1, \vec{\delta}_2, \vec{\delta}_3 \right) \begin{pmatrix} 3 & 2 & -1 \\ 2 & 2 & 1 \\ -1 & 1 & 0 \end{pmatrix} \begin{pmatrix} 5 \\ 3 \\ 7 \end{pmatrix} = 14\vec{\delta}_1 + 23\vec{\delta}_2 - 2\vec{\delta}_3 \, \langle \text{a vector} \rangle$

where $\vec{\delta}_i \hat{=} $ unit vector in the i-direction, with i=1, 2, 3.

(b) $\vec{v} \cdot \vec{\vec{\tau}} = \vec{\vec{\tau}} \cdot \vec{v}$ since $\vec{\vec{\tau}}$ is symmetric.

(c) $\vec{v}\vec{v} = \Sigma_i \Sigma_j \vec{\delta}_i \vec{\delta}_j v_i v_j = 25\vec{\delta}_1\vec{\delta}_1 + 15\vec{\delta}_1\vec{\delta}_2 + 35\vec{\delta}_1\vec{\delta}_3$

$$+15\vec{\delta}_2\vec{\delta}_1 + 9\vec{\delta}_2\vec{\delta}_2 + 21\vec{\delta}_2\vec{\delta}_3$$

$$+35\vec{\delta}_3\vec{\delta}_1 + 21\vec{\delta}_3\vec{\delta}_2 + 49\vec{\delta}_3\vec{\delta}_3$$

(d) $\vec{\vec{\tau}} \cdot \vec{\vec{\delta}} = \Sigma_i \Sigma_l \vec{\delta}_i \vec{\delta}_l \left(\Sigma_j \tau_{ij} \delta_{jl} \right) = 3\vec{\delta}_1\vec{\delta}_1 + 2\vec{\delta}_1\vec{\delta}_2 - 1\vec{\delta}_1\vec{\delta}_3$

$$+2\vec{\delta}_2\vec{\delta}_1 + 2\vec{\delta}_2\vec{\delta}_2 + 1\vec{\delta}_2\vec{\delta}_3$$

$$-1\vec{\delta}_3\vec{\delta}_1 + 1\vec{\delta}_3\vec{\delta}_2 + 0\vec{\delta}_3\vec{\delta}_3$$

A.1.3 Some Tensor Identities

$$\nabla rs = r\nabla s + s\nabla r \tag{A.1.10}$$

$$\left(\nabla \cdot s\vec{v} \right) = \left(\nabla s \cdot \vec{v} \right) + s\left(\nabla \cdot \vec{v} \right) \tag{A.1.11}$$

$$\left(\nabla \cdot [\vec{v} \times \vec{w}] \right) = \left(\vec{w} \cdot [\nabla \times \vec{v}] \right) - \left(\vec{v} \cdot [\nabla \times \vec{w}] \right) \tag{A.1.12}$$

$$[\nabla \times s\vec{v}] = [\nabla s \times \vec{v}] + s[\nabla \times \vec{v}] \tag{A.1.13}$$

$$[\nabla \cdot \nabla \vec{v}] = \nabla(\nabla \cdot \vec{v}) - [\nabla \times [\nabla \times \vec{v}]] \tag{A.1.14}$$

$$[\vec{v} \cdot \nabla \vec{v}] = \tfrac{1}{2}\nabla(\vec{v} \cdot \vec{v}) - [\vec{v} \times [\nabla \times \vec{v}]] \tag{A.1.15}$$

$$[\nabla \cdot \vec{v}\vec{w}] = [\vec{v} \cdot \nabla \vec{w}] + \vec{w}(\nabla \cdot \vec{v}) \tag{A.1.16}$$

$$\left[\nabla \cdot s\vec{\vec{\delta}} \right] = \nabla s \tag{A.1.17}$$

$$\left[\nabla \cdot s\vec{\vec{\tau}} \right] = \left[\nabla s \cdot \vec{\vec{\tau}} \right] + s\left[\nabla \cdot \vec{\vec{\tau}} \right] \tag{A.1.18}$$

$$\nabla(\vec{v} \cdot \vec{w}) = \left[(\nabla \vec{v}) \cdot \vec{w} \right] + \left[(\nabla \vec{w}) \cdot \vec{v} \right] \tag{A.1.19}$$

A.2 DIFFERENTIATION

A.2.1 Differential Time Operators

In order to understand and solve fluid mechanics problems, the basic skills in linear algebra, as well as differentiating and integrating functions, graphing, analyzing functions, and curve fitting, are definitely <u>prerequisites</u>. If a review is necessary, the reader may want to consult Spiegel (1971), Schaum's Outline series, or Greenberg (1998), among many other texts.

The different notations and the physical meanings of various time derivatives (i.e., differential operators) are presented as follows:

- Partial time derivative: $\partial\#/\partial t \,\hat{=}\,$ Changes in variable "#" with time observed from a fixed position in space, i.e., stationary observer.
- Substantial or material time derivate: $D\#/Dt \,\hat{=}\,$ Changes of variable "#" with time following the fluid/material element in motion. The Stokes, or material, derivative is defined as:

$$\frac{D\#}{Dt} \equiv \frac{\partial\#}{\partial t} + (\vec{v}\cdot\nabla)\# \tag{A.2.1}$$

In Eq. (A.2.1) the Lagrangian time rate of change is expressed in terms of Eulerian derivatives. For example, c being a species concentration $\left[M/L^3\right]$, the material time derivative is

$$\frac{Dc}{Dt} \equiv \frac{\partial c}{\partial t} + (\vec{v}\cdot\nabla)c \tag{A.2.2a}$$

In rectangular coordinates

$$\frac{Dc}{Dt} = \frac{\partial c}{\partial t} + u\frac{\partial c}{\partial x} + v\frac{\partial c}{\partial y} + w\frac{\partial c}{\partial z} \tag{A2.2b}$$

while in tensor notation

$$\frac{Dc}{Dt} = \frac{\partial c}{\partial t} + v_k\frac{\partial c}{\partial x_k}; \; k=1,\,2,\,3 \tag{A.2.2c}$$

where $\partial c/\partial t \,\hat{=}\,$ local time derivative (i.e., accumulation of species c) and $v_k\,\partial c/\partial x_k \,\hat{=}\,$ convective derivatives (i.e., mass transfer by convection).

Note: Repeated indices imply summation of these terms, here $k = 1, 2, 3$ <Einstein convention>.

Total time derivative: $d\#/dt \triangleq$ Changes of $\#$ with respect to time observed from a point *moving differently* than the flow field. For example:

$$\frac{dc}{dt} = \frac{\partial c}{\partial t} + \frac{dx}{dt}\frac{\partial c}{\partial x} + \frac{dy}{dt}\frac{\partial c}{\partial y} + \frac{dz}{dt}\frac{\partial c}{\partial z} \tag{A.2.3}$$

where $dx/dt = u$, $dy/dt = v$, and $dz/dt = w$ are the velocity components of the moving observer.

A.2.2 The Total Differential

Dependent variables describing transport phenomena, such as fluid velocity, pressure, and species concentration, are often a function of more than one independent variable. For example, the fluid velocity is a function of three spatial coordinates, say, x, y, and z and, if the flow is unsteady, time t. Thus, $\vec{v} = \vec{v}(x,y,z,t)$. The total differential is defined as

$$d\vec{v} = \frac{\partial \vec{v}}{\partial x}dx + \frac{\partial \vec{v}}{\partial y}dy + \frac{\partial \vec{v}}{\partial z}dz + \frac{\partial \vec{v}}{\partial t}dt \tag{A.2.4}$$

If the spatial coordinates x, y, and z are also functions of time, then the total (particle) time derivative is:

$$\frac{d\vec{v}}{dt} = \frac{\partial \vec{v}}{\partial x}\frac{dx}{dt} + \frac{\partial \vec{v}}{\partial y}\frac{dy}{dt} + \frac{\partial \vec{v}}{\partial z}\frac{dz}{dt} + \frac{\partial \vec{v}}{\partial t} \tag{A.2.5}$$

Such differentiation can be extended to the calculation of fluid acceleration and mass transport where the local quantities change with time. For a scalar, say, the pressure, we have

$$dp = \frac{\partial p}{\partial x}dx + \frac{\partial p}{\partial y}dy + \frac{\partial p}{\partial z}dz \tag{A.2.6}$$

A.2.3 Truncated Taylor Series Expansions and Binomial Theorem

In order to approximate a function, say, $y(x)$, around some point $x = x_0$, we employ two or three terms of the Taylor series. For one independent variable,

$$y(x) = y\Big|_{x=x_0} + \frac{dy}{dx}\Big|_{x=x_0}(x-x_0) + \frac{1}{2}\frac{d^2 y}{dx^2}\Big|_{x=x_0}(x-x_0)^2 + $$
$$\frac{1}{6}\frac{d^3 y}{dx^3}\Big|_{x=x_0}(x-x_0)^3 + \cdots \qquad (A.2.7)$$

Clearly, the first two terms provide a straight-line fit and the first three a parabolic fit of $y(x)$ at $x = x_0$.

If we want to estimate the value of y a very small distance away from the known $y(x)$, i.e., what is $y(x+\varepsilon)$ where $\varepsilon \ll 1$, we can write with Eq. (A.2.7):

$$y(x+\varepsilon) \approx y(x) + \frac{dy}{dx}\Big|_{x=\varepsilon}\varepsilon \qquad (A.2.8)$$

As the graph indicates, the step size ε and local curvature of $y(x)$ determine the accuracy of Eq. (A.2.8). Equation (A.2.8) is employed in Sect. 1.3 to derive equations in differential form.

For functions of two variables, e.g., $f(x,y)$, we write

$$f(x+\varepsilon, y+\delta) \approx f(x,y) + \frac{\partial f}{\partial x}\Big|_{x=\varepsilon}\varepsilon + \frac{\partial f}{\partial y}\Big|_{y=\delta}\delta \quad (A.2.9)$$

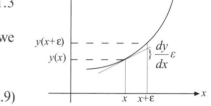

When dealing with rational *fractional* functions, it is often advantageous to express them in terms of partial fractions and then expand them, using the *binomial theorem*. For example, the expansion

$$(c+\varepsilon)^n = c^n + nc^{n-1}\varepsilon + \frac{n(n-1)}{2!}c^{n-2}\varepsilon^2$$
$$+ \frac{n(n-1)(n-2)}{3!}c^{n-3}\varepsilon^3 + \cdots \qquad (A.2.10)$$

is valid for all values of n if $|\varepsilon| < |c|$. If $|c| < |\varepsilon|$, the expansion is valid only if n is a nonnegative integer.

A.2.4 Hyperbolic Functions

Next to the circular trigonometric functions, the hyperbolic functions $\sinh x$, $\cosh x$, and $\tanh x$ appear frequently in science and engineering and hence in the present text.

Graph:

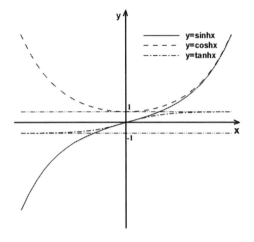

Relations:

$$\sinh x = \frac{1}{2}(e^x - e^{-x})$$

$$\cosh x = \frac{1}{2}(e^x + e^{-x})$$

$$\cosh^2 x - \sinh^2 x = 1$$

$$\sinh(x \pm y) = \sinh x \cosh y$$
$$\pm \cosh x \sinh y$$

$$\cosh(x \pm y) = \cosh x \cosh y$$
$$\pm \sinh x \sinh y$$

A.3 INTEGRAL TRANSFORMATIONS

A.3.1 The Divergence Theorem

As established by Gauss, the divergence theorem states that the integration over the dot product of a vector field, $\vec{v}$, with a closed regular area element, $d\vec{A}$, is equal to the integration of the divergence of $\vec{v}$, i.e., $\nabla \cdot \vec{v}$, over the interior volume, $\forall$:

$$\iint_A \vec{v} \cdot d\vec{A} = \iiint_\forall (\nabla \cdot \vec{v})\, d\forall \tag{A.3.1}$$

Equation (A.3.1) is being used in Sect. 1.3 when converting all surface integrals in the Reynolds Transport Theorem into volume integrals in order to express the conservation laws of mass, momentum, and energy in *differential form*.

Sample Problem: Given $\vec{v} = 4xz\hat{i} - y^2\hat{j} + yz\hat{k}$, i.e., $u = 4xz$, $v = -y^2$, and $w = yz$, in a unit cube, i.e., $0 \le x \le 1$, $0 \le y \le 1$, and $0 \le z \le 1$, show that Eq. (A.3.1) holds.

Solution:

(a) Six-sided surface integral: $\displaystyle\iint_A \vec{v} \cdot d\vec{A} = \int_0^1\int_0^1 \vec{v} \cdot \hat{n}\, dA$

where

$$\hat{n} = \hat{i} \text{ at } x = 1 \text{ and } \hat{n} = -\hat{i} \text{ at } x = 0 \text{ with } dA = dydz$$

$$\hat{n} = \hat{j} \text{ at } y = 1 \text{ and } \hat{n} = -\hat{j} \text{ at } y = 0 \text{ with } dA = dxdz$$

and

$$\hat{n} = \hat{k} \text{ at } z = 1 \text{ and } \hat{n} = -\hat{k} \text{ at } z = 0 \text{ with } dA = dxdy$$

Thus,

$$\int_0^1\int_0^1 \vec{v} \cdot \hat{i}\, dA = \int_0^1\int_0^1 4xz\Big|_{x=1}\, dydz = 4y\Big|_0^1\left(\frac{1}{2}z^2\Big|_0^1\right) = 2$$

Similarly,

$$\int_0^1\int_0^1 \vec{v}\cdot\hat{j}\,dA=-1;\ \int_0^1\int_0^1 \vec{v}\cdot\hat{k}\,dA=\frac{1}{2}$$

and the other three negative surface integrals are zero. Hence,

$$\iint_A \vec{v}\cdot\hat{n}\,dA=2-1+\frac{1}{2}+0+0+0=\frac{3}{2}$$

(b) Volume integral: $\iiint_\forall \nabla\cdot\vec{v}\,d\forall=\int_0^1\int_0^1\int_0^1\left(\frac{\partial u}{\partial x}+\frac{\partial v}{\partial y}+\frac{\partial w}{\partial z}\right)dx\,dy\,dz$

$$\int_0^1\int_0^1\int_0^1(4z-2y+y)\,dx\,dy\,dz=\left|\frac{4z^2}{2}-\frac{z}{2}\right|_0^1=\frac{3}{2}$$

Comment:

It is evident that such integral operations over _variable_ vector fields result in scalars (i.e., numbers); this implies that the Reynolds Transport Theorem generates scalar quantities, i.e., numerical values for flow rates, averaged velocities, forces, wall stresses, pressure drops, etc.

A.3.2 Leibniz Rule

A switch in the order of operation is justified with Leibniz's rule: If $F(t)=\int_a^b f(x,t)\,dx$ and a, b are constants, then

$$\frac{dF}{dt}\equiv\frac{d}{dt}\left[\int_a^b f(x,t)\,dx\right]=\int_a^b\frac{\partial f}{\partial t}\,dx \tag{A.3.2}$$

Equation (A.3.2) is occasionally applied when dealing with the transient term <volume integral> in the RTT. In general, Leibniz's rule reads

$$\frac{d}{dt}\int_{a(t)}^{b(t)} f(x,t)\,dx=\int_a^b\frac{\partial f}{\partial t}\,dx+b'f\big[b(t),t\big]-a'f\big[a(t),t\big]$$

A.3.3 Error Function

Numerous natural phenomena follow exponential functions. For example,

$$\text{erf}(x) = \frac{2}{\sqrt{\pi}} \int_0^x e^{-\xi^2} \, d\xi \tag{A.3.3a}$$

where

$$\text{erf}(0) = 0 \text{ and } \text{erf}(\infty) = 1 \tag{A.3.3b,c}$$

The integrand $\exp(-\xi^2)$ is a normal probability distribution. Thus, Eq. (A.3.3) is a solution part of processes governed by Gaussian-type distributions. The "complementary error function" is defined as

$$1 - \text{erf}(x) \equiv \text{erfc}(x) \tag{A.3.4}$$

A.3.4 Integral Methods

Two solution techniques dealing with integral equations are briefly discussed. The first method starts with the integration of a given set of partial differential equations that describe a given flow system, known as the integral method. The second approach starts with balance equations in integral form, i.e., the Reynolds Transport Theorem, which assures the conservation laws for a control volume.

Von Karman Integral Method   In contrast to separation of variables and similarity theory, the integral method is an approximation method. The von Karman integral method is the most famous member of the family of integral relations, which in turn is a special case of the method of weighted residuals (MWR). Specifically, a transport equation in normal form can be written as [cf. Eq. (1.7)]

$$L(\phi) \equiv \frac{\partial \phi}{\partial t} + \nabla \cdot (\vec{v}\phi) - \nu \nabla^2 \phi - S = 0 \tag{A.3.5}$$

where $L(\bullet)$ is a (nonlinear) operator, ϕ is a dependent variable, and S represents sink/source terms. Now, the unknown ϕ-function is replaced by an *approximate* expression, i.e., a "profile" or functional $\tilde{\phi}$ that satisfies the boundary conditions but contains a number of unknown coefficients or parameters. As can be expected,

$$L(\tilde{\phi}) \neq 0, \text{ i.e., } L(\tilde{\phi}) \equiv R \tag{A.3.6a,b}$$

where R is the residual. In requiring that

$$\int_{\Omega} W R \, d\Omega = 0 \qquad\qquad\qquad (A.3.6c)$$

we force the weighted residual over the computational domain Ω to be zero and thereby determine the unknown coefficients or parameters in the assumed ϕ-function. The type of weighing function W determines the special case of the MWR, e.g., integral method, collocation method, Galerkin finite element method, control volume method, etc. (cf. Finlayson, 1972).

The von Karman method is best applicable to laminar/turbulent similar or nonsimilar *boundary-layer-type* flows for which appropriate velocity, concentration, and temperature profiles are known, i.e., thin and thick wall shear layers as well as plumes, jets, and wakes. Solutions of such problems yield global or integral system parameters, such as flow rates, fluxes, forces, boundary-layer thicknesses, shape factors, drag coefficients, etc.

In general, a two-dimensional partial differential equation is integrated in one direction, typically normal to the main flow and thereby transformed into an ordinary differential equation, which is then solved analytically or numerically. Implementation of the integral method rests on two general characteristics of boundary-layer-type problems: (i) the boundary conditions for a particular system simplify the integration process significantly so that a simpler differential equation is obtained and (ii) all extra unknown functions, or parameters, remaining in the governing differential equation are approximated on physical grounds or by empirical relationships. Thus, closure is gained using, for example, the entrainment concept for plumes, jets, and wakes or by expressing velocity and temperature profiles with power expansions for high-Reynolds-number flows past submerged bodies.

A.4 ORDINARY DIFFERENTIAL EQUATIONS

For most applications in biofluids, the key differential equations are the equation of continuity, i.e., conservation of fluid mass, and the Navier-Stokes equations, i.e., conservation of linear momentum for constant fluid properties, as well as the scalar transport equations for species mass and heat transfer. They reflect the conservation laws in terms of differential balances for fluid mass, momentum, species concentration, and energy.

If the dependent variable, say, the velocity, is a function of more than one independent variable (e.g., x, y, z, t or r, q, z, t), then the describing equation is a partial differential equation (PDE); otherwise, it is an ordinary differential equation (ODE). Clearly, solving PDEs requires usually elaborate transformations or numerical algorithms (see Ozisik, 1993 and Hoffman, 2001). For that reason and to gain direct physical insight, simplified base-case problems are discussed (see Chapter 1) where the continuity equation is fully satisfied and the Navier-Stokes equations are reduced to ODEs. Exact solutions for ODEs are listed in Polyamin & Zaitsev (1995). Numerical ODE solutions may be obtained with commercial software such as Matlab and Mathcad, which for their underlying finite-difference approximations rely on selected terms of the Taylor series (see Sect. A.2.3).

After developing a mathematical model describing approximately the fluid flow problem at hand, the resulting ODE (or system of coupled ODEs) has to be classified. One has to determine if the ODE, say, for $y(x)$ is:

- linear or nonlinear (e.g., y^2, yy', $\sqrt{y''}$, etc.);
- with constant coefficients or not;
- homogeneous or inhomogeneous;
- of first, second, or nth order;
- an initial-value problem (IVP) or a boundary-value problem (BVP).

The last two types of ODEs can be solved numerically with the Runge-Kutta method (IVPs) or a shooting method (BVPs) as available from www.netlib.org/odepack.

Fortunately, most introductory fluid flow problems are governed by ODEs of the form

$$\frac{d^n}{dy^n}[f(y)] = g(y) \qquad (A.4.1)$$

which can be solved by direct integration, subject to n boundary conditions (BCs).

Typically, $n = 2$ and $g(y) \equiv K$, a constant, so that Eq. (A.4.1) can be rewritten as

$$\frac{d}{dy}\left[\frac{df}{dy}\right] = K \qquad\qquad (\text{A.4.2a})$$

or

$$d\left[\frac{df}{dy}\right] = Kdy \qquad\qquad (\text{A.4.2b})$$

Hence,

$$\frac{df}{dy} = Ky + C_1 \qquad\qquad (\text{A.4.2c})$$

and integrating again yields

$$f(y) = \frac{K}{2}y^2 + C_1 y + C_2 \qquad\qquad (\text{A.4.3})$$

where the integration constants are determined with two given boundary conditions for $f(y)$.

In cylindrical coordinates, we may encounter an ODE somewhat similar to Eq. (A.4.1) of the form

$$\frac{1}{r}\frac{df(r)}{dr} + \frac{d^2 f(r)}{dr^2} = g(r) \qquad\qquad (\text{A.4.4a})$$

which can be rewritten for direct integration as:

$$\frac{1}{r}\frac{d}{dr}\left(r\frac{df(r)}{dr}\right) = g(r) \qquad\qquad (\text{A.4.4b})$$

For example, with $g(r) \equiv K = $ constant and $f \equiv v$, say the fully developed axial velocity in a tube of radius R, we have

$$d\left(r\frac{dv(r)}{dr}\right) = Krdr \qquad\qquad (\text{A.4.5a})$$

so that after integration:

$$\frac{dv(r)}{dr} = \frac{K}{2}r + \frac{C_1}{r} \qquad (A.4.5b)$$

Integrating again yields

$$v(r) = \frac{K}{4}r^2 + C_1 \ln r + C_2 \qquad (A.4.6)$$

The differences between the solutions of (A.4.3) and (A.4.6) as well as the impact of different BCs on Eq. (A.4.6) were discussed in Sect. 1.3.

Another interesting case where term contraction allows for direct integration is as an ODE of the form

$$\frac{df(x)}{dx} + \frac{n}{x}f(x) \equiv \frac{1}{x^n}\frac{d}{dx}\left[x^n f(x)\right] = g(x); \qquad n = 1,2 \qquad (A.4.7a)$$

which yields after the first integration:

$$x^n f(x) = \int x^n g(x) dx + C_1 \qquad (A.4.7b)$$

The term $\int x^n g(x) dx$ could be solved via integration by parts, i.e.,

$$\int_a^b u\,dv = (uv)\Big|_a^b - \int_a^b v\,du \qquad (A.4.8)$$

Numerous natural processes can be described by a linear, inhomogeneous second-order ODE with constant coefficients, i.e.,

$$f'' + Af' + Bf = F(x) \qquad (A.4.9)$$

where $F(x)$ is a prescribed (forcing) function. Typically,

$$f(x) = f_{\text{homog.}} + f_{\text{inhomog.}} \qquad (A.4.10a)$$

In general, $f_{\text{homog.}}$ admits exponential solutions, e.g.,

$$f(x) \sim e^{\lambda x} \tag{A.4.10b}$$

where λ can be obtained from the quadratic equation

$$\lambda^2 + A\lambda + B = 0 \tag{A.4.10c}$$

so that

$$f(x) = C_1 e^{\lambda_1 x} + C_2 e^{\lambda_2 x} \tag{A.4.10d}$$

Table A.4.1 summarizes typical ODEs describing transport phenomena, where f and g are functions of x and the quantities a, b, and c are real constants.

Table A4.1   Typical ODEs and Their General Solutions

Equation	General Solution
$\dfrac{dy}{dx} = \dfrac{f(x)}{g(y)}$	$\displaystyle\int g\,dy = \int f\,dx + C_1$
	$y = e^{-\int f\,dx}\left(\displaystyle\int e^{\int f\,dx} g\,dx + C_1\right)$
$\dfrac{dy}{dx} + f(x)y = g(x)$	$y = C_1 \cos ax + C_2 \sin ax$
	$y = C_1 \cosh ax + C_2 \sinh ax$
$\dfrac{d^2 y}{dx^2} + a^2 y = 0$	$y = C_3 e^{+ax} + C_4 e^{-ax}$
$\dfrac{d^2 y}{dx^2} - a^2 y = 0$	$y = \dfrac{C_1}{x}\cos ax + \dfrac{C_2}{x}\sin ax$
$\dfrac{1}{x}\dfrac{d}{dx}\left(x^2 \dfrac{dy}{dx}\right) + a^2 y = 0$	

A.5 TRANSPORT EQUATIONS (CONTINUITY, MOMENTUM, AND HEAT TRANSFER)

A.5.1 Continuity Equation

$$\frac{\partial \rho}{\partial t} + \nabla \cdot \left(\rho \vec{v} \right) = 0$$

Note: For $\rho = \cancel{c}$ $\Rightarrow$ $\nabla \cdot \vec{v} = 0$

- Rectangular coordinates: $\dfrac{\partial \rho}{\partial t} + \dfrac{\partial}{\partial x}\left(\rho u \right) + \dfrac{\partial}{\partial y}\left(\rho v \right) + \dfrac{\partial}{\partial z}\left(\rho w \right) = 0$

- Cylindrical coordinates: $\dfrac{\partial \rho}{\partial t} + \dfrac{1}{r}\dfrac{\partial}{\partial r}\left(\rho r v_r \right) + \dfrac{1}{r}\dfrac{\partial}{\partial \theta}\left(\rho v_\theta \right) + \dfrac{\partial}{\partial z}\left(\rho v_z \right) = 0$

A.5.2 Equation of Motion (or Linear Momentum Equation)

CAUCHY Equation:

$$\rho \frac{D\vec{v}}{Dt} = -\nabla p + \nabla \cdot \vec{\vec{\tau}} + \rho \vec{g}$$

where

$$\frac{D\#}{Dt} \equiv \frac{\partial \#}{\partial t} + (\vec{v} \cdot \nabla)\#$$

- Rectangular Coordinates:

$$\rho \left(\frac{\partial u}{\partial t} + u\frac{\partial u}{\partial x} + v\frac{\partial u}{\partial y} + w\frac{\partial u}{\partial z} \right) = -\frac{\partial p}{\partial x} + \left[\frac{\partial}{\partial x}\tau_{xx} + \frac{\partial}{\partial y}\tau_{yx} + \frac{\partial}{\partial z}\tau_{zx} \right] + \rho g_x$$

$$\rho \left(\frac{\partial v}{\partial t} + u\frac{\partial v}{\partial x} + v\frac{\partial v}{\partial y} + w\frac{\partial v}{\partial z} \right) = -\frac{\partial p}{\partial y} + \left[\frac{\partial}{\partial x}\tau_{xy} + \frac{\partial}{\partial y}\tau_{yy} + \frac{\partial}{\partial z}\tau_{zy} \right] + \rho g_y$$

$$\rho \left(\frac{\partial w}{\partial t} + u\frac{\partial w}{\partial x} + v\frac{\partial w}{\partial y} + w\frac{\partial w}{\partial z} \right) = -\frac{\partial p}{\partial z} + \left[\frac{\partial}{\partial x}\tau_{xz} + \frac{\partial}{\partial y}\tau_{yz} + \frac{\partial}{\partial z}\tau_{zz} \right] + \rho g_z$$

- Cylindrical Coordinates:

$$\rho\left(\frac{\partial v_r}{\partial t}+v_r\frac{\partial v_r}{\partial r}+\frac{v_\theta}{r}\frac{\partial v_r}{\partial \theta}+v_z\frac{\partial v_r}{\partial z}-\frac{v_\theta^2}{r}\right)=-\frac{\partial p}{\partial r}$$

$$+\left[\frac{1}{r}\frac{\partial}{\partial r}\left(r\tau_{rr}\right)+\frac{1}{r}\frac{\partial}{\partial \theta}\tau_{\theta r}+\frac{\partial}{\partial z}\tau_{zr}-\frac{\tau_{\theta\theta}}{r}\right]+\rho g_r$$

$$\rho\left(\frac{\partial v_\theta}{\partial t}+v_r\frac{\partial v_\theta}{\partial r}+\frac{v_\theta}{r}\frac{\partial v_\theta}{\partial \theta}+v_z\frac{\partial v_\theta}{\partial z}+\frac{v_r v_\theta}{r}\right)=-\frac{1}{r}\frac{\partial p}{\partial \theta}$$

$$+\left[\frac{1}{r^2}\frac{\partial}{\partial r}\left(r^2\tau_{r\theta}\right)+\frac{1}{r}\frac{\partial}{\partial \theta}\tau_{\theta\theta}+\frac{\partial}{\partial z}\tau_{z\theta}+\frac{\tau_{\theta r}-\tau_{r\theta}}{r}\right]+\rho g_\theta$$

$$\rho\left(\frac{\partial v_z}{\partial t}+v_r\frac{\partial v_z}{\partial r}+\frac{v_\theta}{r}\frac{\partial v_z}{\partial \theta}+v_z\frac{\partial v_z}{\partial z}\right)=-\frac{\partial p}{\partial z}$$

$$+\left[\frac{1}{r}\frac{\partial}{\partial r}\left(r\tau_{rz}\right)+\frac{1}{r}\frac{\partial}{\partial \theta}\tau_{\theta z}+\frac{\partial}{\partial z}\tau_{zz}\right]+\rho g_z$$

A.5.3 Momentum Equation for Constant-Property Fluids
NAVIER-STOKES Equation:

$$\rho\frac{D\vec{v}}{Dt}=-\nabla p+\mu\nabla^2\vec{v}+\rho\vec{g}$$

- Rectangular Coordinates:

$$\rho\left(\frac{\partial u}{\partial t}+u\frac{\partial u}{\partial x}+v\frac{\partial u}{\partial y}+w\frac{\partial u}{\partial z}\right)=-\frac{\partial p}{\partial x}+\mu\left[\frac{\partial^2 u}{\partial x^2}+\frac{\partial^2 u}{\partial y^2}+\frac{\partial^2 u}{\partial z^2}\right]+\rho g_x$$

$$\rho\left(\frac{\partial v}{\partial t}+u\frac{\partial v}{\partial x}+v\frac{\partial v}{\partial y}+w\frac{\partial v}{\partial z}\right)=-\frac{\partial p}{\partial y}+\mu\left[\frac{\partial^2 v}{\partial x^2}+\frac{\partial^2 v}{\partial y^2}+\frac{\partial^2 v}{\partial z^2}\right]+\rho g_y$$

$$\rho\left(\frac{\partial w}{\partial t}+u\frac{\partial w}{\partial x}+v\frac{\partial w}{\partial y}+w\frac{\partial w}{\partial z}\right)=-\frac{\partial p}{\partial z}+\mu\left[\frac{\partial^2 w}{\partial x^2}+\frac{\partial^2 w}{\partial y^2}+\frac{\partial^2 w}{\partial z^2}\right]+\rho g_z$$

- Cylindrical Coordinates:

$$\rho\left(\frac{\partial v_r}{\partial t} + v_r\frac{\partial v_r}{\partial r} + \frac{v_\theta}{r}\frac{\partial v_r}{\partial \theta} + v_z\frac{\partial v_r}{\partial z} - \frac{v_\theta^2}{r}\right) = -\frac{\partial p}{\partial r}$$

$$+ \mu\left[\frac{\partial}{\partial r}\left(\frac{1}{r}\frac{\partial}{\partial r}(rv_r)\right) + \frac{1}{r^2}\frac{\partial^2 v_r}{\partial \theta^2} + \frac{\partial^2 v_r}{\partial z^2} - \frac{2}{r^2}\frac{\partial v_\theta}{\partial \theta}\right] + \rho g_r$$

$$\rho\left(\frac{\partial v_\theta}{\partial t} + v_r\frac{\partial v_\theta}{\partial r} + \frac{v_\theta}{r}\frac{\partial v_\theta}{\partial \theta} + v_z\frac{\partial v_\theta}{\partial z} + \frac{v_r v_\theta}{r}\right) = -\frac{1}{r}\frac{\partial p}{\partial \theta}$$

$$+ \mu\left[\frac{\partial}{\partial r}\left(\frac{1}{r}\frac{\partial}{\partial r}(rv_\theta)\right) + \frac{1}{r^2}\frac{\partial^2 v_\theta}{\partial \theta^2} + \frac{\partial^2 v_\theta}{\partial z^2} + \frac{2}{r^2}\frac{\partial v_r}{\partial \theta}\right] + \rho g_\theta$$

$$\rho\left(\frac{\partial v_z}{\partial t} + v_r\frac{\partial v_z}{\partial r} + \frac{v_\theta}{r}\frac{\partial v_z}{\partial \theta} + v_z\frac{\partial v_z}{\partial z}\right) = -\frac{\partial p}{\partial z}$$

$$+ \mu\left[\frac{1}{r}\frac{\partial}{\partial r}\left(r\frac{\partial v_z}{\partial r}\right) + \frac{1}{r^2}\frac{\partial^2 v_z}{\partial \theta^2} + \frac{\partial^2 v_z}{\partial z^2}\right] + \rho g_z$$

A.5.4 Heat Transfer Equation for Constant-Property Fluids

$$\rho c_p\frac{DT}{Dt} = k\nabla^2 T + \mu\Phi; \quad \frac{k}{\rho c_p} \equiv \alpha \langle\text{thermal diffusivity}\rangle$$

Note: For species mass transport of concentration $c\left[M/L^3\right]$:

$$\frac{Dc}{Dt} = \mathcal{D}_{AB}\nabla^2 c + S_c; \,\hat{=}\, \mathcal{D}_{AB} \text{ binary mass diffusivity}$$

Note: The equations of Sect. A.5.3 in conjunction with A.5.1 plus A.5.4 are nowadays summarized as the Navier-Stokes equation system.

- Rectangular Coordinates (see Sect. A.5.6 for viscous dissipation function Φ):

$$\frac{\partial T}{\partial t} + u\frac{\partial T}{\partial x} + v\frac{\partial T}{\partial y} + w\frac{\partial T}{\partial z} = \alpha\left[\frac{\partial^2 T}{\partial x^2} + \frac{\partial^2 T}{\partial y^2} + \frac{\partial^2 T}{\partial z^2}\right] + \frac{\mu}{\rho c_p}\Phi$$

- Cylindrical Coordinates:

$$\frac{\partial T}{\partial t} + v_r \frac{\partial T}{\partial r} + \frac{v_\theta}{r} \frac{\partial T}{\partial \theta} + v_z \frac{\partial T}{\partial z} = \alpha \left[\frac{1}{r} \frac{\partial}{\partial r} \left(r \frac{\partial T}{\partial r} \right) + \frac{1}{r^2} \frac{\partial^2 T}{\partial \theta^2} + \frac{\partial^2 T}{\partial z^2} \right] + \frac{\mu}{\rho c_p} \Phi$$

A.5.5 Stresses: $\vec{\vec{\tau}} = \mu \left[\nabla \vec{v} + \left(\nabla \vec{v} \right)^{tr} \right]$ and Fluxes: $\vec{q}_{cond} = -k \nabla T$

Note: Incompressible fluids.

- Rectangular Coordinates:

$$\tau_{xx} = 2\mu \frac{\partial u}{\partial x}; \quad \tau_{yy} = 2\mu \frac{\partial v}{\partial y}; \quad \tau_{zz} = 2\mu \frac{\partial w}{\partial z}$$

$$\tau_{xy} = \tau_{yx} = \mu \left[\frac{\partial v}{\partial x} + \frac{\partial u}{\partial y} \right]$$

$$\tau_{yz} = \tau_{zy} = \mu \left[\frac{\partial w}{\partial y} + \frac{\partial v}{\partial z} \right]$$

$$\tau_{zx} = \tau_{xz} = \mu \left[\frac{\partial u}{\partial z} + \frac{\partial w}{\partial x} \right]$$

$$q_x = -k \frac{\partial T}{\partial x}; \quad q_y = -k \frac{\partial T}{\partial y}; \quad q_z = -k \frac{\partial T}{\partial z}$$

- Cylindrical Coordinates:

$$\tau_{rr} = 2\mu \frac{\partial v_r}{\partial r}; \quad \tau_{\theta\theta} = 2\mu \left(\frac{1}{r} \frac{\partial v_\theta}{\partial \theta} + \frac{v_r}{r} \right); \quad \tau_{zz} = 2\mu \frac{\partial v_z}{\partial z}$$

$$\tau_{r\theta} = \tau_{\theta r} = \mu \left[r \frac{\partial}{\partial r} \left(\frac{v_\theta}{r} \right) + \frac{1}{r} \frac{\partial v_r}{\partial \theta} \right]$$

$$\tau_{\theta z} = \tau_{z\theta} = \mu \left[\frac{1}{r} \frac{\partial v_z}{\partial \theta} + \frac{\partial v_\theta}{\partial z} \right]$$

$$\tau_{zr} = \tau_{rz} = \mu \left[\frac{\partial v_r}{\partial z} + \frac{\partial v_z}{\partial r} \right]$$

$$q_r = -k \frac{\partial T}{\partial r}; \quad q_\theta = -k \frac{1}{r} \frac{\partial T}{\partial \theta}; \quad q_z = -k \frac{\partial T}{\partial z}$$

A.5.6 Dissipation Function for Newtonian Fluids

- Rectangular Coordinates:

$$\Phi = 2\left[\left(\frac{\partial u}{\partial x}\right)^2 + \left(\frac{\partial v}{\partial y}\right)^2 + \left(\frac{\partial w}{\partial z}\right)^2\right] + \left[\frac{\partial v}{\partial x} + \frac{\partial u}{\partial y}\right]^2 + \left[\frac{\partial w}{\partial y} + \frac{\partial v}{\partial z}\right]^2$$

$$+ \left[\frac{\partial u}{\partial z} + \frac{\partial w}{\partial x}\right]^2$$

- Cylindrical Coordinates:

$$\Phi = 2\left[\left(\frac{\partial v_r}{\partial r}\right)^2 + \left(\frac{1}{r}\frac{\partial v_\theta}{\partial \theta} + \frac{v_r}{r}\right)^2 + \left(\frac{\partial v_z}{\partial z}\right)^2\right]$$

$$+ \left[r\frac{\partial}{\partial r}\left(\frac{v_\theta}{r}\right) + \frac{1}{r}\frac{\partial v_r}{\partial \theta}\right]^2 + \left[\frac{1}{r}\frac{\partial v_z}{\partial \theta} + \frac{\partial v_\theta}{\partial z}\right]^2 + \left[\frac{\partial v_r}{\partial z} + \frac{\partial v_z}{\partial r}\right]^2$$

APPENDIX B

B.1 CONVERSION FACTORS

DIMENSION	METRIC	METRIC/ENGLISH
Acceleration	$1 \text{ m/s}^2 = 100 \text{ cm/s}^2$	$1 \text{ m/s}^2 = 3.2808 \text{ ft/s}^2$ $1 \text{ ft/s}^2 = 0.3048^* \text{ m/s}^2$
Area	$1 \text{ m}^2 = 10^4 \text{ cm}^2 = 10^6 \text{ mm}^2 = 10^{-6} \text{ km}^2$	$1 \text{ m}^2 = 1550 \text{ in}^2 = 10.764 \text{ ft}^2$ $1 \text{ ft}^2 = 144 \text{ in}^2 = 0.09290304^* \text{ m}^2$
Density	$1 \text{ g/cm}^3 = 1 \text{ kg/L} = 1000 \text{ kg/m}^3$	$1 \text{ g/cm}^3 = 62.428 \text{ lbm/ft}^3 = 0.036127 \text{ lbm/in}^3$ $1 \text{ lbm/in}^3 = 1728 \text{ lbm/ft}^3$ $1 \text{ kg/m}^3 = 0.062428 \text{ lbm/ft}^3$
Energy, heat, work, internal energy, enthalpy	$1 \text{ kJ} = 1000 \text{ J} = 1000 \text{ N} \cdot \text{m} = 1 \text{ kPa} \cdot \text{m}^3$ $1 \text{ kJ/kg} = 1000 \text{ m}^2/\text{s}^2$ $1 \text{ kWh} = 3600 \text{ kJ}$ $1 \text{ cal}^\dagger = 4.184 \text{ J}$ $1 \text{ IT cal}^\dagger = 4.1868 \text{ J}$ $1 \text{ Cal}^\dagger = 4.1868 \text{ kJ}$	$1 \text{ kJ} = 0.94782 \text{ Btu}$ $1 \text{ Btu} = 1.055056 \text{ kJ}$ $\quad = 5.40395 \text{ psia} \cdot \text{ft}^3 = 778.169 \text{ lbf} \cdot \text{ft}$ $1 \text{ Btu/lbm} = 25{,}037 \text{ ft}^2/\text{s}^2 = 2.326^* \text{ kJ/kg}$ $1 \text{ kJ/kg} = 0.430 \text{ Btu/lbm}$ $1 \text{ kWh} = 3412.14 \text{ Btu}$ $1 \text{ therm} = 10^5 \text{ Btu} = 1.055 \times 10^5 \text{ kJ}$ $\quad$ (natural gas)
Force	$1 \text{ N} = 1 \text{ kg} \cdot \text{m/s}^2 = 10^5 \text{ dyne}$ $1 \text{ kgf} = 9.80665 \text{ N}$	$1 \text{ N} = 0.22481 \text{ lbf}$ $1 \text{ lbf} = 32.174 \text{ lbm} \cdot \text{ft/s}^2 = 4.44822 \text{ N}$
Heat flux	$1 \text{ W/cm}^2 = 10^4 \text{ W/m}^2$	$1 \text{ W/m}^2 = 0.3171 \text{ Btu/h} \cdot \text{ft}^2$
Heat transfer coefficient	$1 \text{ W/m}^2 \cdot {}^\circ\text{C} = 1 \text{ W/m}^2 \cdot \text{K}$	$1 \text{ W/m}^2 \cdot {}^\circ\text{C} = 0.17612 \text{ Btu/h} \cdot \text{ft}^2 \cdot {}^\circ\text{F}$
Length	$1 \text{ m} = 100 \text{ cm} = 1000 \text{ mm} = 10^6 \text{ }\mu\text{m}$ $1 \text{ km} = 1000 \text{ m}$	$1 \text{ m} = 39.370 \text{ in} = 3.2808 \text{ ft} = 1.0926 \text{ yd}$ $1 \text{ ft} = 12 \text{ in} = 0.3048^* \text{ m}$ $1 \text{ mile} = 5280 \text{ ft} = 1.6093 \text{ km}$ $1 \text{ in} = 2.54^* \text{ cm}$
Mass	$1 \text{ kg} = 1000 \text{ g}$ $1 \text{ metric ton} = 1000 \text{ kg}$	$1 \text{ kg} = 2.2046226 \text{ lbm}$ $1 \text{ lbm} = 0.45359237^* \text{ kg}$ $1 \text{ ounce} = 28.3495 \text{ g}$ $1 \text{ slug} = 32.174 \text{ lbm} = 14.5939 \text{ kg}$ $1 \text{ short ton} = 2000 \text{ lbm} = 907.1847 \text{ kg}$
Power, heat transfer rate	$1 \text{ W} = 1 \text{ J/s}$ $1 \text{ kW} = 1000 \text{ W} = 1.341 \text{ hp}$ $1 \text{ hp}^\dagger = 745.7 \text{ W}$	$1 \text{ kW} = 3412.14 \text{ Btu/h}$ $\quad = 737.56 \text{ lbf} \cdot \text{ft/s}$ $1 \text{ hp} = 550 \text{ lbf} \cdot \text{ft/s} = 0.7068 \text{ Btu/s}$ $\quad = 42.41 \text{ Btu/min} = 2544.5 \text{ Btu/h}$ $\quad = 0.74570 \text{ kW}$ $1 \text{ boiler hp} = 33{,}475 \text{ Btu/h}$ $1 \text{ Btu/h} = 1.055056 \text{ kJ/h}$ $1 \text{ ton of refrigeration} = 200 \text{ Btu/min}$
Pressure	$1 \text{ Pa} = 1 \text{ N/m}^2$ $1 \text{ kPa} = 10^3 \text{ Pa} = 10^{-3} \text{ MPa}$ $1 \text{ atm} = 101.325 \text{ kPa} = 1.01325 \text{ bars}$ $\quad = 760 \text{ mm Hg at } 0{}^\circ\text{C}$ $\quad = 1.03323 \text{ kgf/cm}^2$ $1 \text{ mm Hg} = 0.1333 \text{ kPa}$	$1 \text{ Pa} = 1.4504 \times 10^{-4} \text{ psia}$ $\quad = 0.020886 \text{ lbf/ft}^2$ $1 \text{ psi} = 144 \text{ lbf/ft}^2 = 6.894757 \text{ kPa}$ $1 \text{ atm} = 14.696 \text{ psia} = 29.92 \text{ in Hg at } 30{}^\circ\text{F}$ $1 \text{ in Hg} = 3.387 \text{ kPa}$
Specific heat	$1 \text{ kJ/kg} \cdot {}^\circ\text{C} = 1 \text{ kJ/kg} \cdot \text{K} = 1 \text{ J/g} \cdot {}^\circ\text{C}$	$1 \text{ Btu/lbm} \cdot {}^\circ\text{F} = 4.1868 \text{ kJ/kg} \cdot {}^\circ\text{C}$ $1 \text{ Btu/lbmol} \cdot \text{R} = 4.1868 \text{ kJ/kmol} \cdot \text{K}$ $1 \text{ kJ/kg} \cdot {}^\circ\text{C} = 0.23885 \text{ Btu/lbm} \cdot {}^\circ\text{F}$ $\quad = 0.23885 \text{ Btu/lbm} \cdot \text{R}$

*Exact conversion factor between metric and English units.

†Calorie is originally defined as the amount of heat needed to raise the temperature of 1 g of water by 1°C, but it varies with temperature. The international steam table (IT) calorie (generally preferred by engineers) is exactly 4.1868 J by definition and corresponds to the specific heat of water at 15°C. The thermochemical calorie (generally preferred by physicists) is exactly 4.184 J by definition and corresponds to the specific heat of water at room temperature. The difference between the two is about 0.06 percent, which is negligible. The capitalized Calorie used by nutritionists is actually a kilocalorie (1000 IT calories).

DIMENSION	METRIC	METRIC/ENGLISH
Specific volume	$1 \ m^3/kg = 1000 \ L/kg = 1000 \ cm^3/g$	$1 \ m^3/kg = 16.02 \ ft^3/lbm$ $1 \ ft^3/lbm = 0.062428 \ m^3/kg$
Temperature	$T(K) = T(°C) + 273.15$ $\Delta T(K) = \Delta T(°C)$	$T(R) = T(°F) + 459.67 = 1.8 T(K)$ $T(°F) = 1.8 \ T(°C) + 32$ $\Delta T(°F) = \Delta T(R) = 1.8 \ \Delta T(K)$
Thermal conductivity	$1 \ W/m \cdot °C = 1 \ W/m \cdot K$	$1 \ W/m \cdot °C = 0.57782 \ Btu/h \cdot ft \cdot °F$
Velocity	$1 \ m/s = 3.60 \ km/h$	$1 \ m/s = 3.2808 \ ft/s = 2.237 \ mi/h$ $1 \ mi/h = 1.46667 \ ft/s$ $1 \ mi/h = 1.6093 \ km/h$
Volume	$1 \ m^3 = 1000 \ L = 10^6 \ cm^3 \ (cc)$	$1 \ m^3 = 6.1024 \times 10^4 \ in^3 = 35.315 \ ft^3$ $\quad = 264.17 \ gal \ (U.S.)$ $1 \ U.S. \ gallon = 231 \ in^3 = 3.7854 \ L$ $1 \ fl \ ounce = 29.5735 \ cm^3 = 0.0295735 \ L$ $1 \ U.S. \ gallon = 128 \ fl \ ounces$
Volume flow rate	$1 \ m^3/s = 60,000 \ L/min = 10^6 \ cm^3/s$	$1 \ m^3/s = 15,850 \ gal/min \ (gpm) = 35.315 \ ft^3/s$ $\quad = 2118.9 \ ft^3/min \ (cfm)$

[1]Mechanical horsepower. The electrical horsepower is taken to be exactly 746 W.

Some Physical Constants

Universal gas constant	$R_u = 8.31447 \ kJ/kmol \cdot K$ $\quad = 8.31447 \ kPa \cdot m^3/kmol \cdot K$ $\quad = 0.0831447 \ bar \cdot m^3/kmol \cdot K$ $\quad = 82.05 \ L \cdot atm/kmol \cdot K$ $\quad = 1.9858 \ Btu/lbmol \cdot R$ $\quad = 1545.37 \ ft \cdot lbf/lbmol \cdot R$ $\quad = 10.73 \ psia \cdot ft^3/lbmol \cdot R$
Standard acceleration of gravity	$g = 9.80665 \ m/s^2$ $\quad = 32.174 \ ft/s^2$
Standard atmospheric pressure	$1 \ atm = 101.325 \ kPa$ $\quad = 1.01325 \ bar$ $\quad = 14.696 \ psia$ $\quad = 760 \ mm \ Hg \ (0°C)$ $\quad = 29.9213 \ in \ Hg \ (32°F)$ $\quad = 10.3323 \ m \ H_2O \ (4°C)$
Stefan–Boltzmann constant	$\sigma = 5.6704 \times 10^{-8} \ W/m^2 \cdot K^4$ $\quad = 0.1714 \times 10^{-8} \ Btu/h \cdot ft^2 \cdot R^4$
Boltzmann's constant	$k = 1.380650 \times 10^{-23} \ J/K$
Speed of light in vacuum	$c_o = 2.9979 \times 10^8 \ m/s$ $\quad = 9.836 \times 10^8 \ ft/s$
Speed of sound in dry air at 0°C and 1 atm	$c = 331.36 \ m/s$ $\quad = 1089 \ ft/s$
Heat of fusion of water at 1 atm	$h_{if} = 333.7 \ kJ/kg$ $\quad = 143.5 \ Btu/lbm$
Enthalpy of vaporization of water at 1 atm	$h_{fg} = 2256.5 \ kJ/kg$ $\quad = 970.12 \ Btu/lbm$

B.2 PROPERTIES

TABLE B.2-1

Molar mass, gas constant, and critical-point properties

Substance	Formula	Molar mass, M kg/kmol	Gas constant, R kJ/kg · K*	Critical-point properties		
				Temperature, K	Pressure, MPa	Volume, m³/kmol
Air	—	28.97	0.2870	132.5	3.77	0.0883
Ammonia	NH_3	17.03	0.4882	405.5	11.28	0.0724
Argon	Ar	39.948	0.2081	151	4.86	0.0749
Benzene	C_6H_6	78.115	0.1064	562	4.92	0.2603
Bromine	Br_2	159.808	0.0520	584	10.34	0.1355
n-Butane	C_4H_{10}	58.124	0.1430	425.2	3.80	0.2547
Carbon dioxide	CO_2	44.01	0.1889	304.2	7.39	0.0943
Carbon monoxide	CO	28.011	0.2968	133	3.50	0.0930
Carbon tetrachloride	CCl_4	153.82	0.05405	556.4	4.56	0.2759
Chlorine	Cl_2	70.906	0.1173	417	7.71	0.1242
Chloroform	$CHCl_3$	119.38	0.06964	536.6	5.47	0.2403
Dichlorodifluoromethane (R-12)	CCl_2F_2	120.91	0.06876	384.7	4.01	0.2179
Dichlorofluoromethane (R-21)	$CHCl_2F$	102.92	0.08078	451.7	5.17	0.1973
Ethane	C_2H_6	30.070	0.2765	305.5	4.48	0.1480
Ethyl alcohol	C_2H_5OH	46.07	0.1805	516	6.38	0.1673
Ethylene	C_2H_4	28.054	0.2964	282.4	5.12	0.1242
Helium	He	4.003	2.0769	5.3	0.23	0.0578
n-Hexane	C_6H_{14}	86.179	0.09647	507.9	3.03	0.3677
Hydrogen (normal)	H_2	2.016	4.1240	33.3	1.30	0.0649
Krypton	Kr	83.80	0.09921	209.4	5.50	0.0924
Methane	CH_4	16.043	0.5182	191.1	4.64	0.0993
Methyl alcohol	CH_3OH	32.042	0.2595	513.2	7.95	0.1180
Methyl chloride	CH_3Cl	50.488	0.1647	416.3	6.68	0.1430
Neon	Ne	20.183	0.4119	44.5	2.73	0.0417
Nitrogen	N_2	28.013	0.2968	126.2	3.39	0.0899
Nitrous oxide	N_2O	44.013	0.1889	309.7	7.27	0.0961
Oxygen	O_2	31.999	0.2598	154.8	5.08	0.0780
Propane	C_3H_8	44.097	0.1885	370	4.26	0.1998
Propylene	C_3H_6	42.081	0.1976	365	4.62	0.1810
Sulfur dioxide	SO_2	64.063	0.1298	430.7	7.88	0.1217
Tetrafluoroethane (R-134a)	CF_3CH_2F	102.03	0.08149	374.2	4.059	0.1993
Trichlorofluoromethane (R-11)	CCl_3F	137.37	0.06052	471.2	4.38	0.2478
Water	H_2O	18.015	0.4615	647.1	22.06	0.0560
Xenon	Xe	131.30	0.06332	289.8	5.88	0.1186

*The unit kJ/kg · K is equivalent to kPa · m³/kg · K. The gas constant is calculated from $R = R_u/M$, where $R_u = 8.31447$ kJ/kmol · K and M is the molar mass.

Source: K. A. Kobe and R. E. Lynn, Jr., *Chemical Review* 52 (1953), pp. 117–236; and ASHRAE, *Handbook of Fundamentals* (Atlanta, GA: American Society of Heating, Refrigerating and Air-Conditioning Engineers, Inc., 1993), pp. 16.4 and 36.1.

TABLE B.2-2

Ideal-gas specific heats of various common gases

(a) At 300 K

Gas	Formula	Gas constant, R kJ/kg · K	c_p kJ/kg · K	c_v kJ/kg · K	k
Air	—	0.2870	1.005	0.718	1.400
Argon	Ar	0.2081	0.5203	0.3122	1.667
Butane	C_4H_{10}	0.1433	1.7164	1.5734	1.091
Carbon dioxide	CO_2	0.1889	0.846	0.657	1.289
Carbon monoxide	CO	0.2968	1.040	0.744	1.400
Ethane	C_2H_6	0.2765	1.7662	1.4897	1.186
Ethylene	C_2H_4	0.2964	1.5482	1.2518	1.237
Helium	He	2.0769	5.1926	3.1156	1.667
Hydrogen	H_2	4.1240	14.307	10.183	1.405
Methane	CH_4	0.5182	2.2537	1.7354	1.299
Neon	Ne	0.4119	1.0299	0.6179	1.667
Nitrogen	N_2	0.2968	1.039	0.743	1.400
Octane	C_8H_{18}	0.0729	1.7113	1.6385	1.044
Oxygen	O_2	0.2598	0.918	0.658	1.395
Propane	C_3H_8	0.1885	1.6794	1.4909	1.126
Steam	H_2O	0.4615	1.8723	1.4108	1.327

Note: The unit kJ/kg · K is equivalent to kJ/kg · °C.
Source: Chemical and Process Thermodynamics 3/E by Kyle, B. G., © 2000. Adapted by permission of Pearson Education, Inc., Upper Saddle River, NJ.

(b) At various temperatures

Temperature, K	c_p kJ/kg · K	c_v kJ/kg · K	k	c_p kJ/kg · K	c_v kJ/kg · K	k	c_p kJ/kg · K	c_v kJ/kg · K	k
	Air			Carbon dioxide, CO_2			Carbon monoxide, CO		
250	1.003	0.716	1.401	0.791	0.602	1.314	1.039	0.743	1.400
300	1.005	0.718	1.400	0.846	0.657	1.288	1.040	0.744	1.399
350	1.008	0.721	1.398	0.895	0.706	1.268	1.043	0.746	1.398
400	1.013	0.726	1.395	0.939	0.750	1.252	1.047	0.751	1.395
450	1.020	0.733	1.391	0.978	0.790	1.239	1.054	0.757	1.392
500	1.029	0.742	1.387	1.014	0.825	1.229	1.063	0.767	1.387
550	1.040	0.753	1.381	1.046	0.857	1.220	1.075	0.778	1.382
600	1.051	0.764	1.376	1.075	0.886	1.213	1.087	0.790	1.376
650	1.063	0.776	1.370	1.102	0.913	1.207	1.100	0.803	1.370
700	1.075	0.788	1.364	1.126	0.937	1.202	1.113	0.816	1.364
750	1.087	0.800	1.359	1.148	0.959	1.197	1.126	0.829	1.358
800	1.099	0.812	1.354	1.169	0.980	1.193	1.139	0.842	1.353
900	1.121	0.834	1.344	1.204	1.015	1.186	1.163	0.866	1.343
1000	1.142	0.855	1.336	1.234	1.045	1.181	1.185	0.888	1.335
	Hydrogen, H_2			Nitrogen, N_2			Oxygen, O_2		
250	14.051	9.927	1.416	1.039	0.742	1.400	0.913	0.653	1.398
300	14.307	10.183	1.405	1.039	0.743	1.400	0.918	0.658	1.395
350	14.427	10.302	1.400	1.041	0.744	1.399	0.928	0.668	1.389
400	14.476	10.352	1.398	1.044	0.747	1.397	0.941	0.681	1.382
450	14.501	10.377	1.398	1.049	0.752	1.395	0.956	0.696	1.373
500	14.513	10.389	1.397	1.056	0.759	1.391	0.972	0.712	1.365
550	14.530	10.405	1.396	1.065	0.768	1.387	0.988	0.728	1.358
600	14.546	10.422	1.396	1.075	0.778	1.382	1.003	0.743	1.350
650	14.571	10.447	1.395	1.086	0.789	1.376	1.017	0.758	1.343
700	14.604	10.480	1.394	1.098	0.801	1.371	1.031	0.771	1.337
750	14.645	10.521	1.392	1.110	0.813	1.365	1.043	0.783	1.332
800	14.695	10.570	1.390	1.121	0.825	1.360	1.054	0.794	1.327
900	14.822	10.698	1.385	1.145	0.849	1.349	1.074	0.814	1.319
1000	14.983	10.859	1.380	1.167	0.870	1.341	1.090	0.830	1.313

Source: Kenneth Wark, *Thermodynamics,* 4th ed. (New York: McGraw-Hill, 1983), p. 783, Table A–4M. Originally published in *Tables of Thermal Properties of Gases,* NBS Circular 564, 1955.

TABLE B.2-3

Properties of common liquids, solids, and foods

(a) Liquids

Substance	Boiling data at 1 atm		Freezing data		Liquid properties		
	Normal boiling point, °C	Latent heat of vaporization h_{fg}, kJ/kg	Freezing point, °C	Latent heat of fusion h_{if}, kJ/kg	Temperature, °C	Density ρ, kg/m³	Specific heat c_p, kJ/kg · K
Ammonia	−33.3	1357	−77.7	322.4	−33.3	682	4.43
					−20	665	4.52
					0	639	4.60
					25	602	4.80
Argon	−185.9	161.6	−189.3	28	−185.6	1394	1.14
Benzene	80.2	394	5.5	126	20	879	1.72
Brine (20% sodium chloride by mass)	103.9	—	−17.4	—	20	1150	3.11
n-Butane	−0.5	385.2	−138.5	80.3	−0.5	601	2.31
Carbon dioxide	−78.4*	230.5 (at 0°C)	−56.6		0	298	0.59
Ethanol	78.2	838.3	−114.2	109	25	783	2.46
Ethyl alcohol	78.6	855	−156	108	20	789	2.84
Ethylene glycol	198.1	800.1	−10.8	181.1	20	1109	2.84
Glycerine	179.9	974	18.9	200.6	20	1261	2.32
Helium	−268.9	22.8	—	—	−268.9	146.2	22.8
Hydrogen	−252.8	445.7	−259.2	59.5	−252.8	70.7	10.0
Isobutane	−11.7	367.1	−160	105.7	−11.7	593.8	2.28
Kerosene	204–293	251	−24.9	—	20	820	2.00
Mercury	356.7	294.7	−38.9	11.4	25	13,560	0.139
Methane	−161.5	510.4	−182.2	58.4	−161.5	423	3.49
					−100	301	5.79
Methanol	64.5	1100	−97.7	99.2	25	787	2.55
Nitrogen	−195.8	198.6	−210	25.3	−195.8	809	2.06
					−160	596	2.97
Octane	124.8	306.3	−57.5	180.7	20	703	2.10
Oil (light)					25	910	1.80
Oxygen	−183	212.7	−218.8	13.7	−183	1141	1.71
Petroleum	—	230–384			20	640	2.0
Propane	−42.1	427.8	−187.7	80.0	−42.1	581	2.25
					0	529	2.53
					50	449	3.13
Refrigerant-134a	−26.1	217.0	−96.6	—	−50	1443	1.23
					−26.1	1374	1.27
					0	1295	1.34
					25	1207	1.43
Water	100	2257	0.0	333.7	0	1000	4.22
					25	997	4.18
					50	988	4.18
					75	975	4.19
					100	958	4.22

* Sublimation temperature. (At pressures below the triple-point pressure of 518 kPa, carbon dioxide exists as a solid or gas. Also, the freezing-point temperature of carbon dioxide is the triple-point temperature of −56.5°C.)

TABLE B.2-3

Properties of common liquids, solids, and foods (*Concluded*)

(*b*) Solids (values are for room temperature unless indicated otherwise)

Substance	Density, ρ kg/m^3	Specific heat, c_p kJ/kg · K	Substance	Density, ρ kg/m^3	Specific heat, c_p kJ/kg · K
Metals			Nonmetals		
Aluminum			Asphalt	2110	0.920
200 K		0.797	Brick, common	1922	0.79
250 K		0.859	Brick, fireclay (500°C)	2300	0.960
300 K	2,700	0.902	Concrete	2300	0.653
350 K		0.929	Clay	1000	0.920
400 K		0.949	Diamond	2420	0.616
450 K		0.973	Glass, window	2700	0.800
500 K		0.997	Glass, pyrex	2230	0.840
Bronze (76% Cu, 2% Zn, 2% Al)	8,280	0.400	Graphite	2500	0.711
			Granite	2700	1.017
Brass, yellow (65% Cu, 35% Zn)	8,310	0.400	Gypsum or plaster board	800	1.09
Copper			Ice		
−173°C		0.254	200 K		1.56
−100°C		0.342	220 K		1.71
−50°C		0.367	240 K		1.86
0°C		0.381	260 K		2.01
27°C	8,900	0.386	273 K	921	2.11
100°C		0.393	Limestone	1650	0.909
200°C		0.403	Marble	2600	0.880
Iron	7,840	0.45	Plywood (Douglas Fir)	545	1.21
Lead	11,310	0.128	Rubber (soft)	1100	1.840
Magnesium	1,730	1.000	Rubber (hard)	1150	2.009
Nickel	8,890	0.440	Sand	1520	0.800
Silver	10,470	0.235	Stone	1500	0.800
Steel, mild	7,830	0.500	Woods, hard (maple, oak, etc.)	721	1.26
Tungsten	19,400	0.130	Woods, soft (fir, pine, etc.)	513	1.38

(*c*) Foods

Food	Water content, % (mass)	Freezing point, °C	Specific heat, kJ/kg · K Above freezing	Specific heat, kJ/kg · K Below freezing	Latent heat of fusion, kJ/kg	Food	Water content, % (mass)	Freezing point, °C	Specific heat, kJ/kg · K Above freezing	Specific heat, kJ/kg · K Below freezing	Latent heat of fusion, kJ/kg
Apples	84	−1.1	3.65	1.90	281	Lettuce	95	−0.2	4.02	2.04	317
Bananas	75	−0.8	3.35	1.78	251	Milk, whole	88	−0.6	3.79	1.95	294
Beef round	67	—	3.08	1.68	224	Oranges	87	−0.8	3.75	1.94	291
Broccoli	90	−0.6	3.86	1.97	301	Potatoes	78	−0.6	3.45	1.82	261
Butter	16	—	—	1.04	53	Salmon fish	64	−2.2	2.98	1.65	214
Cheese, swiss	39	−10.0	2.15	1.33	130	Shrimp	83	−2.2	3.62	1.89	277
Cherries	80	−1.8	3.52	1.85	267	Spinach	93	−0.3	3.96	2.01	311
Chicken	74	−2.8	3.32	1.77	247	Strawberries	90	−0.8	3.86	1.97	301
Corn, sweet	74	−0.6	3.32	1.77	247	Tomatoes, ripe	94	−0.5	3.99	2.02	314
Eggs, whole	74	−0.6	3.32	1.77	247	Turkey	64	—	2.98	1.65	214
Ice cream	63	−5.6	2.95	1.63	210	Watermelon	93	−0.4	3.96	2.01	311

Source: Values are obtained from various handbooks and other sources or are calculated. Water content and freezing-point data of foods are from *ASHRAE, Handbook of Fundamentals*, SI version (Atlanta, GA: American Society of Heating, Refrigerating and Air-Conditioning Engineers, Inc., 1993), Chapter 30, Table 1. Freezing point is the temperature at which freezing starts for fruits and vegetables, and the average freezing temperature for other foods.

B.3 DRAG COEFFICIENT: (A) Smooth Sphere and (B) An Infinite Cylinder as a Function of Reynolds Number

(A)

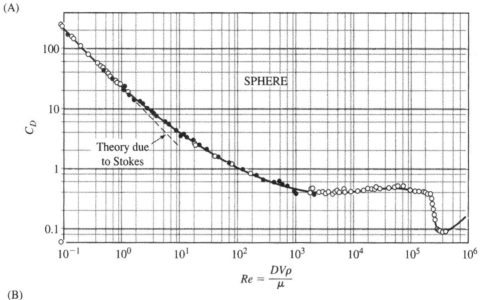

(B)

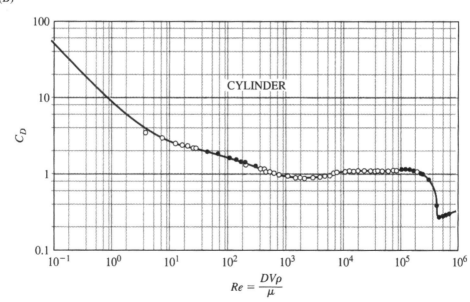

B.4 MOODY CHART

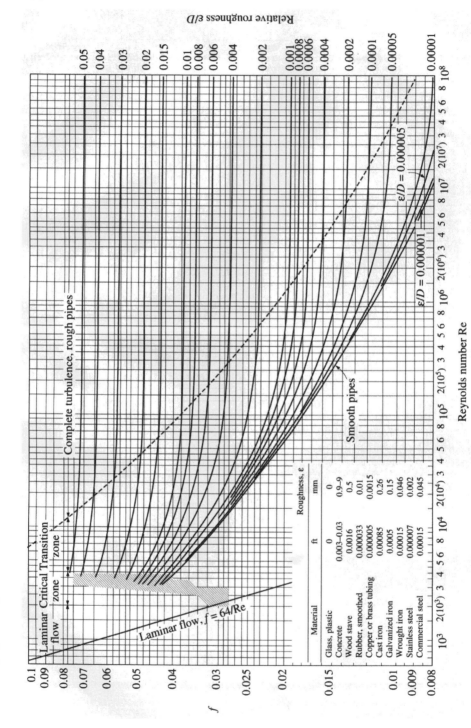

References (Appendices)

Aris, R. 1989, Vectors, *Tensors, and the Basic Equations of Fluid Mechanics*, Dover Publications, New York.

Bird, R.B., Stewart, W.E., Lightfoot, E.N., 2002, *Transport Phenomena*, 2nd ed., Wiley, New York.

Finlayson, B.A., 1972, *The Method of Weighted Residuals and Variational Principles: with Application in Fluid Mechanics, Heat and Mass Transfer*, Academic Press, New York.

Greenberg, M.D., 1998, *Advanced Engineering Mathematics*, 2nd ed, Prentice-Hall,

Hoffman, J.D., 2001, *Numerical Methods for Engineers and Scientists*, 2nd ed, Marcel Dekker, New York.

Ozisik, M.N., 1993, *Heat Conduction*, 2nd ed, Wiley Interscience, New York.

Polyamin, A.D., Zaitsev, V.F., 1995, *Handbook of Exact Solution for Ordinary Differential Equations*. CRC Press, New York.

Schey, H.M., 1973, *Div, Grad, Curl, and All That*, Norton, New York.

Spiegel, M., 1971, *Schaum's Outline of Advanced Mathematics for Scientists and Engineers*, McGraw-Hill, New York.

INDEX